AF750375

Springer Handbook of Auditory Research

For further volumes:
http://www.springer.com/series/2506

Yale E. Cohen • Arthur N. Popper • Richard R. Fay
Editors

Neural Correlates of Auditory Cognition

Springer

Editors
Yale E. Cohen
Department of Otorhinolaryngology
Perelman School of Medicine
University of Pennsylvania
Philadelphia, PA, USA

Arthur N. Popper
Department of Biology
University of Maryland
College Park, MD, USA

Richard R. Fay
Marine Biological Laboratory
Woods Hole, MA, USA

ISSN 0947-2657
ISBN 978-1-4614-2349-2 ISBN 978-1-4614-2350-8 (eBook)
DOI 10.1007/978-1-4614-2350-8
Springer New York Heidelberg Dordrecht London

Library of Congress Control Number: 2012945007

© Springer Science+Business Media New York 2013
This work is subject to copyright. All rights are reserved by the Publisher, whether the whole or part of the material is concerned, specifically the rights of translation, reprinting, reuse of illustrations, recitation, broadcasting, reproduction on microfilms or in any other physical way, and transmission or information storage and retrieval, electronic adaptation, computer software, or by similar or dissimilar methodology now known or hereafter developed. Exempted from this legal reservation are brief excerpts in connection with reviews or scholarly analysis or material supplied specifically for the purpose of being entered and executed on a computer system, for exclusive use by the purchaser of the work. Duplication of this publication or parts thereof is permitted only under the provisions of the Copyright Law of the Publisher's location, in its current version, and permission for use must always be obtained from Springer. Permissions for use may be obtained through RightsLink at the Copyright Clearance Center. Violations are liable to prosecution under the respective Copyright Law.
The use of general descriptive names, registered names, trademarks, service marks, etc. in this publication does not imply, even in the absence of a specific statement, that such names are exempt from the relevant protective laws and regulations and therefore free for general use.
While the advice and information in this book are believed to be true and accurate at the date of publication, neither the authors nor the editors nor the publisher can accept any legal responsibility for any errors or omissions that may be made. The publisher makes no warranty, express or implied, with respect to the material contained herein.

Printed on acid-free paper

Springer is part of Springer Science+Business Media (www.springer.com)

We dedicate this volume to our colleague and friend, James C. Saunders, an outstanding teacher and mentor. Jim's enthusiasm for science and the scientific process was, and continues to be, contagious to all.

Series Preface

The preface that follows is the one that we published in Volume 1 of the Springer Handbook of Auditory Research back in 1992. Thus, 2012 marks the 20th year of SHAR. As anyone reading the original preface, or the many users of the series, will note, we have far exceeded our original expectation of eight volumes. Indeed, with books published to date, and those in the pipeline, we are now set for more than 50 volumes in SHAR, and we are still open to new and exciting ideas for additional books.

We are very proud that there seems to be consensus, at least among our friends and colleagues, that SHAR has become an important and influential part of the auditory literature. While we have worked hard to develop and maintain the quality and value of SHAR, the real value of the books is very much attributable to the numerous authors who have given their time to write outstanding chapters and to our many coeditors who have provided the intellectual leadership to the individual volumes. We have worked with a remarkable and wonderful group of people, many of whom have become great personal friends of both of us. We also continue to work with a spectacular group of editors at Springer, currently Ann Avouris. Indeed, several of our past editors have moved on in the publishing world to become senior executives. To our delight, this includes the current president of Springer US, Dr. William Curtis.

But the truth is that the series would not and could not be possible without the support of our families, and we want to take this opportunity to dedicate all of the SHAR books, past and future, to them. Our wives, Catherine Fay and Helen Popper, and our children, Michelle Popper Levit, Melissa Popper Levinsohn, Christian Fay, and Amanda Fay, have been immensely patient as we developed and worked on this series. We thank them, and state, without doubt, that this series could not have happened without them.

Preface 1992

The Springer Handbook of Auditory Research presents a series of comprehensive and synthetic reviews of the fundamental topics in modern auditory research. The volumes are aimed at all individuals with interests in hearing research including advanced graduate students, post-doctoral researchers, and clinical investigators. The volumes are intended to introduce new investigators to important aspects of hearing science and to help established investigators to better understand the fundamental theories and data in fields of hearing that they may not normally follow closely.

Each volume presents a particular topic comprehensively, and each serves as a synthetic overview and guide to the literature. As such, the chapters present neither exhaustive data reviews nor original research that has not yet appeared in peer-reviewed journals. The volumes focus on topics that have developed a solid data and conceptual foundation rather than on those for which a literature is only beginning to develop. New research areas will be covered on a timely basis in the series as they begin to mature.

Each volume in the series consists of a few substantial chapters on a particular topic. In some cases, the topics will be ones of traditional interest for which there is a substantial body of data and theory, such as auditory neuroanatomy (Vol. 1) and neurophysiology (Vol. 2). Other volumes in the series deal with topics that have begun to mature more recently, such as development, plasticity, and computational models of neural processing. In many cases, the series editors are joined by a coeditor having special expertise in the topic of the volume.

Richard R. Fay, Falmouth, MA
Arthur N. Popper, College Park, MD

Volume Preface

Hearing and communication present a variety of challenges to the nervous system. To be heard and to be understood, a communication signal must be transformed from a time-varying acoustic waveform to a perceptual representation and then to an abstract representation that combines the perceptual representation with memory stores and semantic/referential information. Finally, this abstract representation must be interpreted to form categorical decisions that guide behavior. Did I hear the stimulus? From where and whom did it come? What does it tell me? How can I use this information to plan an action? All of these issues and questions underlie auditory cognition.

This volume emphasizes those literatures that *combine* measurements of behavioral (psychophysical) responses with measurements of the responses of single neurons. Since the auditory cortex is modulated by cognitive factors (e.g., attention, memory, and reward values) and ongoing task demands, their effect on the neural code can be tested only when behavioral responses and single-neuron activity are simultaneously recorded. This relationship between behavior and single-neuron activity is one of the most significant issues in auditory neuroscience and, more generally, in sensory and cognitive neuroscience.

Chapters 2 and 3 set the stage for the remaining chapters in the book. In Chapter 2, Rauschecker reviews the hypothesis that the auditory cortex contains two functional processing streams: one pathway devoted to the neural computations underlying spatial processing (where is the sound?) and the other pathway devoted to computations underlying auditory-object processing (what is the sound?). In Chapter 3, Scheich and Brosch review the contribution of the auditory cortex to cognitive components of auditory behavior. The next section of this volume expands on the topics raised by Rauschecker in Chapter 2. In Chapter 4, Recanzone discusses the role of the auditory cortex in spatial processing and localization, while in Chapter 5, Schnupp, Honey, and Willmore focus on auditory-object perception and the neural correlates underlying this form of perception.

Chapters 6 and 7 focus on the neural processing and perception underlying a specific type of auditory object: vocal-communication signals. Chapter 6 by Steinschneider focuses on how listeners extract phonemes—the smallest unit of human speech—from the acoustic environment. In our daily lives, this ability seems automatic and even trivial. In Chapter 7, Ghazanfar and Chandrasekaran overview the multimodal nature of communication and the neural signals mediating multi-modal communication.

The final three chapters of the book focus on attention, memory, and plasticity, themes that are initially introduced by Scheich and Brosch in Chapter 3. The inter-dependencies of attention and memory and the neural codes that underlie auditory attention and memory are discussed by Poremba and Bigelow in Chapter 8. Chapter 9 by Fritz, David, and Shamma raises the possibility that this idea may need refining, and the authors review a body of literature indicating that the tuning properties of a neuron are plastic and are functionally dependent on the immediate needs of the listener. Finally, in Chapter 10, Shepard, Kilgard, and Liu continue with the theme of plasticity and how a listener's experience has a profound impact on a neuron's response properties and how auditory information is organized and represented in the brain.

As is always the case with SHAR volumes, this volume builds on and complements many of the outstanding previous volumes in the SHAR series. In particular, the material in this volume is complemented by material in several volumes that focus on psychophysics and neural activity in the human auditory cortex including *Human Psychophysics* (Vol. 3, edited by Yost, Popper, and Fay), *Auditory Perception of Sound Sources* (Vol. 29, edited by Yost, Popper, and Fay), and *Human Auditory Cortex* (Vol. 43, edited by Poeppel, Overath, Fay, and Popper).

Yale E. Cohen, Philadelphia, PA
Arthur N. Popper, College Park, MD
Richard R. Fay, Falmouth, MA

Contents

Contributors

James Bigelow Department of Psychology, University of Iowa, E11 SSH, Iowa City, IA 52242, USA

Michael Brosch Leibniz-Institut für Neurobiologie and Center for Behavioral Brain Sciences, Brenneckestraße 6, 39118 Magdeburg, Germany

Chandramouli Chandrasekaran Department of Psychology, Neuroscience Institute, Princeton University, Princeton, NJ USA

Yale E. Cohen Department of Otorhinolaryngology, Perelman School of Medicine, University of Pennsylvania, 3400 Spruce St – 5 Ravdin, Philadelphia, PA 19104, USA

Stephen David Oregon Hearing Research Center, Oregon Health and Science University, 3181 S.W. Sam Jackson Park Road, Portland, OR 97239-3098, USA

Jonathan B. Fritz Neural Systems Lab, Institute for Systems Research, Electrical and Computer Engineering, University of Maryland, College Park, MD 20742, USA

Asif A. Ghazanfar Departments of Psychology and Ecology & Evolutionary Biology, Neuroscience Institute, Princeton University, Princeton, NJ 08540, USA

Christian Honey Department of Physiology, Anatomy and Genetics, University of Oxford, Parks Road, Oxford OX1 3PT, UK

Michael P. Kilgard School of Behavioral and Brain Sciences, University of Texas at Dallas, GR 41, Richardson, TX 75083, USA

Robert C. Liu Department of Biology, Emory University, 1510 Clifton Rd. NE Atlanta, GA 30322, USA; Center for Behavioral Neuroscience, Georgia State University, PO Box 3999, Atlanta, GA 30302, USA and Center for Translational Social Neuroscience, Yerkes National Primate Research Center, Emory University, 954 Gatewood Rd., Atlanta, GA 30322, USA

Amy Poremba Department of Psychology, University of Iowa, E11 SSH, Iowa City, IA 52242, USA

Josef P. Rauschecker Department of Neuroscience, Georgetown University Medical Center, New Research Building, Room WP19, Box 571460, Washington, DC 20057, USA

Gregg H. Recanzone Center for Neuroscience and Department of Neurobiology, Physiology and Behavior, University of California at Davis, 1544 Newton Ct., Davis, CA 95618, USA

Henning Scheich Leibniz-Institut für Neurobiologie and Center for Behavioral Brain Sciences, Brenneckestraße 6, 39118 Magdeburg, Germany

Jan W.H. Schnupp Department of Physiology, Anatomy and Genetics, University of Oxford, Parks Road, Oxford OX1 3PT, UK

Shihab Shamma Neural Systems Lab, Institute for Systems Research, Electrical and Computer Engineering, University of Maryland, College Park, MD 20742, USA

Kathryn N. Shepard Neuroscience Graduate Program and Department of Biology, Emory University, 1510 Clifton Rd. NE; and Center for Behavioral Neuroscience, Atlanta, GA 30322, USA

Mitchell Steinschneider Department of Neurology, Albert Einstein College of Medicine, Rose F. Kennedy Center, 1300 Morris Park Avenue, Bronx, NY 10461, USA

Ben D.B. Willmore Department of Physiology, Anatomy and Genetics, University of Oxford, Parks Road, Oxford OX1 3PT, UK

Chapter 1
Auditory Cognition: The Integration of Psychophysics with Neurophysiology

Yale E. Cohen

Keywords Auditory identity • Auditory object • Auditory scene • Communication signals • Decision-making • Learning • Memory • Multi-modal • Neurometric • parallel processing • Sound localization • Speech

1.1 Introduction

Hearing and communication present a variety of challenges to the nervous system. To be heard and to be understood, an auditory signal must be transformed from a time-varying acoustic waveform to a perceptual representation and then to an abstract representation that combines this perceptual representation with memory stores and semantic/referential information (Griffiths & Warren, 2004). Finally, this abstract representation must be interpreted to form a categorical decision that guides behavior. Did I hear the stimulus? From where and whom did it come? What does it tell me? How can I use this information to plan an action? All of these issues and questions underlie auditory cognition.

Since the early 1990s, there has been a rebirth of studies that test the neural correlates of auditory cognition, with an emerging emphasis on recording neural activity in behaving animals. Continuing today, a great deal of scientific energy is devoted to elucidating the neural computations that underlie auditory cognition. Although the growth of the field is exciting, it is still in its quite nascent stages. In the last decade, only 5% of the PubMed-cited manuscripts that examined audition or hearing focused on the auditory cortex. This proportion falls substantially for studies that tested the relationship between the auditory cortex and behavior (1%) and even more dramatically for studies that tested the interaction between the auditory cortex, behavior, and

Y.E. Cohen (✉)
Department of Otorhinolaryngology, Perelman School of Medicine,
University of Pennsylvania, 3400 Spruce St – 5 Ravdin, Philadelphia, PA 19104, USA
e-mail: ycohen@mail.med.upenn.edu

Y.E. Cohen et al. (eds.), *Neural Correlates of Auditory Cognition*, Springer Handbook of Auditory Research 45, DOI 10.1007/978-1-4614-2350-8_1,
© Springer Science+Business Media New York 2013

neural activity (0.3%). Thus, it is timely to produce a volume that brings together this exciting literature on the neural correlates of auditory cognition.

This volume builds on and complements many of the outstanding previous volumes in the SHAR series, in particular Volumes 9 and 43 that focus on psychophysics and neural activity in the human auditory cortex, respectively (Yost et al., 1993; Poeppel et al., 2012). In this volume, we emphasize those literatures that *combine* measurements of behavioral (psychophysical) responses with measurements neural activity. Because the auditory cortex is modulated by cognitive factors (e.g., attention, memory, and reward values) and ongoing task demands, their effect on the neural code can be tested only when behavioral responses and neural activity are simultaneously recorded. This relationship between behavior and neural activity is one of the most significant issues in auditory neuroscience (Micheyl et al., 2007; Schnupp & Bizley, 2010) and, more generally, in sensory and cognitive neuroscience (Parker & Newsome, 1998).

1.2 Overview of Chapters in the Volume

Chapters 2 and 3 by Rauschecker and by Scheich and Brosch set the stage for the remaining chapters in the book. In Chapter 2, Rauschecker reviews the hypothesis that the auditory cortex contains two functional processing streams: one pathway devoted to the neural computations underlying spatial processing (where is the sound?) and the other pathway devoted to computations underlying auditory-object processing (what is the sound?). Since their description in the late 1990s, these "what–where" pathways have become a theoretical framework for many studies of auditory cognition. Rauschecker also presents a more nuanced theory of the "where" pathway. This new theory posits that the "where" pathway segregates and groups acoustic stimuli into auditory objects and then maps actions onto these identified auditory objects. In other words, the dorsal pathway mediates audiomotor processing.

In Chapter 3, Scheich and Brosch review the contribution of the auditory cortex to cognitive components of auditory behavior: that is, the modulation of auditory neurons by non-acoustic events such as response selection, reward value, and behavioral choice. Scheich and Brosch also lay out a learning-theory framework that forms a basis for understanding how these non-acoustic events are represented in early sensory areas. These ideas are particularly fascinating when they are contrasted with other studies that have not identified cognitive-related activity in the auditory cortex (Tsunada et al., 2011). Do different task demands and different forms of learning differentially engage the auditory cortex? Finally, the ideas discussed by Scheich and Brosch are also interesting when compared with the visual system (Gold & Shadlen, 2007). In the visual system, response selection is associated with computations that occur in the parietal cortex. Do these differences reflect hierarchical or functional differences between the sensory cortices? Or is there something unique about the temporal dynamics of auditory stimuli?

The next section of this volume expands on the topics raised by Rauschecker in Chapter 2. In Chapter 4, Recanzone discusses the role of the auditory cortex in spatial processing and localization. Much of sound localization is thought to occur in the midbrain. However, as discussed by Recanzone, the cortex is vital for the perception of a sound's location (as opposed to reflex spatial orienting that is mediated by the midbrain). Recanzone also highlights the hierarchical nature of auditory-spatial processing that occurs in different fields of the auditory cortex: between the core and belt regions of the auditory cortex, neural activity becomes more correlated with the psychophysical thresholds of a listener.

Chapter 5 by Schnupp, Honey, and Willmore focuses on auditory-object perception and the neural correlates underlying this form of perception. In reference to Bregman's seminal book (Bregman, 1990), the study of auditory objects is called "auditory-scene analysis." Auditory-scene analysis refers to the auditory system's capacity to first detect regularities in the acoustic environment (e.g., the onset and offset of auditory stimuli or the harmonicity of a multi-tone stimulus) and then use these regularities as a basis for grouping and segregating auditory stimuli into perceptual representations (i.e., auditory objects). In other words, auditory objects are the computational result of the auditory system's ability to detect, extract, segregate, and group the spatiotemporal regularities in the acoustic environment.

Chapter 6 by Steinschneider and Chapter 7 by Ghazanfar and Chandrasekaran focus on the neural processing and perception underlying a specific type of auditory object: vocal-communication signals. Because vocal communication is a fundamental component of both human and nonhuman animal behavior, it is likely that these acoustic signals had a significant influence on the evolution of the auditory system (Ghazanfar & Hauser, 1999). Indeed, the evolutionary forces that act on the cortex for vocalization processing are thought to be primarily related to the information that is transmitted by these vocalizations (Hauser, 1997). However, because vocalizations are also auditory objects. it is very likely that the computations that transform the acoustic representation of a vocalization into a perceptual auditory object have also been important in the evolution of the auditory cortex (Miller & Cohen, 2010).

Steinschneider's chapter focuses on how listeners extract phonemes—the smallest unit of human speech—from the acoustic environment. In our daily lives, this ability seems automatic and even trivial. However, on reflection, it is clear that this ability is not trivial. We are able to comprehend phonemes despite the inherent variability in different speakers' accents and rate of speech, a hallmark of the brain's remarkable ability to categorize information. Moreover, an acoustic analysis of a phoneme indicates that, whereas its perception is stable, its acoustic content is fundamentally dependent on the acoustic signals that precede and follow it. Clearly then, underlying this apparently trivial ability is a rich and sophisticated set of neural computations.

In Chapter 7, Ghazanfar and Chandrasekaran overview the multimodal nature of communication and the neural signals mediating multimodal communication. One of the best examples of the multimodal nature of communication is the McGurk effect (McGurk & MacDonald, 1976). In the McGurk effect, an observer hears one communication signal and views an individual mouthing a different signal.

The percept, though, is neither the auditory signal nor the visual signal. Instead, it is a morphed version of both. Chapter 7 reviews how primates perceive and respond to these auditory and visual signals as a coherent communication signal, how extra-communicative information (e.g., body size) can be extracted from these signals, and how the brain codes and integrates these dynamic signals.

The final three chapters of the book focus on attention, memory, and plasticity, themes that are initially introduced by Scheich and Brosch in Chapter 3. Whereas memory has an important role in all sensory systems, its role in audition is unique. Unlike visual stimuli, the primary feature of auditory stimuli is that they change over time. Consequently, it is imperative for a listener to keep track of past auditory events to predict the future stimulus accurately. Indeed, it has been posited that this ability to remember past auditory events and predict future auditory events is the primary basis by which the auditory system creates perceptual auditory objects (Winkler et al., 2009).

Attention and memory are often intertwined. A person's ability to orient his or her attention to a sudden loud sound (e.g., the sound of a police siren) does not require memory resources because this ability is a "bottom-up" automatic process. However, to tag that sound with a name requires that you compare that sound with your memory of sounds and semantic memories. In contrast, when searching for a lost child, the memory of the child's voice needs to be maintained on-line (so-called working memory) while shifting attention from one voice to another in search of the child. These interdependencies of attention and memory and the neural codes that underlie auditory attention and memory are discussed by Poremba and Bigelow in Chapter 8.

As discussed in Poremba and Bigelow's chapter, most studies of attention and memory have shown that these cognitive processes simply modulate the tuning properties of a neuron but do not fundamentally alter them. For example, when a listener attends to a 9-kHz tone, the firing rates of neurons in the auditory cortex that are tuned for 9 kHz increase. However, Chapter 9 raises the possibility that this idea may need refining. Fritz, David, and Shamma review a body of literature indicating that the tuning properties of a neuron are plastic and are functionally dependent on the immediate needs of a listener. In other words, as the statistical properties of the environment change and the attentional demands of the listener change, a neuron's tuning properties (e.g., selectivity for a particular frequency, sensitivity to a frequency range, etc.) rapidly change to meet these ongoing task and contextual demands.

The final chapter (Chapter 10) by Shepard, Kilgard, and Liu continues with the theme of plasticity and how a listener's experience has a profound impact on neural-response properties and how auditory information is organized and represented in the brain. Unlike the chapter by Fritz et al., this chapter focuses on long-term plastic changes in auditory representations and the role that the environment plays in shaping these representations. For example, an enriched environment that contains numerous types and kinds of sounds induces one form of plastic change, whereas an environment with only one kind of sound or unstructured sounds can induce a different form of plastic change. Chapter 10 also touches on the importance of ethological stimuli and how the interplay between ethological stimuli and non-neural signaling (e.g., hormones) can profoundly impact plasticity. Finally, Chapter 10 discusses the role of mapped representations of auditory information (e.g., frequency). Is more better? Does an expanded representation of a particular frequency imply that an animal can detect that frequency better? Are these changes permanent or do they only reflect intermediate stages in memory or learning?

1.3 Conclusion and Future Work

The ideas, hypotheses, and data that are reviewed in this volume raise more questions than they answer. What is the causal relationship between auditory activity and auditory perception and cognition? How do the "what–where" pathways interact to form coherent representations that can guide executive behavior? What factors determine a neuron will be modulated by cognitive factors? Ultimately, a perceptual decision is formed: Was that sound *bing* or was it *bong*? Although some studies have touched on decision-related activity, the auditory community does not have any computational or neural models that relate psychophysical decisions to neural activity. A challenge for the community is to develop and test such models. Finally, most people live in a noisy environment. What are the neural mechanisms that allow a listener to segregate, detect, and hear a target sound that is embedded in a noisy background? Together, answers to these questions and those raised in the chapters of this volume will help to delineate the neural mechanisms underlying hearing and auditory cognition.

References

Bregman, A. S. (1990). *Auditory scene analysis*. Boston, MA: MIT Press.

Ghazanfar, A. A., & Hauser, M. D. (1999). The neuroethology of primate vocal communication: Substrates for the evolution of speech. *Trends in Cognitive Science*, 3, 377–384.

Gold, J. I., & Shadlen, M. N. (2007). The neural basis of decision making. *Annual Review of Neuroscience*, 30, 535–574.

Griffiths, T. D., & Warren, J. D. (2004). What is an auditory object? *Nature Reviews Neuroscience*, 5, 887–892.

Hauser, M. D. (1997). *The evolution of communication*. Cambridge, MA: MIT Press.

McGurk, H., & MacDonald, J. (1976). Hearing lips and seeing voices. *Nature*, 264(5588), 746–748.

Micheyl, C., Carlyon, R. P., Gutschalk, A., Melcher, J. R., Oxenham, A. J., Rauchecker, J. P., et al. (2007). The role of auditory cortex in the formation of auditory streams. *Hearing Research*, 229, 116–131.

Miller, C. T., & Cohen, Y. E. (2010). Vocalization processing. In A. Ghazanfar & M. L. Platt (Eds.), *Primate neuroethology* (pp. 237–255). Oxford: Oxford University Press.

Parker, A. J., & Newsome, W. T. (1998). Sense and the single neuron: Probing the physiology of perception. *Annual Review of Neuroscience*, 21, 227–277.

Poeppel, D., Overath, T., Popper, A. N., & Fay, R. R., Eds. (2012). *The human auditory cortex*. New York: Springer.

Schnupp, J. W., & Bizley, J. K. (2010). On pitch, the ear and the brain of the beholder. *Journal of Neurophysiology*, 103, 1708–1711.
Tsunada, J., Lee, J. H., & Cohen, Y. E. (2011). Representation of speech categories in the primate auditory cortex. *Journal of Neurophysiology*, 105, 2634–2646.
Winkler, I., Denham, S. L., & Nelken, I. (2009). Modeling the auditory scene: Predictive regularity representations and perceptual objects. *Trends in Cognitive Science*, 13, 532–540.
Yost, W. A., Popper, A. N., & Fay, R. R., Eds. (1993). *Human psychophysics*. New York: Springer.

Chapter 2
Processing Streams in Auditory Cortex

Josef P. Rauschecker

Keywords Anterior ectosylvian • Bandpass noise • Combination sensitivity • Frequency modulation • Functional MRI • Inferior frontal • Inferior parietal lobule • Lateral belt • Magnetoencephalography • Medial geniculate • Monkey calls • PET imaging • Prefrontal cortex • Premotor cortex • Superior temporal

Abbreviations

A1	primary auditory cortex
AEA	anterior ectosylvian auditory area
AES	anterior ectosylvian sulcus
AL	anterolateral area
aST	anterior superior temporal
BA	Brodmann area
BOLD	blood oxygen level–dependent
BP	band-passed
BPN	band-passed noise
CD	compact disk
CL	caudolateral area
CM	caudomedial belt field
cs	central sulcus
CS	combination sensitivity
DCN	dorsal cochlear nucleus
DLPFC	dorsolateral prefrontal cortex
dPMC, vPMC	dorsal and ventral premotor cortex

J.P. Rauschecker (✉)
Department of Neuroscience, Georgetown University Medical Center,
New Research Building, Room WP19, Box 571460, Washington, DC 20057, USA
e-mail: rauschej@georgetown.edu

Y.E. Cohen et al. (eds.), *Neural Correlates of Auditory Cognition*, Springer Handbook of Auditory Research 45, DOI 10.1007/978-1-4614-2350-8_2,

© Springer Science+Business Media New York 2013

FM	frequency-modulated
fMRI	functional magnetic resonance imaging
IFC	inferior frontal cortex
IFG, SFG	inferior and superior frontal gyrus
IPL	inferior parietal lobule
ITDILD	interaural time and level differences
LB, MB	lateral and medial belt
LIP	lateral intraparietal area
MCs	monkey calls
MCPI	monkey call preference index
MEG	magnetoencephalography
MGd	dorsal nucleus of the medial geniculate
MGm	medial nucleus of the medial geniculate
ML	middle lateral area
MMN	mismatch negativity
MSO, LSO	medial and lateral superior olive
NSF	National Science Foundation
PAF	posterior auditory field
PET	positron emission tomography
PFC	prefrontal cortex
PMC	premotor cortex
PPC	posterior parietal cortex
pre-SMA	presupplementary motor area
pST	posterior superior temporal
PT	planum temporale
R, RL	rostral (=rostrolateral) field
SC	superior colliculus
STG	superior temporal gyrus
STP	supratemporal plane

2.1 Hierarchical Processing of Communication Sounds in the Auditory Ventral Stream

2.1.1 Auditory Communication as a Pattern Recognition Problem

Auditory communication involves the decoding of complex sounds and the assignment of specific sounds to behaviorally relevant meanings. While the latter problem has long been discussed in the realms of ethology and linguistics, the former should be solvable more immediately by neurophysiology. But even the phonetic decoding problem is a difficult one, comparable in complexity with visual pattern recognition. Like in vision, it involves the combination of a multitude of features that occur simultaneously and in temporal sequence. The unique combination of features in the spectral and temporal domain, therefore, characterizes a specific communication sound. This alone, however, is not enough. The feature representation has to be robust and invariant against spurious changes and distortions caused by unpredictable influences.

The way such feature representations are generated in the brain is equally unclear. It could be that neurons in the auditory pathways become more and more specific for frequency, as one ascends the different levels of processing from the cochlea to auditory cortex and beyond. Such highly frequency-specific neurons could then be simultaneously active in concert when a complex sound, such as a communication sound, is present in the animal's environment. The opposite, however, is conceivable as well: Neurons at higher levels of the auditory pathways could become more and more broadly tuned for frequency, as they combine more and more inputs across the frequency domain. As a consequence, they would become more and more specific for certain sounds and unresponsive to others.

It appears that both processes are in fact happening. While the tuning of neurons for frequency is still relatively broad at the level of the auditory nerve, it becomes sharpened in the lemniscal pathways of brain stem and thalamus, with neurons in primary auditory cortex (A1) very narrowly tuned for frequency. As one travels beyond A1, however, the trend reverses dramatically. One of the discoveries of Merzenich and Brugge in their early mapping studies of auditory cortex in the rhesus monkey (*Macaca mulatta*) (Fig. 2.1a, b) was that neurons in the nonprimary, lateral part of the supratemporal plane (STP) responded only poorly to pure tones and were, therefore, hard to characterize (Brugge & Merzenich, 1973; Merzenich & Brugge, 1973). Merzenich and Brugge's assumption was that the neurons in these lateral fields might respond to more complex sounds. As was discovered later, this is indeed the case (Rauschecker et al., 1995): Neurons in the lateral belt (LB) clearly prefer band-passed (BP) noise bursts to tone pips and also respond well to other types of complex sounds, as reviewed here.

Therefore, the coding problem of auditory pattern recognition does not end at the level of primary auditory cortex; it actually begins there. The realization of the existence of a multitude of auditory cortical representations (Merzenich & Brugge, 1973; Reale & Imig, 1980) on one hand adds to the complexity of the problem but on the other hand may be an important part of the solution. Could the various cortical fields be specialized for certain aspects of the auditory world and represent different levels of analysis? This is the answer suggested by research on visual pattern and object recognition over the last 30 years, in which it has become abundantly clear that different cortical areas do indeed specialize in certain aspects of the visual world. The specialization is, of course, not absolute—in fact, there exists substantial overlap—but there is clear evidence for "families" of specialized areas that are organized into processing streams emphasizing distinct aspects of our natural surroundings.

2.1.2 Early Parallel Processing in Auditory Cortex

Parallel processing streams in auditory cortex start as early as the core areas: Area A1 and the rostral auditory area (R) are both koniocortical areas with neurons sharply tuned for frequency and tonotopic maps that are mirror-symmetric. Combined lesion and tracer studies (Rauschecker et al., 1997) have shown that both

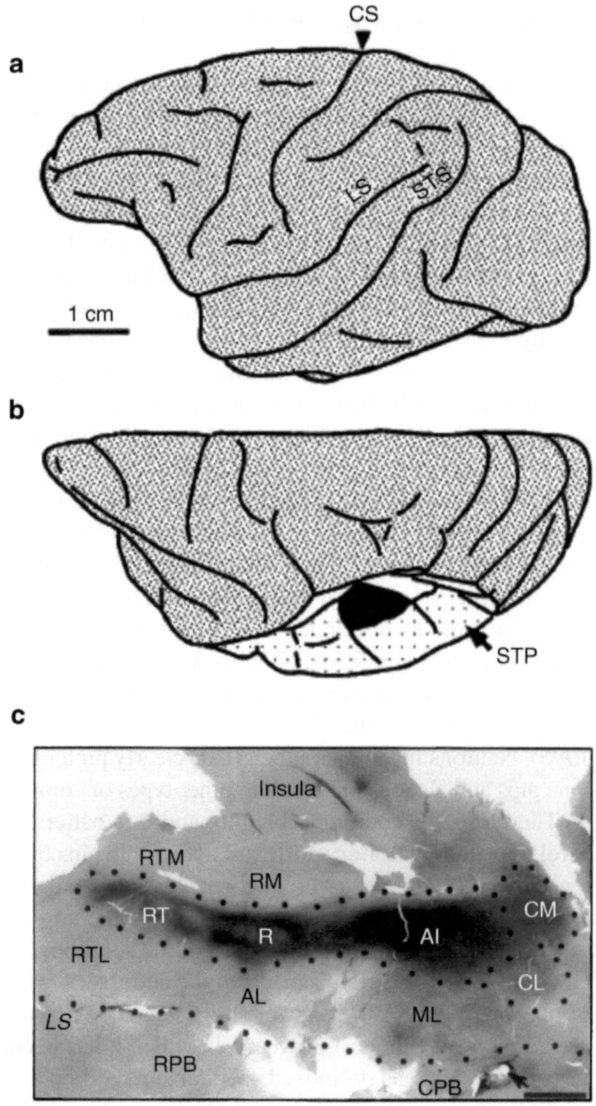

Fig. 2.1 Location of auditory cortical areas in the macaque brain. Left hemisphere is displayed. (**b**) Primary auditory cortex is shown marked in black on the supratemporal plane (STP) inside the lateral sulcus (LS). STS, superior temporal sulcus; CS, central sulcus. (**a**, **b** adapted from Merzenich & Brugge, 1973). (**c**) Histochemical identification of core and belt areas by parvalbumin staining in a flatmount of rhesus monkey auditory cortex. (**c** from Hackett et al., 1998)

cortical core areas receive input from the principal relay nucleus of the auditory thalamus, the ventral nucleus of the medial geniculate (MGv). By contrast, the other prominent area on the supratemporal plane of the macaque, the caudomedial area (CM), does not receive input from MGv but only from the medial and dorsal subnuclei of the medial geniculate (MGd and MGm). As a consequence, lesions of A1 lead to unresponsiveness of neurons in CM to tonal stimulation, but not of neurons in area R, which receive independent input from MGv. To be sure, the parallel input to areas of the supratemporal plane may start even more peripherally than the thalamus. Studies of the auditory brain stem indicate that the ventral and dorsal cochlear nuclei (VCN, DCN) have very different response characteristics and may ultimately give rise to different functions of hearing, including auditory pattern and space processing, respectively (Yu & Young, 2000).

2.1.3 Processing of Sounds with Intermediate Complexity in the Auditory Belt

As has been recognized early on the basis of cytoarchitectonics (Pandya & Sanides, 1972), the auditory region in the superior temporal cortex consists of a "core" with a koniocortical appearance surrounded by a "belt." The cytoarchitecture is matched by distinct histochemical differences (Kaas & Hackett, 2000; Hackett, 2011) that make the core stand out by dark staining compared to the belt (with intermediate staining) and another zone termed "parabelt" (PB) (Morel et al., 1993; Hackett et al., 1998) with very light staining (Fig. 2.1c) (see also Jones et al., 1995). Some processing characteristics of the belt are discussed first and are then compared with those of the core.

2.1.3.1 Selectivity for BP Noise

One fundamental finding that was secured in initial studies (Rauschecker et al., 1995) and confirmed in detail later (Rauschecker & Tian, 2004) was the enhanced response of LB neurons to BP noise compared to pure tones (Fig. 2.2a, b). This demonstrated the ability of LB neurons to integrate over a finite frequency spectrum in a facilitatory fashion. By comparison, this integrative ability is largely absent in A1 neurons, a significant difference that we return to later.

The finding of robust auditory responses to band-passed noise (BPN) stimuli in LB neurons also was of great practical value, however, because it permitted systematic mapping of the LB. BPN bursts have a clearly defined center frequency as well as a defined bandwidth. Mapping of the LB along the rostrocaudal dimension reveals a smooth gradient for best center frequency with two reversals (Rauschecker et al., 1995; Rauschecker & Tian, 2004). This means that there exist three cochleotopically organized areas within the LB, which were termed the anterolateral, middle lateral, and caudolateral areas (AL, ML, and CL; Fig. 2.2c).

Not only do LB neurons integrate over frequency, but they do so in a rather specific way, which produces the best response at a specific "best bandwidth" (BBW) (Fig. 2.3a). Presumably, this is the result of intricate interactions between excitatory and inhibitory inputs. BBWs in the LB are about equally distributed over the whole bandwidth spectrum, whereas A1 neurons clearly prefer pure tones to BP noise. Thus, there was a clear trend for BBW to increase from core toward belt (Rauschecker & Tian, 2004). The same was later found for medial belt (Kusmierek & Rauschecker, 2009).

Neurons with selectivity for the center frequency and bandwidth of BPN bursts are ideally suited to participate in the decoding of communication sounds. Such sounds contain many instances of BPN bursts (Fig. 2.4), not only in rhesus monkeys, but also in many other species (Wang, 2000), including humans. BPN detectors

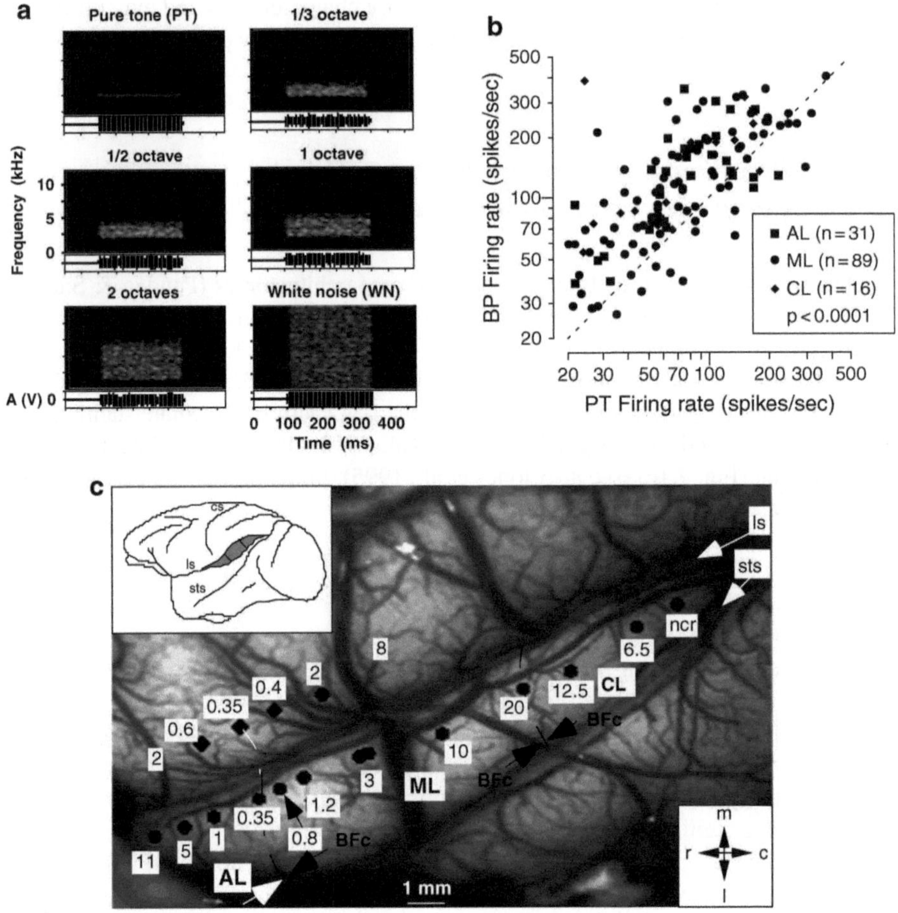

Fig. 2.2 Characterization of lateral belt (LB) areas in rhesus monkey auditory cortex. (**a, b**) Preference of neurons in the three LB areas (AL, ML, and CL) for band-passed (BP) noise bursts over tone pips (PT). Almost invariably BP stimuli elicit a higher spike rate in the same neurons than PT stimuli. (**c**) Cochleotopic organization of LB areas. ls, lateral sulcus; cs, central sulcus; sts, superior temporal sulcus. (From Rauschecker et al., 1995)

Fig. 2.3 Selectivity of LB neurons for sounds of intermediate complexity. (**a**) Responses (in spikes per second) of two auditory cortical neurons to BP noise bursts with different bandwidth. Maximum firing rate is plotted diagrammatically against bandwidth. Filled symbols show net firing rate after subtracting spontaneous (base) firing rate (shown in open symbols). (From Rauschecker & Tian, 2004.) (**b, c**) Response of a typical LB neuron to FM sweeps of different rate and direction. Peristimulus time histograms and raster dot displays are shown above a schematic display of the respective stimuli. Upward FM directions are shown in (**b**), downward directions in (**c**). FM rates are displayed on top of each column. A clear preference for a highly specific FM rate and for FM sweeps in the upward direction is found. (From Tian & Rauschecker, 2004)

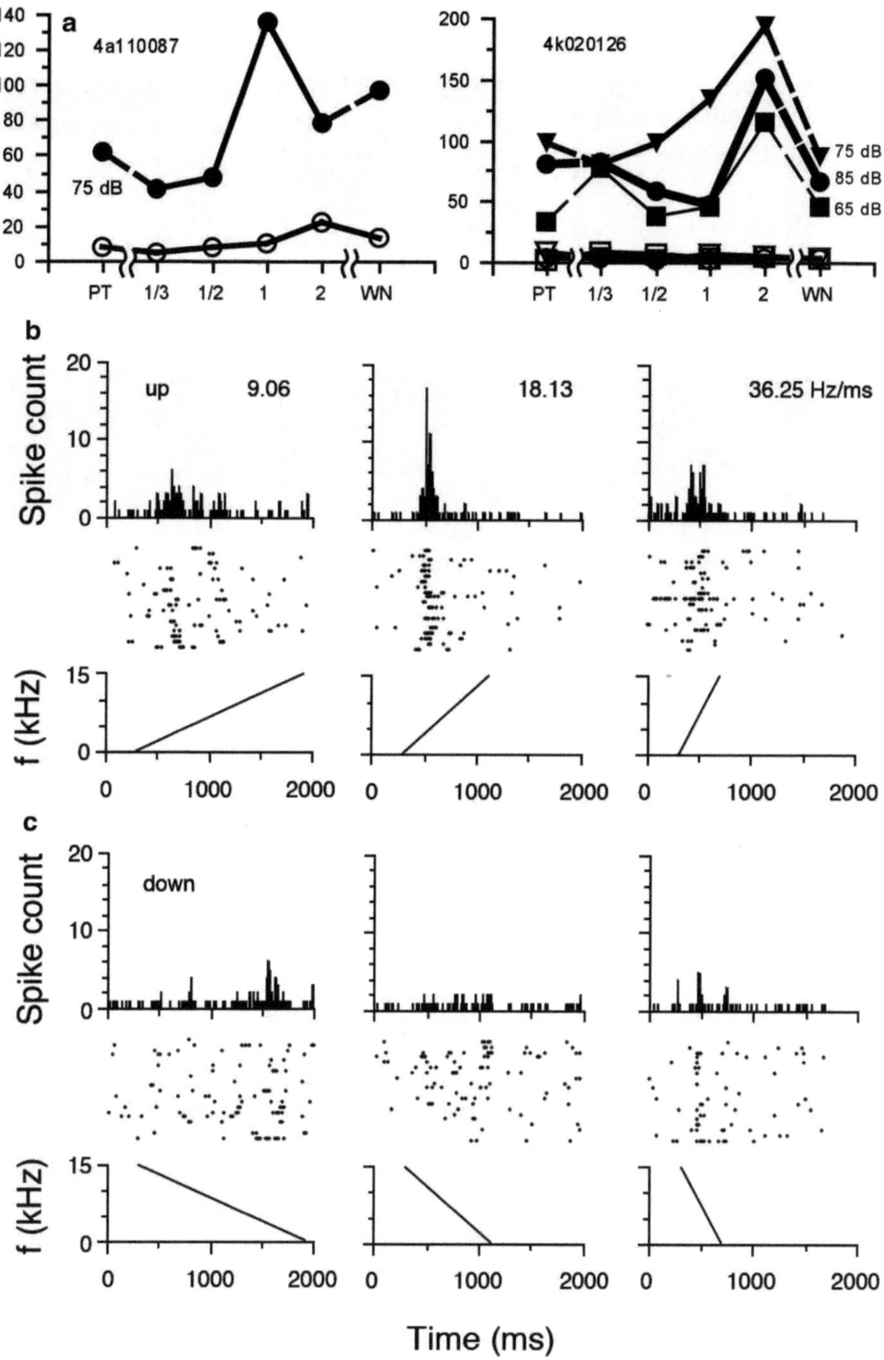

Fig. 2.3 (continued)

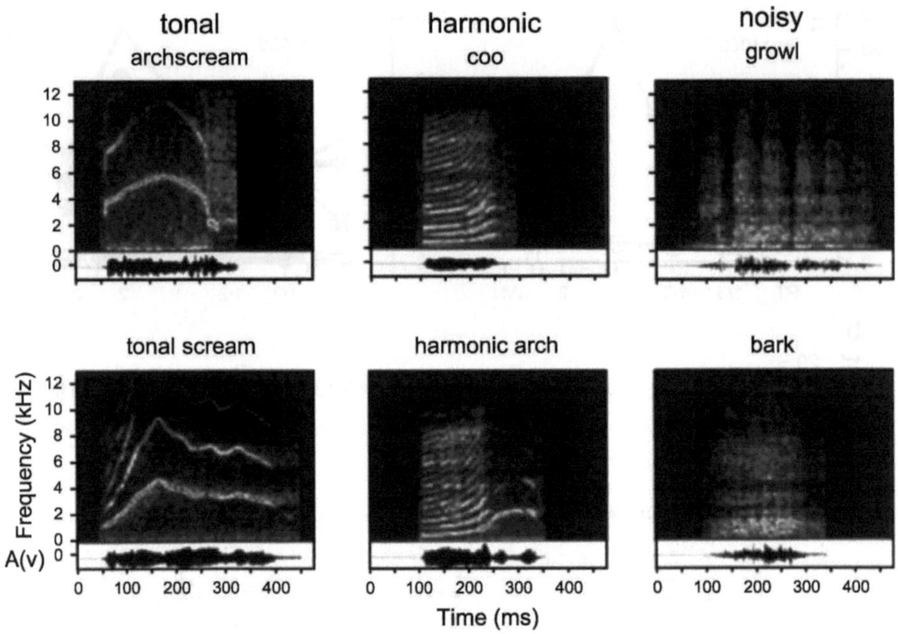

Fig. 2.4 Six examples of rhesus monkey calls in three phonetic-acoustic categories. First column: tonal calls; second column: harmonic calls; third column: noisy calls. Spectrograms are shown with corresponding time signals underneath. (From Rauschecker, 1998b)

would, therefore, almost have to be included in the repertoire of feature detectors dealing with communication sounds. To perform such a task adequately, however, the feature detectors would have to preserve their selectivity regardless of sound intensity. Indeed, as one of the examples in Fig. 2.3a shows, LB neurons generally do prefer the same center frequency at different intensities.

2.1.3.2 Selectivity for Frequency-Modulated Sweeps

Other features that are highly typical for communication sounds in most species are changes in frequency over time ("frequency-modulated [FM] sweeps"), also sometimes referred to as chirps or glides. FM sweeps are characterized by two parameters: FM rate and direction. Neurons in the LB are highly selective for both parameters (Tian & Rauschecker, 2004). First, 94% of LB neurons responded to FM stimuli in at least one direction. To characterize FM direction selectivity (DS) quantitatively, a DS index was calculated. A neuron was considered direction selective when the response in one FM direction for one or more FM rate was at least twice as large as that in the other direction (Mendelson & Cynader, 1985). An example is shown in Fig. 2.3b and c. About 60% of LB neurons were classified as direction-selective on the basis of this criterion, with roughly equal proportions of neurons preferring upward and downward directions (Tian & Rauschecker, 2004).

Even more striking was the selectivity of LB neurons for FM rate. Various types of FM-rate tuning can be discerned in the LB, including high-pass, low-pass, and band-pass tuning (Tian & Rauschecker, 2004). Neurons tuned to both FM direction and FM rate, like the one in Fig. 2.3, would be ideal candidates for the extraction of communication-sound features, such as formant transitions in human speech. Preferred FM rate differed markedly between the three LB areas. AL neurons preferred lower FM rates: More than half of AL neurons had their preferred FM rates below 64 Hz/ms, with medians of 25 and 50 Hz/ms for upward and downward sweeps, respectively. CL neurons, in contrast, preferred higher FM rates: About 70% of CL neurons preferred FM rates above 64 Hz/ms, with medians of 160 Hz/ms for both directions. ML neurons preferred FM rates in between.

According to these differences, AL neurons would be very well suited to participate in the decoding of species-specific vocalizations, which range mostly between 8 and 50 Hz/ms (Hauser, 1996; Rauschecker, 1998b). The various harmonics in the widely occurring "coo" calls fall between 10 and 40 Hz/ms. Only some of the "screams" contain FM rates above 100 Hz/ms (tonal scream: 103 Hz/ms; arch scream: 314 and 826 Hz/ms for the downward portion). Some of the neurons in AL do include responses to these faster sweeps. It is noteworthy that screams play an important role as alarm calls, which have to be well localizable by members of the same species.

In its role, area AL can be likened to visual area V4, which contains neurons selective for the size of visual objects (Desimone & Schein, 1987) and plays a pivotal role in the ventral visual "what" stream. Just as inferotemporal cortex, which receives input from V4, constitutes the later stages of visual object recognition (Lueschow et al., 1994), neurons in the rostral parabelt (and further anterior in the STG) are expected to rely on input from AL, compute invariances against distortions, and ensure perceptual constancy.

2.1.4 Selectivity for Species-Specific Calls

Neurons in the LB responded more vigorously to time-variant FM sweeps than to tones of constant frequency, and FM sweeps were also generally more effective than BPN bursts. LB neurons were also tested directly with whole monkey calls (MC; Fig. 2.4) or components thereof (Fig. 2.5). Just as BPN bursts and FM sweeps, MC stimuli elicited more vigorous responses in LB than pure tones. MC stimuli were also generally more effective than BPN bursts but not necessarily more so than FM sweeps, which often remained the best stimuli.

2.1.4.1 Nonlinear Integration Mechanisms

LB neurons responded differentially to different types of MC. Although calls often had the same or comparable bandwidths, neuronal responses differed. Response selectivity, therefore, must be based on features contained in the phonetic fine

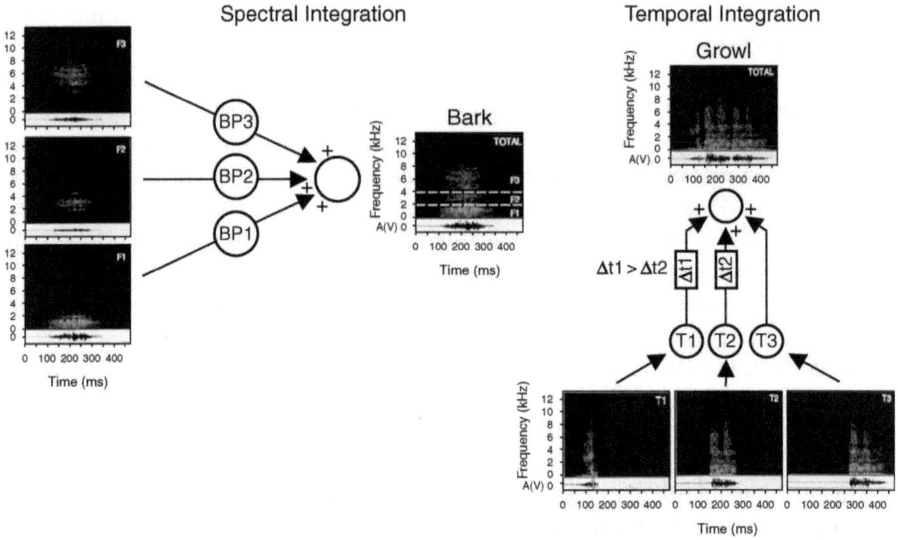

Fig. 2.5 Models of nonlinear spectral and temporal integration ("combination sensitivity") in neurons of the LB (and presumably PB). Selectivity for complex sounds, such as specific communication calls, is created by combining inputs from lower-order neurons that are BP-selective and/or FM-rate and FM-direction-selective. Delay lines need to be implemented, as explained in the text, to create temporal combination sensitivity. (Extracted from Rauschecker & Scott, 2009)

structure of the calls. The conclusion is not far-fetched that it is the combination of features that causes a cell to respond to a specific type of call and not to others. Indeed, two fundamental mechanisms were identified as causing neuronal selectivity: nonlinear summation (a) in the spectral domain ("spectral facilitation," SFA) and (b) in the temporal domain ("temporal facilitation," TFA). This corresponds to spectral and temporal combination sensitivity (CS), respectively, as it has been described previously in other species, such as bats, frogs, and songbirds (Suga et al., 1978; Narins & Capranica, 1980; Margoliash & Fortune, 1992).

In spectral CS or SFA, inputs from lower-order neurons, such as band-pass–selective neurons, are combined in the frequency domain (Fig. 2.5, top). In temporal CS or TFA, inputs are combined in the time domain (Fig. 2.5, bottom). However, both mechanisms are based on the same principle: coincidence detection by cell membranes with a relatively high threshold, that is, a logical AND-gate principle. Only with all inputs present simultaneously a response is evoked; with one input alone, no response follows. This explains why single components or syllables within a call usually are not sufficient to elicit a response. Temporal summation is accomplished by introducing staggered delays in the input pathways transmitting the early components, so all inputs eventually arrive simultaneously at the higher-order target neuron.

In some cases, however, the opposite is found: LB neurons respond decently to single components, but the response is suppressed by presenting the whole call. This is referred to as spectral or temporal suppression (SSU or TSU), respectively. We currently do not fully understand the significance of such units but assume that they are part of the logical alphabet implemented at that level.

One of the most striking differences between core and belt areas identified so far is the difference in their ability to nonlinearly integrate information both in the spectral and the time domain. Whereas more than half of the neurons in LB show some form of nonlinear interaction (SFA, SSU; TFA, TSU), only approximately 10% (or less) of the neurons in A1 or R display the same form of behavior. This demonstrates a quantum leap in the processing characteristics of auditory cortex and is one of the strongest arguments for a hierarchical organization in auditory cortex.

2.1.4.2 MC and Spatial Selectivity

To quantify MC selectivity in different LB areas, a monkey call preference index (MCPI) was calculated depending on the number of calls to which the neuron responds (Tian et al., 2001). In most cases, a standard battery of seven of the most frequently occurring calls was used. An MCPI of seven, therefore, means that the cell responded to all the calls presented. An MCPI of three or less corresponds to a cell that responded to fewer than half of the calls and can be termed "MC-selective," whereas cells that responded to five or more of the calls are termed "MC-nonselective." The LB areas differed in their degree of MC selectivity, as quantified on this basis (Tian et al., 2001). Area AL had the greatest percentage of highly selective neurons (MCPI ≤ 2), followed by ML, whereas CL had the smallest percentage of highly selective neurons. Naturally, for the most nonselective neurons (MCPI ≥ 6), the opposite was found: CL had the greatest percentage of such nonselective neurons, AL the least, with ML somewhere between those two extremes. Comparison of MC selectivity showed AL to be more selective than both ML and CL ($p = 0.0006$ and $p = 0.0287$, respectively, Mann–Whitney *U*-test). This difference was also highly significant when all three areas were compared together ($p = 0.0026$, Kruskal-Wallis, df = 2).

Together with the finding that spatial tuning in neurons of the LB shows the opposite areal distribution—highest selectivity is found in CL and lowest in AL (see Section 2.2)—this has led to the hypothesis that AL and CL, which lie on opposite ends of the LB along its rostrocaudal extent, form the beginning of two pathways for the processing of auditory object and space information (Rauschecker & Tian, 2000; Tian et al., 2001). Selectivity along the anteroventral stream increases further toward more anterior locations (Kikuchi et al., 2010). This trend extends all the way to the temporal pole, which is auditorily activated in the macaque and shows a hemispheric difference for species-specific communication sounds (Poremba et al., 2003, 2004).

2.1.5 Auditory Belt Projections to Prefrontal Cortex

An anatomical study in rhesus monkeys, combined with physiological mapping of LB areas, has demonstrated the existence of largely separate pathways originating in the LB and projecting to different target regions in the prefrontal cortex (Romanski

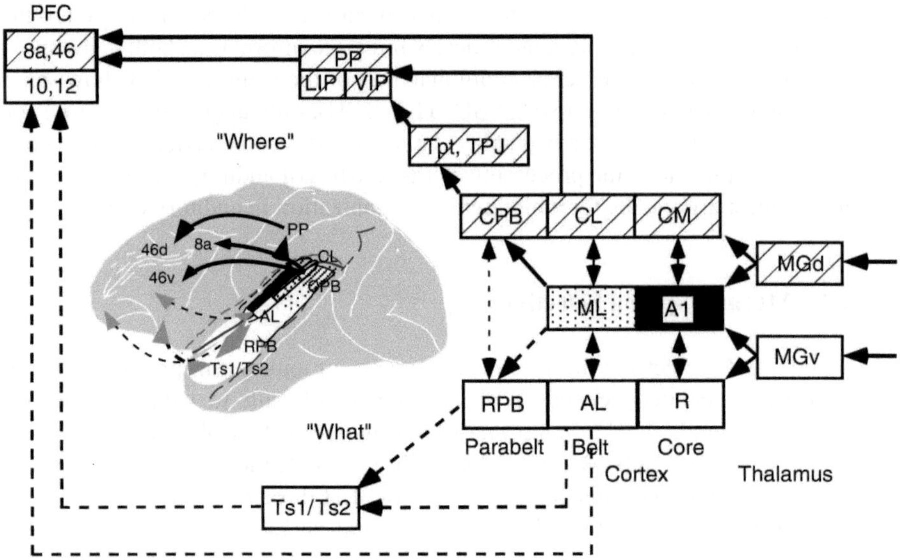

Fig. 2.6 Schematic diagram of dual auditory cortical pathways in primates representing auditory object/pattern ("what") processing in an anteroventral projection and auditory space ("where") processing in a posterodorsal projection. (Modified and expanded from Rauschecker, 1998a; Rauschecker & Tian, 2000; Rauschecker, 2007). The projections of the posterodorsal stream are highlighted in solid lines; participating cortical areas are marked with oblique lines. The anteroventral pathway is shown in dashed lines. Areas that are not uniquely participating in either pathway are shown in dark blocks (primary auditory cortex, A1) or stippled (middle lateral belt area, ML). Prefrontal connections of the LB are also shown directly on a lateral view of a rhesus monkey brain. (From Romanski et al., 1999.) MGd, medial geniculate nucleus, dorsal division; MGv, medial geniculate nucleus, ventral division; CM, caudomedial area; R, rostral area; CL, caudolateral area; CPB, caudal parabelt area; RPB, rostral parabelt area; Tpt, temporoparietal area; TPJ, temporoparietal junction; PP, posterior parietal cortex; LIP, lateral intraparietal area; VIP, ventral intraparietal area; Ts1, Ts2, rostral temporal areas of Pandya and Sanides (1972); PFC, prefrontal cortex. Brodmann areas are abbreviated with their respective numbers

et al., 1999) (see Fig. 2.6). In this study, three different fluorescent tracers were injected into matched frequency regions of the three LB areas after these had been physiologically mapped. Injections into area AL produced label in ventrolateral and orbital regions of prefrontal cortex, whereas CL injections led to labeling of dorsolateral prefrontal cortex (DLPFC). The latter is known for its involvement in spatial working memory, whereas the former regions are assumed to participate in object working memory (Goldman-Rakic, 1996).

These projection patterns conform to the physiological response properties found in the aforementioned study of Tian et al. (2001), which assigned superior selectivity for auditory patterns and space to areas AL and CL, respectively. The studies by Tian et al. (2001) and Romanski et al. (1999), therefore, form the cornerstones of the theory according to which dual processing streams in nonprimary auditory cortex underlie the perception of auditory objects and auditory space (Rauschecker & Tian,

2000): According to the tracer results, the anteroventral pathway originates from area AL of the LB. Recent physiological data indicate, however, that this pathway may have its origin already in the rostral auditory core area R (Kusmierek et al., 2012). The anteroventral stream projects further from AL via the rostral STG and STS into ventrolateral prefrontal cortex (VLPFC). The anteroventral pathway forms the main substrate for auditory pattern recognition and object identification. An auditory domain is found in VLPFC, in which neurons show responses to complex, nonspatial sounds, including animal and human vocalizations (Romanski & Goldman-Rakic, 2002; Cohen et al., 2009). By contrast, another pathway projecting caudodorsally into posterior parietal cortex and DLPFC is thought to be involved in auditory spatial processing (as discussed in Section 2.2).

2.1.6 Human Imaging Studies

Human neuroimaging studies have confirmed the organization of auditory cortex into core and belt areas by using the same types of stimuli as in the monkey studies (Wessinger et al., 2001; Chevillet et al., 2011). A core region, robustly activated by pure-tone stimuli, is found along Heschl's gyrus. This pure-tone responsive region in human auditory cortex is surrounded by belt regions both medially and laterally, which are activated preferentially by BPN bursts, corresponding to results in nonhuman primates (Rauschecker et al., 1995; Kusmierek & Rauschecker, 2009). Finally, a region activated by vowel sounds was identified more anterolaterally (Chevillet et al., 2011) (Fig. 2.7).

Various findings from human neuroimaging strongly support the dual-stream hypothesis of auditory processing: Anterolateral areas of the superior temporal cortex are activated by intelligible speech (Scott et al., 2000; Binder et al., 2004) or speech-like sounds (Binder et al., 2000; Alain et al., 2001; Maeder et al., 2001), whereas caudal belt and parabelt areas (projecting up dorsally into posterior parietal cortex) are activated by auditory spatial discrimination tasks (see Section 2.2). Some of the areas in anterior human STG do seem to represent species-specific sounds, because they light up only with speech but not with animal calls (Fecteau et al., 2004), whereas others may encode more general auditory object information (Zatorre et al., 2004; Leaver & Rauschecker, 2010).

Thus it becomes increasingly clear that behaviorally relevant auditory patterns are discriminated in an anterior auditory "what"-stream. Especially with regard to speech it had long been assumed that these processes are located posteriorly in a region called the planum temporale or "Wernicke's area." These views were largely based on human stroke studies performed more than a century ago (Galaburda et al., 1978). By contrast, a recent meta-analysis of human speech processing reviewing more than 100 neuroimaging studies of phoneme, word, and phrase recognition confirms an anterior rather than posterior location of "Wernicke's area", as defined in this way (DeWitt & Rauschecker, 2012).

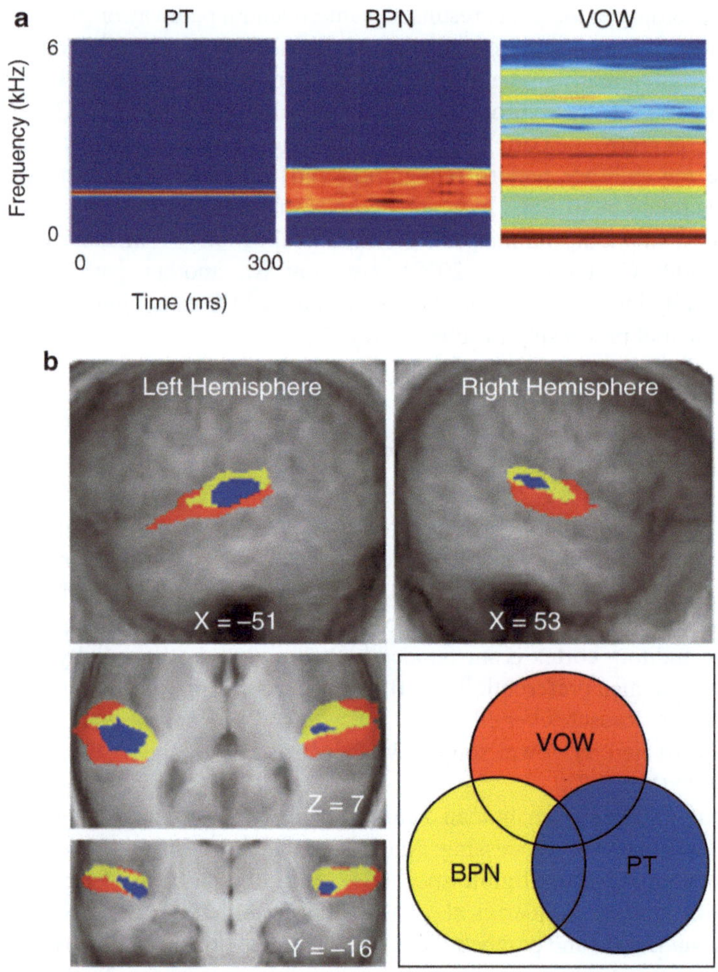

Fig. 2.7 Three functionally discrete regions identified in human auditory cortex using blood oxygen level–dependent (BOLD) imaging and sounds with different spectral complexity. (**a**) Example stimuli used. Subjects were presented with pure tones (PT), band-passed noises (BPN), and species-specific vocalizations (vowel-like speech sounds, VOW), based on the ability to differentiate core, belt and parabelt using these stimulus classes in previous nonhuman primate studies. (**a**) Random-effects group analysis ($n = 13$). A hierarchy of three functionally separable regions is clearly visible in each hemisphere. (From Chevillet et al., 2011)

In conclusion, it appears that, like in the visual system, studies of nonhuman primates can serve as excellent models for human studies. Conversely, imaging studies (in both humans and monkeys) can provide useful guidance for microelectrode studies in nonhuman primates, which permit analyses at much higher spatial and temporal resolution than would be possible in most human studies, with some exceptions (Howard et al., 2000).

2.1.7 Summary

Contrary to common belief, which places speech perception in posterior regions of the superior temporal gyrus (STG), converging evidence from three lines of research suggests that communication sounds in both human and nonhuman primates are processed along an anteroventral axis in the STG: (1) Neurophysiological single-unit studies in the rhesus monkey (*Macaca mulatta*) find increased selectivity for species-specific vocalizations along the auditory ventral stream starting with the rostral area (R). (2) The anterolateral area (AL) of auditory belt cortex sends direct anatomical projections to the ventrolateral prefrontal cortex (VLPFC), which has previously been implicated in working memory for patterns and objects. (3) Neuroimaging studies in humans demonstrate that activation associated with the processing of phonemes and words (as well as other auditory "objects") is consistently localized in areas of the ventral stream.

Thus emerges a picture of an anteriorly directed hierarchical processing stream dedicated to the identification and recognition of behaviorally relevant auditory patterns, which include those used for communication. Activity of neurons in this pathway signals "what" a complex sound represents. Areas R and AL constitute an early stage in this process. Neurons at the level of the belt participate in the decomposition and recombination of auditory features, as they are contained in complex sounds, including conspecific communication calls and human speech. Examples of such features are BP noise bursts and FM sweeps, for which neurons in AL are selective. Neurons in the rostral parabelt (and further anterior in the ventral stream) are expected to rely on input from anterolateral belt, compute invariances against distortions, and ensure perceptual constancy.

2.2 Processing of Space and Motion in the Auditory Dorsal Stream

2.2.1 Brain Stem Processing of Auditory Space

The superior colliculus (SC) is often considered the seat of auditory spatial perception, because it contains a map of auditory space. The SC receives its input from the inferior colliculus (IC), in particular the "external" nuclei of the IC, which in turn receive their input from the dorsal part of the cochlear nuclei (DCN). The medial superior olive (MSO) and the lateral superior olive (LSO) are responsible for encoding interaural time and level differences (ITD and ILD), respectively. For a more complete review of brainstem mechanisms of auditory space processing see Irvine (1992).

2.2.2 *Spatial Selectivity in Primary and Nonprimary Auditory Cortex*

Early studies have suggested a role for auditory cortex in sound localization (Diamond et al., 1956; Ravizza & Masterton, 1972; Heffner & Masterton, 1975). The first study, however, to unequivocally demonstrate that a lesion of primary auditory cortex (A1) in cats causes a deficit in sound localization was performed by Jenkins and Merzenich (1984). Particularly convincing was the fact that the sound localization deficits after small A1 lesions were frequency-specific. These findings were confirmed in later studies using different tasks (Heffner & Heffner, 1990; Beitel & Kaas, 1993). In these studies, A1 appeared to be the only region of auditory cortex whose ablation caused a localization deficit. However, cats have an auditory cortical region that is hidden deep in the anterior ectosylvian sulcus (AES), which was later implicated in sound localization: the anterior ectosylvian auditory area (AEA) (Korte & Rauschecker, 1993; Rauschecker & Korte, 1993; Middlebrooks et al., 1994) or "field AES" (Meredith & Clemo, 1989). It is the main source of auditory cortical input to the SC in cats. In addition, on the basis of cortical cooling studies, the posterior auditory field (PAF) has also been shown to play a role in sound localization by cats (Malhotra et al., 2004; Lomber & Malhotra, 2008).

Spatial tuning of single cortical neurons in cat A1 was measured by a number of groups (Imig et al., 1990; Rajan et al., 1990a, b). They all found two types of spatial tuning: single-peak and hemifield. In single-peak neurons, the best response is found at a particular azimuth location; in hemifield neurons, the response is largely restricted to speaker locations in the contra- or ipsilateral hemifield. In rhesus monkeys, spatially tuned neurons are also found in A1. However, such neurons are found at an even higher rate in the caudal belt (Rauschecker et al., 1997; Recanzone, 2000; Tian et al., 2001). Further, when monkeys are trained in an auditory localization task, the firing rate of caudal belt neurons correlates more tightly with behavioral performance than that of neurons in A1, which is a strong indication that the caudal belt plays an important role in sound localization (Recanzone et al., 2000).

2.2.3 *Spatial and Pattern Selectivity in the Lateral Belt*

In order to compare the spatial selectivity of neurons in the rostral and caudal LB directly in the same animals, broadband species-specific communication calls (MCs) were presented in free field from different locations (Tian et al., 2001) (see Section 2.1). LB areas differed in their degree of MC selectivity, as quantified on the basis of a monkey call preference index (MCPI). The anterolateral area (AL) had the greatest percentage of highly selective neurons (MCPI ≤ 2), whereas the caudolateral area (CL) had the smallest percentage of highly MC-selective neurons. Spatial tuning, as measured by the number of speaker locations from which a significant response could be elicited in a free-field experiment, showed the opposite areal distribution in LB than MC: the highest selectivity was found in CL and the lowest in AL (Fig. 2.8).

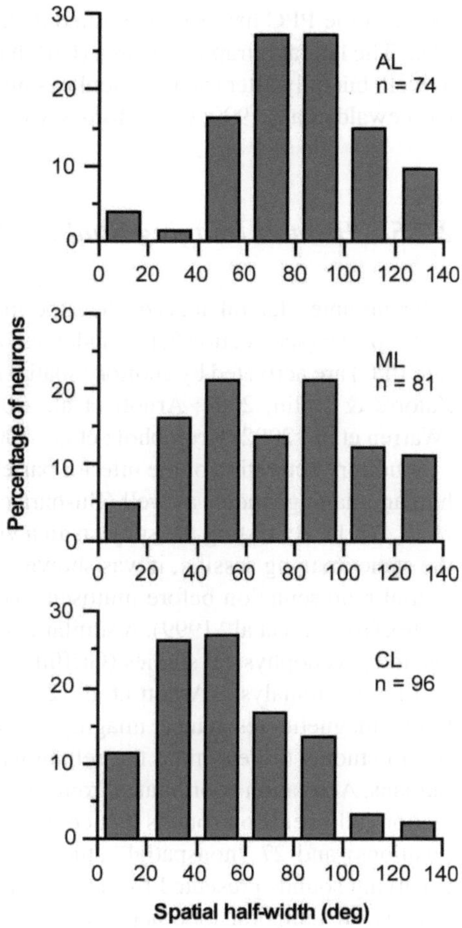

Fig. 2.8 Distribution of spatial half-width in areas AL, ML, and CL. Summary data from four monkeys are shown in histogram form. Number of units recorded in each area is given on the right. Neurons in CL show significantly greater spatial selectivity than neurons in AL or ML. By contrast, neurons in AL are more selective for monkey calls than neurons in either of the other areas (see Section 2.1). (From Tian et al., 2001)

2.2.4 *Caudal Belt Projections to Parietal and Prefrontal Cortex*

The anatomical tracer study of Romanski et al. (1999) demonstrated the existence of largely segregated pathways from the LB to prefrontal cortex (see Section 2.1 and Fig. 2.6). Injections into area AL produced label in ventrolateral and orbital regions of prefrontal cortex (areas 10, 12), whereas CL injections led to labeling of DLPFC (areas 8a, 46). These projection patterns conformed to the physiological response properties in LB found in the study of Tian et al. (2001), which assigned superior selectivity for auditory patterns and space to areas AL and CL, respectively. The pathway projecting caudodorsally from A1 to the caudal belt (areas CM and CL) and parabelt (areas Tpt and CPB) into posterior parietal cortex (PPC) and DLPFC is thought to be involved in auditory spatial processing.

A projection from posterior STG to PPC in monkeys has been found independently by Lewis and Van Essen (2000). Specifically, the ventral intraparietal area

(VIP) in the PPC has been identified as the primary recipient of auditory input to PPC. The lateral intraparietal area (LIP) has been found to contain auditory neurons as well, but only after training monkeys on auditory saccades (Stricanne et al., 1996; Grunewald et al., 1999; cf. Gifford & Cohen, 2004).

2.2.5 *Human Imaging Studies of the Auditory Dorsal Stream*

Whereas anterolateral areas of the superior temporal cortex are activated by intelligible speech (see Section 2.1), caudal belt and parabelt areas (projecting up dorsally into PPC) are activated by auditory spatial discrimination tasks (Maeder et al., 2001; Zatorre & Belin, 2001; Arnott et al., 2004) or tasks involving auditory motion (Warren et al., 2002; Krumbholz et al., 2005) (Fig. 2.9).

Auditory activation of the inferior parietal lobule (IPL) has been demonstrated in human imaging studies as well (Bushara et al., 1999; Weeks et al., 1999; Bremmer et al., 2001). By testing the subjects in a visual as well as in an auditory task during the same imaging session, it was shown that IPL does contain a unimodal auditory spatial representation before multisensory convergence occurs in superior parietal cortex (Bushara et al., 1999). A similar conclusion was reached on the basis of clinical and psychophysical studies (Griffiths et al., 1997, 1998).

In a meta-analysis, Arnott et al. (2004) reviewed evidence from auditory functional magnetic resonance imaging (fMRI) and positron emission tomography (PET) studies to determine the reliability of the auditory dual-pathway model in humans. Activation coordinates from 11 "spatial" studies (i.e., listeners made localization judgments on sounds that could occur at two or more perceptually different positions) and 27 "nonspatial" studies (i.e., listeners completed nonspatial tasks involving sounds presented from the same location) were entered into the analysis. Almost all temporal lobe activity observed during spatial tasks was confined to posterior areas. In addition, all but one of the spatial studies reported activation within the IPL as opposed to only 41% of the nonspatial studies. Finally, inferior frontal activity (Brodmann areas [BA] 45 and 47) was reported in only 9% of the spatial studies, but in 56% of the nonspatial studies. These results support an auditory dual-pathway model in humans in which nonspatial sound information (e.g., sound identity) is processed primarily along an anteroventral stream whereas sound location is processed along a posterodorsal stream, that is, within areas posterior to primary auditory cortex.

In a PET study by Zatorre et al. (2002) posterior auditory cortex responded to sounds that varied in their spatial distribution, but only when multiple complex stimuli were presented simultaneously. Consistent with other studies, these authors also found that the right inferior parietal cortex was specifically recruited in localization tasks.

An fMRI study by Krumbholz and co-workers (2005) found that interaural time differences were represented along a posterior pathway comprising the planum temporale (PT) and IPL of the respective contralateral hemisphere. The response

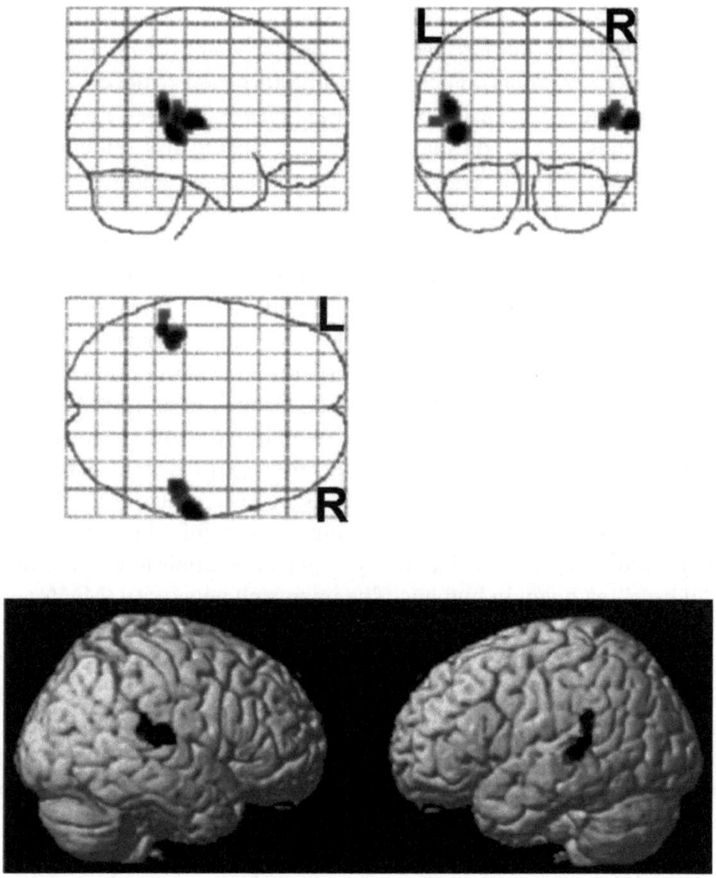

Fig. 2.9 Human imaging of motion in virtual auditory space (all-motion minus stationary-sound). Statistical parametric maps of fMRI group data are shown as sagittal, coronal, and axial projections (above) and rendered onto a canonical brain surface template (below). All voxels significant at the $p < 0.05$ level (corrected for multiple comparisons) are displayed. (From Warren et al., 2002)

was stronger and extended further into adjacent regions of the IPL when the sound was moving than when it was stationary, a finding that confirmed earlier results by Warren et al. (2002) (Fig. 2.9). In contrast to Zatorre et al. (2002), the study by Krumbholz et al. (2005) found that stationary lateralized sounds did produce a significant activation increase in the PT of the respective contralateral hemisphere compared to a centrally presented sound. This discrepancy may be due to the inferior sensitivity of PET relative to fMRI, or to the fact that the spatial ranges of the sounds used by Zatorre et al. (2002) were centered around the midline, and thus always comprised equal parts of both hemifields. This suggests that Zatorre et al. (2002) were unable to detect the contralateral tuning that was observed in the study of Krumbholz et al. (2005).

Timing differences between the two ears can be used to localize sounds in space only when the inputs to the two ears have similar spectrotemporal profiles (high binaural coherence). Zimmer and Macaluso (2005) used fMRI to investigate any modulation of auditory responses by binaural coherence. They assessed how processing of these cues depends on spatial information being task-relevant and whether brain activity correlates with subjects' localization performance. They found that activity in Heschl's gyrus increased with increasing coherence, irrespective of localization being task-relevant. Posterior auditory regions also showed increased activity for high coherence, but only when sound localization was required and subjects successfully localized sounds. The authors concluded that binaural coherence cues are processed throughout auditory cortex, but that these cues are used in posterior regions of the STG for successful auditory localization (Zimmer & Macaluso, 2005). In another series of fMRI experiments, Deouell et al. (2007) showed that a region in the human medial PT is sensitive to auditory spatial changes, even when subjects are not engaged in a sound localization task, that is, when the spatial changes are occurring in the background. Thus, acoustic space is firmly represented in the human PT even when sound processing is not required by the ongoing task.

Tata and Ward (2005a,b) used auditory evoked potentials to explore the putative auditory "where"-pathway in humans. The mismatch negativity (MMN) elicited by deviations in sound location is comprised of two temporally and anatomically distinct phases: an early phase with a generator posterior to primary auditory cortex and contralateral to the deviant stimulus, and a later phase with generators that are more frontal and bilaterally symmetric. The posterior location of the early-phase generator suggests the engagement of neurons within a posterior "where"-pathway for processing spatial auditory information (Tata & Ward, 2005a). Transient attention oriented in cue-target paradigms results in several modulations of the auditory event-related potential. Its earliest component (the Nd1) also reflects modulation of neurons posterior to primary auditory cortex within or near the temporoparietal junction (TPJ) (Tata & Ward, 2005b).

Selective attention was also used to differentiate the effects of sound location and pitch of an auditory stimulus in an fMRI study (Degerman et al., 2006). Attention to either sound feature produced activation in areas of the superior temporal cortex and in prefrontal and inferior parietal regions. However, during attention to location these activations were located more posterior on the STG than during attention to pitch.

In a study combining fMRI and MEG, Brunetti and co-workers found that the processing of sound coming from different locations activates a neural circuit similar to the auditory "where"-pathway described in monkeys (Brunetti et al., 2005). This system included Heschl's gyrus, the posterior STG, and the IPL. Their MEG analysis allowed assessment of the timing of this circuit: Activation of Heschl's gyrus was observed 139 ms after the auditory stimulus, the peak latency of the source located in the posterior STG was at 156 ms, and the IPL and the supramarginal gyrus peaked at 162 ms. Both hemispheres were found to be involved in the processing of sounds coming from different locations, but a stronger activation was observed in the right hemisphere (Brunetti et al., 2005).

A similar study combining fMRI and MEG was conducted by Ahveninen et al. (2006). They found a double dissociation in response adaptation to sound pairs with phonetic versus spatial sound changes, demonstrating that the human nonprimary auditory cortex processes speech-sound identity and location in parallel anterior "what" (in anterolateral Heschl's gyrus, anterior superior temporal gyrus, and posterior planum polare) and posterior "where" (in PT and posterior STG) pathways as early as approximately 70–150 ms after stimulus onset. These data further showed that the "where"-pathway is activated approximately 30 ms earlier than the "what"-pathway.

2.2.6 *Summary*

Contrary to popular belief, which places auditory space processing wholly in the brain stem, several lines of evidence suggest that auditory cortex plays an important role in spatial perception. Lesion studies in animals and humans demonstrate severe deficits in sound localization after damage to auditory cortex. Single-unit recording studies find neurons tuned to spatial location in auditory cortical areas. While these neurons exist already in primary auditory cortex, their prevalence and sharpness of spatial tuning increases in nonprimary areas of the caudal belt, as defined in nonhuman primates. The firing of neurons in the caudal belt also shows a tighter correlation with the behavioral performance of alert monkeys engaged in sound localization behavior. Caudal belt and parabelt project to posterior parietal cortex and to areas of dorsolateral prefrontal cortex, such as the frontal eye and pinna fields, known to be involved in spatial attention and control of eye and pinna movements. This has led to the notion that a posterodorsal processing stream is intimately involved in aspects of auditory spatial processing.

The existence of an auditory "where"-stream is also suggested by functional neuroimaging studies in humans in which subjects process stationary or moving sounds in space. Consistently, posterior aspects of the superior temporal cortex and adjoining inferior parietal areas are activated during these tasks. Thus, although brain stem nuclei perform an important service by computing some of the basic parameters that are prerequisites for spatial processing, such as interaural time and intensity differences, these parameters (together with monaural spectral cues that depend on head and pinnae) are integrated at the cortical level. Auditory space processing, including processing of motion in space, is, therefore, ultimately accomplished at the cortical level.

2.3 Sensorimotor Integration and Control in the Dorsal Stream

2.3.1 *Introduction*

The dual-pathway model of auditory cortical processing assumes that two largely segregated processing streams originating in the lateral belt subserve the two main functions of hearing: identification of auditory patterns or "objects," including recognition of speech sounds; and localization of sounds in space (Rauschecker & Tian, 2000).

Evidence has accumulated, chiefly from work in humans and nonhuman primates, that an anteroventral pathway supports the former function, whereas a posterodorsal stream (including posterior superior temporal cortex, pST, and inferior parietal lobule, IPL) supports the latter, that is, processing of space and motion-in-space. A role of the posterodorsal stream in space and motion would seem to be at odds with classic claims of a function of pST and IPL in speech and language in humans. However, as we have argued previously (Rauschecker & Scott, 2009; Rauschecker, 2011), the neural functions related to space and spoken speech, in a computational sense, may not be as incompatible as they seem. Rather, both functions share a common set of properties that actually require a neural system like the dorsal stream, which creates an interface between sensory and motor networks and performs a matching operation between predicted outcomes and actual events. Although the computational algorithms in the brain are far from clear, they must resemble the internal "forward models" that have revolutionized thinking in motor control and robotics (Wolpert et al., 1995; Kawato, 1999). In the case of audition, an efference copy sent from prefrontal and premotor cortex provides the basis for "optimal state estimation" in the IPL and in sensory areas of the posterior auditory cortex.

2.3.2 Role of the Human Dorsal Auditory Pathway in Speech and Language

2.3.2.1 Is the Dorsal Pathway Really Involved in Speech Processing?

Section 2.2 of this chapter summarized the evidence for a role of the posterior ST (pST) region (and the IPL regions connected with it) in processing auditory space and motion-in-space. This function is undeniably present in both monkeys and humans (as well as nonprimate animals). However, another view about the function of pST in humans has classically been even more pervasive: the view that pST is involved in speech or language comprehension (Geschwind, 1965; Damasio & Damasio, 1980). Many textbooks refer to pST and surrounding regions as "Wernicke's area," so it seems as if this view dates back to Carl Wernicke (1874), who described patients with lesions of the ST region having difficulties with various aspects of speech. Closer examination of Wernicke's case studies reveals, however, that the pertinent lesions were not necessarily found in pST alone. A figure in one of his own textbooks (Wernicke, 1881) explicitly marked the whole ST region as speech-related, including its anterior aspects. To reserve the term "Wernicke's area" for the posterior one-third of ST is, therefore, misleading (Wise et al., 2001; Rauschecker & Scott, 2009).

Wernicke did, however, make the insightful claim that auditory ST regions subserving the deciphering of speech sounds must be connected somehow with the motor speech area in the frontal cortex, which had been discovered by Broca (1861) about a decade earlier. Based on gross anatomical studies of aphasic stroke patients, later researchers assumed that this functional connectivity was provided by a fiber bundle that wound its way from the posterior ST region to Broca's area, the "arcuate fascicle" (Geschwind, 1965). Present-day work is being performed with high-resolution structural imaging techniques (Rilling et al., 2008; Bernal & Ardila, 2009; Keller et al., 2009). At least one of these studies has revealed that a direct connection from pST to Broca's area, as in the monkey and its homologous areas (Petrides & Pandya, 2009), is sparse at best (Frey et al., 2008). Instead, fibers projecting to Broca's area from ST originate in its anterior aspects and follow a whole different pathway via the extreme capsule and/or the uncinate fascicle (Ebeling & von Cramon, 1992; Friederici et al., 2006)). In fact, Wernicke himself suspected that the connection from ST to Broca's area went via the anterior insula, a region that has recently been found to play a role in communication sound processing of monkeys (Remedios et al., 2009). All this adds to the support for an anteroventral pathway in auditory speech processing and one might be tempted to reject the claim of a specific pST (and dorsal-stream) involvement in speech processing altogether. However, this would be "throwing the baby out with the bathwater."

To salvage a genuine role for the pST region in speech and language and to reconcile this role with the spatial functions of that region, one merely has to back away from the claim that pST is involved in the "perception" of speech, that is, primarily an acoustic-phonetic decoding of speech sounds. Instead, one needs to analyze the incidents under which pST and IPL are activated by sounds or tasks with other than spatial connotations.

2.3.2.2 Representation of Action Sounds in the Dorsal Stream

Various studies have demonstrated activation of left parietal cortical regions while subjects were listening to sounds generated by actions, such as tool sounds (Lewis et al., 2005; Pizzamiglio et al., 2005; Engel et al., 2009). These activations often include posterior STS and STG regions, especially when contrasted with unrecognizable control sounds. One possibility is that these regions contain representations of "doable" sounds (Rauschecker & Scott, 2009). In particular, it has been suggested that the medial PT region (Warren et al., 2005) contains templates of "doable" articulations (not limited to speech sounds) against which incoming sounds are matched. Studies of silent articulation (Wise et al., 2001) and covert rehearsal of speech (Hickok et al., 2009) have also identified activation in the posterior medial PT region within the posterodorsal stream.

Such findings resonate with the "affordance" model of Gibson (1977), in which objects and events are described in terms of action possibilities. Gibson's views undoubtedly had an influence on the mirror-neuron theory of Rizzolatti and colleagues (2006; see also Arbib et al., 2013).

2.3.2.3 A Multisensory Reference Frame

The posteromedial region of the PT has been identified as a possible key node for the feedback control of speech production (Dhanjal et al., 2008) because it shows a response to somatosensory input from articulators as well as to auditory speech input. Adjacent to pST, the temporoparietal junction (TPJ) has been discussed independently in both auditory and visual contexts, but probably constitutes a multisensory region having to do with temporal order judgment of spatially separate events (Davis et al., 2009).

In relation to these studies, it is fitting that neurophysiological evidence from nonhuman primates shows that caudal belt neurons not only are responsive to auditory input but also reveal multisensory responses (Fu et al., 2003; Brosch et al., 2005; Kayser et al., 2007), although this has been reported about other areas as well (Ghazanfar et al., 2005; Bulkin & Groh, 2006; Lakatos et al., 2007). Neuroanatomical studies demonstrate that both caudal medial and lateral belt fields receive input from somatosensory and multisensory cortex as well as thalamic nuclei (Smiley et al., 2007). In contrast, core and anterior areas show only sparse multisensory connections. Thus, the posterodorsal stream, by bringing together input from different sensory modalities, may create a supramodal reference frame in which any transformations, whether spatial or otherwise, can be conducted. Some studies suggest that this reference frame transformation begins at earlier stages of the auditory pathway (Mullette-Gillman et al., 2005).

2.3.2.4 Encoding and Retrieval of Sound Sequences

One of the unsolved puzzles in auditory neuroscience is how the brain encodes and stores sequences of sound (Schubotz et al., 2000; Rauschecker, 2005). Unlike tape recorders and compact disk (CD) players the brain does not have any moving parts that could translate temporal order of a sound sequence into location on a physical medium for storage and retrieval. Digital music players, on the other hand, use specific file formats to preserve the spectrotemporal integrity of, for instance, a piece of music. If we look for structures in the brain that may be suitable for storage and reproduction of temporal sequences, we are quickly reminded of the fact that motor areas must be able to do just that: a simple motor act or gesture requires the production of sequences of nerve signals sent to specific muscles (or motor neurons) controlling the various limbs involved in that gesture in a particular order. The act of speaking or singing is an example of a motor performance during which a multitude of fine-grained muscles have to be controlled in a highly time-order specific fashion to keep both rhythm and pitch exactly right. While the motor cortex provides the origin of axons projecting to the spinal cord for control of muscles, it is commonly assumed that subcortical entities such as the basal ganglia or the cerebellum set up the patterns reflecting temporal sequential structure of motor acts.

Indeed, singing or speaking, like other motor acts, light up cortical motor areas as well as subcortical structures (Perry et al., 1999). Singing also activates auditory areas, which would not be surprising (because the subjects hear their own voice) if the activation did not persist even after subtracting out auditory perceptual activation. Interestingly, the remaining auditory activation appears in pST. Even more interestingly, listening to music also activates motor areas

(Wilson et al., 2004; Zatorre et al., 2007; Chen et al., 2008). It thus appears as if we are looking at a sensorimotor loop, wherein both afferent and efferent branches are active in either situation.

Finally, even imagery of music (Halpern & Zatorre, 1999) and anticipation of familiar melodies after playing the preceding melody (Leaver et al., 2009) leads to activation of both auditory and motor structures (Fig. 2.10), cortical and subcortical (cerebellum and basal ganglia). The amount of basal ganglia versus frontal cortical activation depends on the state of familiarity of the sequence, basal ganglia being more active during the learning period (Leaver et al., 2009).

There is also strong psychophysical evidence suggesting that auditory–motor processing dissociates from auditory–perceptual processing (Repp, 2005; Rauschecker & Scott, 2009): Listeners can accurately tap along to auditory sequences, and their motor responses can track changes in the rates of these sequences. This tracking of sequences could occur in the dorsal stream. Functional imaging evidence does indeed suggest that the intraparietal sulcus plays a role in streaming, sequence detection, and dissociation of figure from ground (Cusack, 2005). These results from human psychophysical and imaging studies would merit further examination in monkey single-unit studies to determine the exact neurophysiological mechanisms of auditory sequence processing and stream segregation (Micheyl et al., 2005).

2.3.2.5 Auditory Perception/Production Links in Voice and Speech

Monkey studies have shown that neurons in auditory cortex are suppressed during vocalization (Müller-Preuss & Ploog, 1981; Eliades & Wang, 2003). This finding is consistent with results from humans, which indicate that superior temporal areas are suppressed during speech production (Numminen et al., 1999; Curio et al., 2000; Houde et al., 2002). This suppression or attenuation of auditory cortex is found even with covert articulation and lipreading, suggesting the existence of an efference-copy pathway from premotor regions to auditory cortex (Kauramäki et al., 2010).

It has been argued that mechanisms of this kind may exist to help distinguish the effects of actions caused by oneself from those caused by the actions of others (Blakemore et al., 1998), specifically differentiating between one's own voice and the voices of others (Rauschecker & Scott, 2009). However in nonhuman primate studies, auditory neurons that are suppressed during actual vocalizations are often more activated by distorted vocalizations (Eliades & Wang, 2008). This suggests a role for these neurons in the comparison of information from the auditory and motor systems during speech production (Guenther, 2006). Work in humans using distorted feedback of speech production has indeed shown enhanced bilateral activation in pST to distorted feedback, even if it is below the threshold for explicit awareness (Tourville et al., 2008).

There have also been persistent claims for a role of the IPL, that is, the angular and supramarginal gyri, in phonology (Caplan et al., 1992), particularly an involvement in the "phonological/articulatory loop" (Baddeley et al., 1984; Aboitiz et al., 2006). This has been confirmed in several functional imaging studies, though the

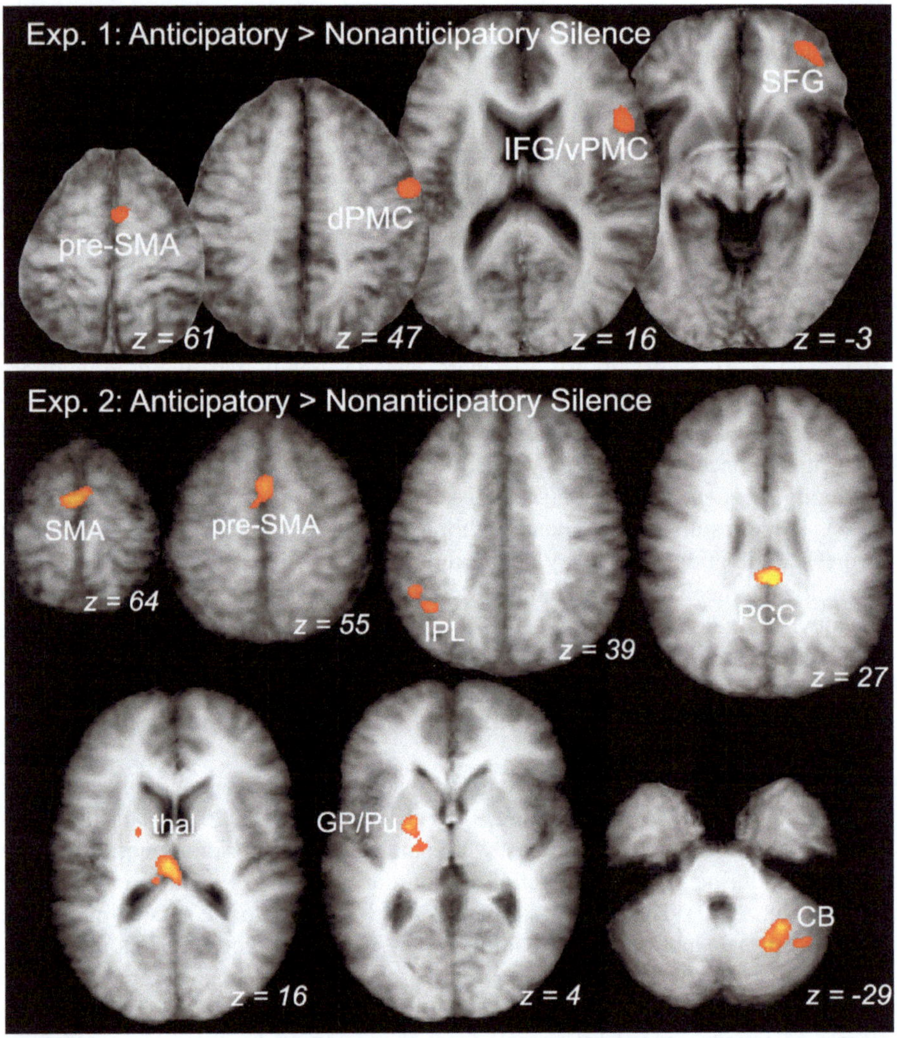

Fig. 2.10 Brain areas active during anticipatory imagery of familiar music. Two experiments were conducted. In Experiment 1 (Exp. 1), stimuli consisted of the final seconds of familiar or unfamiliar tracks from a compact disk (CD), followed by 8 s of silence. During the silence following familiar tracks from their favorite CD (anticipatory silence, AS, following familiar music, FM), subjects (Ss) reported experiencing anticipatory imagery for each subsequent track. Stimuli presented during unfamiliar trials consisted of music that the Ss had never heard before (unfamiliar music, UM). Thus, during this condition, Ss could not anticipate the onset of the following track (nonanticipatory silence, NS). While in the MRI scanner, Ss were instructed to attend to the stimulus being presented and to imagine, but not vocalize, the subsequent melody where appropriate. In Experiment 2 (Exp. 2), Ss were trained in associating newly composed melodies with each other. In both experiments, the outcome was similar: Activated brain regions were found in frontal and premotor regions, including inferior and superior frontal gyrus (IFG, SFG), presupplementary motor area (pre-SMA), as well as dorsal and ventral premotor cortex (dPMC, vPMC). (From Leaver et al., 2009)

precise localization of activity does vary with the type of task used (Gelfand & Bookheimer, 2003; Buchsbaum & D'Esposito, 2008). What seems clear is that the IPL, like pST, is not driven by acoustic–phonetic factors in speech processing but is associated with more domain-general factors (Friederici et al., 2006; Rauschecker & Scott, 2009).

New work using diffusion tensor imaging in humans demonstrates that there are direct connections between the pars opercularis of Broca's area (BA44) and the IPL (Frey et al., 2008; Saur et al., 2008; Bernal & Ardila, 2009), but hardly at all with pST, calling into question the notion of a direct connection between "Broca's" and "Wernicke's" area, as postulated in most textbooks. In addition, there is the known projection from ventral premotor (vPM) cortex to the IPL (Petrides & Pandya, 1984, 2009), and connections between parietal cortex and pST are also well known (Seltzer & Pandya, 1994); together, this could form the basis for a feed-forward network between speech production areas and posterior temporal auditory areas (Fig. 2.11).

2.3.3 Unified Function of the Dorsal Stream: Anticipatory Control of Sensorimotor Events

As documented in Section 2.2, posterior ST regions and the IPL participate in the processing of auditory space and motion. At the same time, pST and IPL in humans are also involved in the processing and imagery of auditory sequences, including speech and music. Both regions receive input from premotor areas in the dorsal and ventral premotor cortex (PMC). PMC is also activated during listening to music (Lahav et al., 2007; Chen et al., 2008) and even during musical imagery and anticipation (Leaver et al., 2009). One conclusion is that premotor areas are responsible for assembling the motor patterns for the production of musical sequences (by singing or playing a musical instrument). The sounds being produced activate neuronal assemblies in auditory cortex, which in turn are matched with the corresponding premotor neurons that helped produce the sounds. Thus, specific sensorimotor networks are established that, together, represent the musical melodies in a quasi-motor code. During learning of musical melodies, which occurs in the same way as learning of motor sequences (Hikosaka et al., 1999), subcortical structures such as the basal ganglia and the cerebellum are also active in binding the correct sets of sensory and motor neurons together (Leaver et al., 2009). One prediction would be, therefore, that learning to play a new piece on a musical instrument or, for that matter, learning to play a familiar piece on a new instrument, should result in characteristic changes in premotor representations. The same would be expected when passive listening to complex sounds is replaced by producing these sounds ("action sounds").

An analogous process can be assumed to be at work during learning of speech and speech production. Once learned, listening to speech activates the same circuits as during speech production. Although it may not strictly be accurate to talk about a "motor code" for speech perception (Liberman et al., 1967), correct speech does require a closing of the loop between perception and production and will lead to

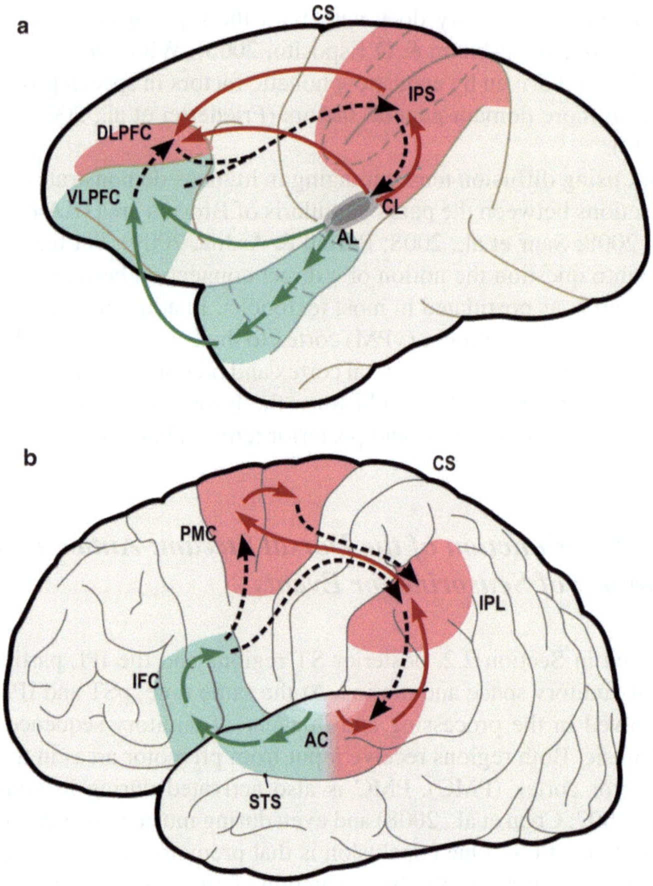

Fig. 2.11 Expanded model of dual auditory processing streams in the primate brain. (**a**) Rhesus monkey. (Modified from Rauschecker & Tian, 2000.) (**b**) Human. (Simplified from Rauschecker & Scott, 2009.) While the role of the anteroventral stream (green) in auditory object recognition, including perception of vocalizations and speech, is now widely accepted, the exact role of the posterodorsal (or just "dorsal") stream (red) is still being debated. Its function clearly includes spatial processing, but a role in human speech and language has also long been postulated. A reinterpretation of these classic studies suggests that the dorsal stream pivots around inferior/posterior parietal cortex, where a quick sketch of sensory event information is compared with an efference copy of motor plans (dashed lines). Thus, the dorsal stream plays a more general role in sensorimotor integration and control. In clockwise fashion, starting out from auditory cortex, the processing loop performs as a forward model: Object information, such as vocalizations and speech, is decoded in the anteroventral stream all the way to category-invariant inferior frontal cortex (IFC, or VLPFC in monkeys) and transformed into articulatory representations (DLPFC or ventral PMC). Frontal activations are transmitted to the IPL and pST, where they are compared with auditory and other sensory information. It is this fronto–parietal–sensory section that turns the dorsal stream on its head and expands its function. AC, auditory cortex; STS, superior temporal sulcus; IFC, inferior frontal cortex; PFC, prefrontal cortex; PMC, premotor cortex; IPL, inferior parietal lobule; IPS, inferior parietal sulcus; CS, central sulcus

coactivation of both networks. The connection between auditory areas in the ST and speech planning areas in the frontal cortex around "Broca's region," as postulated by Wernicke, runs through aST and inferior frontal cortex; the loop is closed through PMC via IPL and back to auditory cortex (Fig. 2.11). Learning to produce new sounds in a foreign language should, therefore, lead to changes in both sensory and motor representations of the corresponding sounds.

This basic structure of the underlying fronto–parietal–sensory loops responsible for sensorimotor planning and control is best described as that of an "internal model" or "emulator," as it is known in motor control theory and robotics (Rauschecker & Scott, 2009). Such models have been used to describe reaching movements or planning of movement trajectories using Kalman filters and Bayesian statistics for optimal state estimation (Kawato, 1999; Desmurget & Grafton, 2000; Sabes, 2000). More recently, these models have been used to model perception and imagery as well (Wolpert et al., 2003; Grush, 2004). The inferior parietal cortex appears to provide an ideal interface for feed-forward information from motor preparatory networks in the PFC and PMC to be matched with feedback signals from sensory areas. The goal of the internal model is to minimize the resulting error signal in this process.

The feed-forward projection from BA 44 and vPM can be considered the pathway carrying an "efference copy" or "corollary discharge" in the classic sense (Sperry, 1950; Von Holst & Mittelstaedt, 1950), informing the sensory system of planned motor articulations that are about to happen. This signal provides a predictive quality to activity running from frontal areas to the IPL, which therefore anticipates the sensory consequences of action. The feedback signal coming to the IPL from posterior ST, on the other hand, can be considered an "afference copy" (Hershberger, 1976) or reafference with relatively short latencies and high temporal precision (Jääskeläinen et al., 2004; Kauramäki et al., 2010). It can be thought of as a sparse but fast primal sketch of ongoing sensory events (Bar et al., 2006) that are compared with the predictive motor signal in the IPL in real time at every instance. In that sense, both spatial processing and real-time processing of speech and music make use of the same general internal model structures that enable the instantiation of smooth sequential motor behaviors, including visuospatial reaching as well as articulation of speech. At the same time, these sensorimotor loops also support the disambiguation of phonological information.

2.4 Summary

This chapter first summarizes the evidence for the existence of dual pathways in auditory cortex (1) for the processing of pattern/object information and (2) for the processing of space and motion. The former pathway follows an anteroventral route, originating in the rostral field (R) and projecting through rostral belt and parabelt into VLPFC (directly as well as through intermediary stations in the rostral STG). The latter pathway follows a posterodorsal route, originating in primary auditory cortex and projecting through caudal belt and parabelt into DLPFC (directly as well as via inferior parietal cortex and premotor cortex). Evidence for these pathways comes from anatomical and physiological studies of nonhuman primates as well as neuroimaging studies in humans. A wider consideration of function of the dorsal stream in a third section then discusses the role of the dorsal stream in sensorimotor integration and control more generally. This view incorporates the theoretical concept of internal models.

Future studies of the ventral stream may concentrate on the formation of invariances in the perception of complex sounds, for instance, against changes in pitch. Future work on the dorsal stream should design more specific experiments to test the concept of forward and inverse models and come up with computational formulations of the underlying neural circuitry. This could be done, for instance, by training animals in the learning of auditory sequences, while monitoring neural activity with fMRI as well as chronic microelectrode recordings. Both of these approaches could help to advance our understanding of cognitive aspects of hearing with particular relevance for understanding the evolution/cognition of language and music.

Acknowledgments The present chapter is an updated synthesis of prior publications by the author (Rauschecker & Tian, 2005; Rauschecker, 2007, 2011). The author's work was supported by grants from the National Institutes of Health (R01 NS052494), the Cognitive Neuroscience Initiative of the National Science Foundation (BCS-0519127), and the NSF PIRE program (OISE-0730255).

References

Aboitiz, F., Garcia, R. R., Bosman, C., & Brunetti, E. (2006). Cortical memory mechanisms and language origins. *Brain and Language*, 98(1), 40–56.

Ahveninen, J., Jääskeläinen, I. P., Raij, T., Bonmassar, G., Devore, S., Hämäläinen, M., et al. (2006). Task-modulated "what" and "where" pathways in human auditory cortex. *Proceedings of the National Academy of Sciences of the USA*, 103(39), 14608–14613.

Alain, C., Arnott, S. R., Hevenor, S., Graham, S., & Grady, C. L. (2001). "What" and "where" in the human auditory system. *Proceedings of the National Academy of Sciences of the USA*, 98(21), 12301–12306.

Arbib, M. A., Verschure, P. F. M. J., & Seifert, U. (2013). Action, Language and Music: Events in Time and Models of the Brain. In M. A. Arbib (Ed.), *Language, music and the brain: A mysterious relationship*. Cambridge, MA: MIT Press.

Arnott, S. R., Binns, M. A., Grady, C. L., & Alain, C. (2004). Assessing the auditory dual-pathway model in humans. *NeuroImage*, 22(1), 401–408.

Baddeley, A., Lewis, V., & Vallar, G. (1984). Exploring the articulatory loop. *The Quarterly Journal of Experimental Psychology*, *A*(36), 233–252.

Bar, M., Kassam, K. S., Ghuman, A. S., Boshyan, J., Schmid, A. M., Dale, A. M., et al. (2006). Top-down facilitation of visual recognition. *Proceedings of the National Academy of Sciences of the USA*, 103(2), 449–454.

Beitel, R. E., & Kaas, J. H. (1993). Effects of bilateral and unilateral ablation of auditory cortex in cats on the unconditioned head orienting response to acoustic stimuli. *Journal of Neurophysiology*, 70(1), 351–369.

Bernal, B., & Ardila, A. (2009). The role of the arcuate fasciculus in conduction aphasia. *Brain*, 132(Pt 9), 2309–2316.

Binder, J. R., Frost, J. A., Hammeke, T. A., Bellgowan, P. S., Springer, J. A., Kaufman, J. N., & Possing, E. T. (2000). Human temporal lobe activation by speech and nonspeech sounds. *Cerebral Cortex*, 10(5), 512–528.

Binder, J. R., Liebenthal, E., Possing, E. T., Medler, D. A., & Ward, B. D. (2004). Neural correlates of sensory and decision processes in auditory object identification. *Nature Neuroscience*, 7(3), 295–301.

Blakemore, S. J., Goodbody, S. J., & Wolpert, D. M. (1998). Predicting the consequences of our own actions: The role of sensorimotor context estimation. *Journal of Neuroscience*, 18(18), 7511–7518.

Bremmer, F., Schlack, A., Shah, N. J., Zafiris, O., Kubischik, M., Hoffmann, K., et al. (2001). Polymodal motion processing in posterior parietal and premotor cortex: A human fMRI study strongly implies equivalencies between humans and monkeys. *Neuron*, 29(1), 287–296.

Broca, P. (1861). Remarques sur le siège de la faculté du language articulé: Suivies d'une observation d'aphémie (perte de la parole). *Bulletin de la Société Anatomique*, 6, 330–357.

Brosch, M., Selezneva, E., & Scheich, H. (2005). Nonauditory events of a behavioral procedure activate auditory cortex of highly trained monkeys. *Journal of Neuroscience*, 25(29), 6797–6806.

Brugge, J. F., & Merzenich, M. M. (1973). Responses of neurons in auditory cortex of the macaque monkey to monaural and binaural stimulation. *Journal of Neurophysiology*, 36, 1138–1158.

Brunetti, M., Belardinelli, P., Caulo, M., Del Gratta, C., Della Penna, S., Ferretti, A., et al. (2005). Human brain activation during passive listening to sounds from different locations: An fMRI and MEG study. *Human Brain Mapping*, 26(4), 251–261.

Buchsbaum, B. R., & D'Esposito, M. (2008). The search for the phonological store: from loop to convolution. *Journal of Cognitive Neuroscience*, 20(5), 762–778.

Bulkin, D. A., & Groh, J. M. (2006). Seeing sounds: visual and auditory interactions in the brain. *Current Opinion in Neurobiology*, 16(4), 415–419.

Bushara, K. O., Weeks, R. A., Ishii, K., Catalan, M.-J., Tian, B., Rauschecker, J. P., & Hallett, M. (1999). Modality-specific frontal and parietal areas for auditory and visual spatial localization in humans. *Nature Neuroscience*, 2(8), 759–766.

Caplan, D., Rochon, E., & Waters, G. S. (1992). Articulatory and phonological determinants of word length effects in span tasks. *The Quarterly Journal of Experimental Psychology*, 45(2), 177–192.

Chen, J. L., Penhune, V. B., & Zatorre, R. J. (2008). Listening to musical rhythms recruits motor regions of the brain. *Cerebral Cortex*, 18(12), 2844–2854.

Chevillet, M., Riesenhuber, M., & Rauschecker, J. P. (2011). Functional correlates of the anterolateral processing hierarchy in human auditory cortex. *Journal of Neuroscience*, 31(25), 9345–9352.

Cohen, Y. E., Russ, B. E., Davis, S. J., Baker, A. E., Ackelson, A. L., & Nitecki, R. (2009). A functional role for the ventrolateral prefrontal cortex in non-spatial auditory cognition. *Proceedings of the National Academy of Sciences of the USA*, 106(47), 20045–20050.

Curio, G., Neuloh, G., Numminen, J., Jousmaki, V., & Hari, R. (2000). Speaking modifies voice-evoked activity in the human auditory cortex. *Human Brain Mapping*, 9(4), 183–191.

Cusack, R. (2005). The intraparietal sulcus and perceptual organization. *Journal of Cognitive Neuroscience*, 17(4), 641–651.

Damasio, H., & Damasio, A. R. (1980). The anatomical basis of conduction aphasia. *Brain*, 103(2), 337–350.

Davis, B., Christie, J., & Rorden, C. (2009). Temporal order judgments activate temporal parietal junction. *Journal of Neuroscience*, 29(10), 3182–3188.

Degerman, A., Rinne, T., Salmi, J., Salonen, O., & Alho, K. (2006). Selective attention to sound location or pitch studied with fMRI. *Brain Research*, 1077(1), 123–134.

Deouell, L. Y., Heller, A. S., Malach, R., D'Esposito, M., & Knight, R. T. (2007). Cerebral responses to change in spatial location of unattended sounds. *Neuron*, 55(6), 985–996.

Desimone, R., & Schein, S. J. (1987). Visual properties of neurons in area V4 of the macaque: Sensitivity to stimulus form. *Journal of Neurophysiology*, 57, 835–868.

Desmurget, M., & Grafton, S. (2000). Forward modeling allows feedback control for fast reaching movements. *Trends in Cognitive Sciences*, 4(11), 423–431.

DeWitt, I., & Rauschecker, J. P. (2012). Phoneme and word recognition in the auditory ventral stream. *Proceedings of the National Academy of Sciences of the USA*, 109(8), E505–E514.

Dhanjal, N. S., Handunnetthi, L., Patel, M. C., & Wise, R. J. (2008). Perceptual systems controlling speech production. *Journal of Neuroscience*, 28(40), 9969–9975.

Diamond, I. T., Fisher, J. F., Neff, W. D., & Yela, M. (1956). Role of auditory cortex in discrimination requiring localization of sound in space. *Journal of Neurophysiology*, 19(6), 500–512.

Ebeling, U., & von Cramon, D. (1992). Topography of the uncinate fascicle and adjacent temporal fiber tracts. *Acta Neurochirurgica*, 115(3–4), 143–148.

Eliades, S. J., & Wang, X. (2003). Sensory-motor interaction in the primate auditory cortex during self-initiated vocalizations. *Journal of Neurophysiology*, 89(4), 2194–2207.

Eliades, S. J., & Wang, X. (2008). Neural substrates of vocalization feedback monitoring in primate auditory cortex. *Nature*, 453(7198), 1102–1106.

Engel, L. R., Frum, C., Puce, A., Walker, N. A., & Lewis, J. W. (2009). Different categories of living and non-living sound-sources activate distinct cortical networks. *NeuroImage*, 47(4), 1778–1791.

Fecteau, S., Armony, J. L., Joanette, Y., & Belin, P. (2004). Is voice processing species-specific in the human brain? An fMRI study. *NeuroImage*, 23(3), 840–848.

Frey, S., Campbell, J. S., Pike, G. B., & Petrides, M. (2008). Dissociating the human language pathways with high angular resolution diffusion fiber tractography. *Journal of Neuroscience*, 28(45), 11435–11444.

Friederici, A. D., Bahlmann, J., Heim, S., Schubotz, R. I., & Anwander, A. (2006). The brain differentiates human and non-human grammars: Functional localization and structural connectivity. *Proceedings of the National Academy of Sciences of the USA*, 103(7), 2458–2463.

Fu, K. G., Shah, A. S., Arnold, L., Garraghty, P. E., Smiley, J., Hackett, T. A., & Schroeder, C. E. (2003). Auditory cortical neurons respond to somatosensory stimulation. *Journal of Neuroscience*, 23, 7510–7515.

Galaburda, A. M., Sanides, F., & Geschwind, N. (1978). Human brain: Cytoarchitectonic left-right asymmetries in the temporal speech region. *Archives of Neurology*, 35(12), 812–817.

Gelfand, J. R., & Bookheimer, S. Y. (2003). Dissociating neural mechanisms of temporal sequencing and processing phonemes. *Neuron*, 38(5), 831–842.

Geschwind, N. (1965). Disconnexion syndromes in animals and man. *Brain*, 88(2), 237–294, 585–644.

Ghazanfar, A. A., Maier, J. X., Hoffman, K. L., & Logothetis, N. K. (2005). Multisensory integration of dynamic faces and voices in rhesus monkey auditory cortex. *Journal of Neuroscience*, 25(20), 5004–5012.

Gibson, J. J. (1977). The theory of affordances. In R. Shaw & J. Bransford (Eds.), *Perceiving, acting, and knowing: Toward an ecological psychology* (pp. 67–82). Hillsdale, NJ: Lawrence Erlbaum.

Gifford, G. W., 3rd, & Cohen, Y. E. (2004). Effect of a central fixation light on auditory spatial responses in area LIP. *Journal of Neurophysiology*, 91(6), 2929–2933.

Goldman-Rakic, P. S. (1996). The prefrontal landscape: Implications of functional architecture for understanding human mentation and the central executive. *Philosophical Transactions of the Royal Society of London B:, Biological Sciences*, 351(1346), 1445–1453.

Griffiths, T. D., Rees, A., Witton, C., Cross, P. M., Shakir, R. A., & Green, G. G. (1997). Spatial and temporal auditory processing deficits following right hemisphere infarction: A psychophysical study. *Brain*, 120(Pt 5), 785–794.

Griffiths, T. D., Rees, G., Rees, A., Green, G. G., Witton, C., Rowe, D., et al. (1998). Right parietal cortex is involved in the perception of sound movement in humans. *Nature Neuroscience*, 1(1), 74–79.

Grunewald, A., Linden, J. F., & Andersen, R. A. (1999). Responses to auditory stimuli in macaque lateral intraparietal area. I. Effects of training. *Journal of Neurophysiology*, 82(1), 330–342.

Grush, R. (2004). The emulation theory of representation: motor control, imagery, and perception. *Behavioral and Brain Sciences*, 27(3), 377–396; discussion 396–442.

Guenther, F. H. (2006). Cortical interactions underlying the production of speech sounds. *Journal of Communication Disorders*, 39(5), 350–365.

Hackett, T. A. (2011). Information flow in the auditory cortical network. *Hearing Research*, 271(1–2), 133–146.

Hackett, T. A., Stepniewska, I., & Kaas, J. H. (1998). Subdivisions of auditory cortex and ipsilateral cortical connections of the parabelt auditory cortex in macaque monkeys. *Journal of Comparative Neurology*, 394(4), 475–495.

Halpern, A. R., & Zatorre, R. J. (1999). When that tune runs through your head: A PET investigation of auditory imagery for familiar melodies. *Cerebral Cortex*, 9(7), 697–704.

Hauser, M. D. (1996). *The evolution of communication*. Cambridge, MA: MIT Press.

Heffner, H., & Masterton, B. (1975). Contribution of auditory cortex to sound localization in the monkey (*Macaca mulatta*). *Journal of Neurophysiology*, 38(6), 1340–1358.

Heffner, H. E., & Heffner, R. S. (1990). Effect of bilateral auditory cortex lesions on sound localization in Japanese macaques. *Journal of Neurophysiology*, 64(3), 915–931.

Hershberger, W. (1976). Afference copy, the closed-loop analogue of von Holst's efference copy. *Cybernetics Forum*, 8, 97–102.

Hickok, G., Okada, K., & Serences, J. T. (2009). Area Spt in the human planum temporale supports sensory-motor integration for speech processing. *Journal of Neurophysiology*, 101(5), 2725–2732.

Hikosaka, O., Nakahara, H., Rand, M. K., Sakai, K., Lu, X., Nakamura, K., et al. (1999). Parallel neural networks for learning sequential procedures. *Trends in Neurosciences*, 22(10), 464–471.

Houde, J. F., Nagarajan, S. S., Sekihara, K., & Merzenich, M. M. (2002). Modulation of the auditory cortex during speech: an MEG study. *Journal of Cognitive Neuroscience*, 14(8), 1125–1138.

Howard, M. A., Volkov, I. O., Mirsky, R., Garell, P. C., Noh, M. D., Granner, M., et al. (2000). Auditory cortex on the human posterior superior temporal gyrus. *Journal of Comparative Neurology*, 416, 79–92.

Imig, T. J., Irons, W. A., & Samson, F. R. (1990). Single-unit selectivity to azimuthal direction and sound pressure level of noise bursts in cat high-frequency primary auditory cortex. *Journal of Neurophysiology*, 63(6), 1448–1466.

Irvine, D. R. F. (1992). Physiology of auditory brainstem pathways. In R. R. Fay & A. A. Popper (Eds.), *The mammalian auditory pathway: Neurophysiology* (pp. 153–231). New York: Springer.

Jääskeläinen, I. P., Ahveninen, J., Bonmassar, G., Dale, A. M., Ilmoniemi, R. J. L., S., Lin, F. H., et al. (2004). Human posterior auditory cortex gates novel sounds to consciousness. *Proceedings of the National Academy of Sciences of the USA*, 101(17), 6809–6814.

Jenkins, W. M., & Merzenich, M. M. (1984). Role of cat primary auditory cortex for sound-localization behavior. *Journal of Neurophysiology*, 52(5), 819–847.

Jones, E. G., Dell'Anna, M. E., Molinari, M., Rausell, E., & Hashikawa, T. (1995). Subdivisions of macaque monkey auditory cortex revealed by calcium- binding protein immunoreactivity. *Journal of Comparative Neurology*, 362(2), 153–170.

Kaas, J. H., & Hackett, T. A. (2000). Subdivisions of auditory cortex and processing streams in primates. *Proceedings of the National Academy of Sciences of the USA*, 97(22), 11793–11799.

Kauramäki, J., Jääskeläinen, I. P., Hari, R., Möttönen, R., Rauschecker, J. P., & Sams, M. (2010). Transient adaptation of auditory cortex organization by lipreading and own speech production. *Journal of Neuroscience*, 30(4), 1314 –1321.

Kawato, M. (1999). Internal models for motor control and trajectory planning. *Current Opinion in Neurobiology*, 9(6), 718–727.

Kayser, C., Petkov, C. I., Augath, M., & Logothetis, N. K. (2007). Functional imaging reveals visual modulation of specific fields in auditory cortex. *Journal of Neuroscience*, 27(8), 1824–1835.

Keller, S. S., Roberts, N., & Hopkins, W. (2009). A comparative magnetic resonance imaging study of the anatomy, variability, and asymmetry of Broca's area in the human and chimpanzee brain. *Journal of Neuroscience*, 29(46), 14607–14616.

Kikuchi, Y., Horwitz, B., & Mishkin, M. (2010). Hierarchical auditory processing directed rostrally along the monkey's supratemporal plane. *Journal of Neuroscience*, 30(39), 13021–13030.

Korte, M., & Rauschecker, J. P. (1993). Auditory spatial tuning of cortical neurons is sharpened in cats with early blindness. *Journal of Neurophysiology*, 70(4), 1717–1721.

Krumbholz, K., Schönwiesner, M., Cramon, D. Y. v., Rübsamen, R., Shah, N. J., Zilles, K., & Fink, G. R. (2005). Representation of interaural temporal information from left and right auditory space in the human planum temporale and inferior parietal lobe. *Cerebral Cortex*, 15(3), 317–324.

Kusmierek, P., & Rauschecker, J. P. (2009). Functional specialization of medial auditory belt cortex in the alert rhesus monkey. *Journal of Neurophysiology*, 102(3), 1606–1622.

Kusmierek, P., Ortiz, M., & Rauschecker, J. P. (2012). Sound-identity processing in early areas of the auditory ventral stream in the macaque. *Journal of Neurophysiology*, 107(4), 1123–1141.

Lahav, A., Saltzman, E., & Schlaug, G. (2007). Action representation of sound: audiomotor recognition network while listening to newly acquired actions. *Journal of Neuroscience*, 27(2), 308–314.

Lakatos, P., Chen, C. M., O'Connell, M. N., Mills, A., & Schroeder, C. E. (2007). Neuronal oscillations and multisensory interaction in primary auditory cortex. *Neuron*, 53(2), 279–292.

Leaver, A., Van Lare, J. E., Zielinski, B. A., Halpern, A., & Rauschecker, J. P. (2009). Brain activation during anticipation of sound sequences. *Journal of Neuroscience*, 29(8), 2477–2485.

Leaver, A. M., & Rauschecker, J. P. (2010). Cortical representation of natural complex sounds: Effects of acoustic features and auditory object category. *Journal of Neuroscience*, 30(22), 7604–7612.

Lewis, J. W., & Van Essen, D. C. (2000). Corticocortical connections of visual, sensorimotor, and multimodal processing areas in the parietal lobe of the macaque monkey. *Journal of Comparative Neurology*, 428(1), 112–137.

Lewis, J. W., Brefczynski, J. A., Phinney, R. E., Janik, J. J., & DeYoe, E. A. (2005). Distinct cortical pathways for processing tool versus animal sounds. *Journal of Neuroscience*, 25(21), 5148–5158.

Liberman, A. M., Cooper, F. S., Shankweiler, D. P., & Studdert-Kennedy, M. (1967). Perception of the speech code. *Psychological Review*, 74, 431–461.

Lomber, S. G., & Malhotra, S. (2008). Double dissociation of 'what' and 'where' processing in auditory cortex. *Nature Neuroscience*, 11(5), 609–616.

Lueschow, A., Miller, E. K., & Desimone, R. (1994). Inferior temporal mechanisms for invariant object recognition. *Cerebral Cortex*, 4(5), 523–531.

Maeder, P. P., Meuli, R. A., Adriani, M., Bellmann, A., Fornari, E., Thiran, J. P., et al. (2001). Distinct pathways involved in sound recognition and localization: A human fMRI study. *NeuroImage*, 14(4), 802–816.

Malhotra, S., Hall, A. J., & Lomber, S. G. (2004). Cortical control of sound localization in the cat: Unilateral cooling deactivation of 19 cerebral areas. *Journal of Neurophysiology*, 92(3), 1625–1643.

Margoliash, D., & Fortune, E. S. (1992). Temporal and harmonic combination-sensitive neurons in the zebra finch's HVc. *Journal of Neuroscience*, 12, 4309–4326.

Mendelson, J. R., & Cynader, M. S. (1985). Sensitivity of cat primary auditory cortex (AI) neurons to the direction and rate of frequency modulation. *Brain Research*, 327, 331–335.

Meredith, M. A., & Clemo, H. R. (1989). Auditory cortical projection from the anterior ectosylvian sulcus (field AES) to the superior colliculus in the cat: An anatomical and electrophysiological study. *Journal of Comparative Neurology*, 289(4), 687–707.

Merzenich, M. M., & Brugge, J. F. (1973). Representation of the cochlear partition on the superior temporal plane of the macaque monkey. *Brain Research*, 50, 275–296.

Micheyl, C., Tian, B., Carlyon, R. P., & Rauschecker, J. P. (2005). Perceptual organization of sound sequences in the auditory cortex of awake macaques. *Neuron*, 48(1), 139–148.

Middlebrooks, J. C., Clock, A. E., Xu, L., & Green, D. M. (1994). A panoramic code for sound location by cortical neurons. *Science*, 264(5160), 842–844.

Morel, A., Garraghty, P. E., & Kaas, J. H. (1993). Tonotopic organization, architectonic fields, and connections of auditory cortex in macaque monkeys. *Journal of Comparative Neurology*, 335(3), 437–459.

Müller-Preuss, P., & Ploog, D. (1981). Inhibition of auditory cortical neurons during phonation. *Brain Research*, 215(1–2), 61–76.

Mullette-Gillman, O. A., Cohen, Y. E., & Groh, J. M. (2005). Eye-centered, head-centered, and complex coding of visual and auditory targets in the intraparietal sulcus. *Journal of Neurophysiology*, 94(4), 2331–2352.

Narins, P. M., & Capranica, R. R. (1980). Neural adaptations for processing the two-note call of the Puerto Rican treefrog, *Eleutherodactylus coqui*. *Brain, Behavior and Evolution*, 17, 48–66.
Numminen, J., Salmelin, R., & Hari, R. (1999). Subject's own speech reduces reactivity of the human auditory cortex. *Neuroscience Letters*, 265(2), 119–122.
Pandya, D. N., & Sanides, F. (1972). Architectonic parcellation of the temporal operculum in rhesus monkey and its projection pattern. *Zeitschrift für Anatomie und Entwicklungsgeschichte*, 139, 127–161.
Perry, D. W., Zatorre, R. J., Petrides, M., Alivisatos, B., Meyer, E., & Evans, A. C. (1999). Localization of cerebral activity during simple singing. *NeuroReport*, 10(18), 3979–3984.
Petrides, M., & Pandya, D. N. (1984). Projections to the frontal cortex from the posterior parietal region in the rhesus monkey. *Journal of Comparative Neurology*, 228(1), 105–116.
Petrides, M., & Pandya, D. N. (2009). Distinct parietal and temporal pathways to the homologues of Broca's area in the monkey. *Public Library of Science Biology*, 7(8), e1000170.
Pizzamiglio, L., Aprile, T., Spitoni, G., Pitzalis, S., Bates, E., D'Amico, S., & Di Russo, F. (2005). Separate neural systems for processing action- or non-action-related sounds. *NeuroImage*, 24(3), 852–861.
Poremba, A., Saunders, R. C., Crane, A. M., Cook, M., Sokoloff, L., & Mishkin, M. (2003). Functional mapping of the primate auditory system. *Science*, 299(5606), 568–572.
Poremba, A., Malloy, M., Saunders, R. C., Carson, R. E., Herscovitch, P., & Mishkin, M. (2004). Species-specific calls evoke asymmetric activity in the monkey's temporal poles. *Nature*, 427(6973), 448–451.
Rajan, R., Aitkin, L. M., & Irvine, D. R. (1990a). Azimuthal sensitivity of neurons in primary auditory cortex of cats. II. Organization along frequency-band strips. *Journal of Neurophysiology*, 64(3), 888–902.
Rajan, R., Aitkin, L. M., Irvine, D. R., & McKay, J. (1990b). Azimuthal sensitivity of neurons in primary auditory cortex of cats. I. Types of sensitivity and the effects of variations in stimulus parameters. *Journal of Neurophysiology*, 64(3), 872–887.
Rauschecker, J. P. (1998a). Cortical processing of complex sounds. *Current Opinion in Neurobiology*, 8, 516–521.
Rauschecker, J. P. (1998b). Parallel processing in the auditory cortex of primates. *Audiology and Neuro-Otology*, 3, 86–103.
Rauschecker, J. P. (2005). Neural encoding and retrieval of sound sequences. *Annals of the New York Academy of Sciences*, 1060, 125–135.
Rauschecker, J. P. (2007). Cortical processing of auditory space: Pathways and plasticity. In F. Mast & L.Jäncke (Eds.), *Spatial processing in navigation, imagery, and perception* (pp. 389–410). New York: Springer.
Rauschecker, J. P. (2011). An expanded role for the dorsal auditory pathway in sensorimotor integration and control. *Hearing Research*, 271, 16–25.
Rauschecker, J. P., & Korte, M. (1993). Auditory compensation for early blindness in cat cerebral cortex. *Journal of Neuroscience*, 13, 4538–4548.
Rauschecker, J. P., & Scott, S. K. (2009). Maps and streams in the auditory cortex: nonhuman primates illuminate human speech processing. *Nature Neuroscience*, 12(6), 718–724.
Rauschecker, J. P., & Tian, B. (2000). Mechanisms and streams for processing of "what" and "where" in auditory cortex. *Proceedings of the National Academy of Sciences of the USA*, 97(22), 11800–11806.
Rauschecker, J. P., & Tian, B. (2004). Processing of band-passed noise in the lateral auditory belt cortex of the rhesus monkey. *Journal of Neurophysiology*, 91(6), 2578–2589.
Rauschecker, J. P., & Tian, B. (2005). Hierarchic processing of communication sounds in primates. In J. S. Kanwal & G. Ehret (Eds.), *Behavior and neurodynamics for auditory communication*. Cambridge, UK: Cambridge University Press.
Rauschecker, J. P., Tian, B., & Hauser, M. (1995). Processing of complex sounds in the macaque nonprimary auditory cortex. *Science*, 268(5207), 111–114.
Rauschecker, J. P., Tian, B., Pons, T., & Mishkin, M. (1997). Serial and parallel processing in rhesus monkey auditory cortex. *Journal of Comparative Neurology*, 382, 89–103.

Ravizza, R. J., & Masterton, B. (1972). Contribution of neocortex to sound localization in opossum (*Didelphis virginiana*). *Journal of Neurophysiology*, 35(3), 344–356.
Reale, R. A., & Imig, T. J. (1980). Tonotopic organization in auditory cortex of the cat. *Journal of Comparative Neurology*, 192, 265–291.
Recanzone, G. H. (2000). Spatial processing in the auditory cortex of the macaque monkey. *Proceedings of the National Academy of Sciences of the USA*, 97(22), 11829–11835.
Recanzone, G. H., Guard, D. C., Phan, M. L., & Su, T. K. (2000). Correlation between the activity of single auditory cortical neurons and sound-localization behavior in the macaque monkey. *Journal of Neurophysiology*, 83(5), 2723–2739.
Remedios, R., Logothetis, N. K., & Kayser, C. (2009). Monkey drumming reveals common networks for perceiving vocal and nonvocal communication sounds. *Proceedings of the National Academy of Sciences of the USA*, 106(42), 18010–18015.
Repp, B. H. (2005). Sensorimotor synchronization: A review of the tapping literature. *Psychonomic Bulletin & Review*, 12(6), 969–992.
Rilling, J. K., Glasser, M. F., Preuss, T. M., Ma, X., Zhao, T., Hu, X., & Behrens, T. E. (2008). The evolution of the arcuate fasciculus revealed with comparative DTI. *Nature Neuroscience*, 11(4), 426–428.
Rizzolatti, G., Ferrari, P. F., Rozzi, S., & Fogassi, L. (2006). The inferior parietal lobule: where action becomes perception. *Novartis Foundation Symposium*, 270, 129–140; discussion 140–125, 164–129.
Romanski, L. M., & Goldman-Rakic, P. S. (2002). An auditory domain in primate prefrontal cortex. *Nature Neuroscience*, 5(1), 15–16.
Romanski, L. M., Tian, B., Fritz, J., Mishkin, M., Goldman-Rakic, P. S., & Rauschecker, J. P. (1999). Dual streams of auditory afferents target multiple domains in the primate prefrontal cortex. *Nature Neuroscience*, 2(12), 1131–1136.
Sabes, P. N. (2000). The planning and control of reaching movements. *Current Opinion in Neurobiology*, 10(6), 740–746.
Saur, D., Kreher, B. W., Schnell, S., Kummerer, D., Kellmeyer, P., Vry, M. S., et al. (2008). Ventral and dorsal pathways for language. *Proceedings of the National Academy of Sciences of the USA*, 105(46), 18035–18040.
Schubotz, R. I., Friederici, A. D., & von Cramon, D. Y. (2000). Time perception and motor timing: A common cortical and subcortical basis revealed by fMRI. *NeuroImage*, 11(1), 1–12.
Scott, S. K., Blank, C. C., Rosen, S., & Wise, R. J. S. (2000). Identification of a pathway for intelligible speech in the left temporal lobe. *Brain*, 123, 2400–2406.
Seltzer, B., & Pandya, D. N. (1994). Parietal, temporal, and occipital projections to cortex of the superior temporal sulcus in the rhesus monkey: A retrograde tracer study. *Journal of Comparative Neurology*, 343(3), 445–463.
Smiley, J. F., Hackett, T. A., Ulbert, I., Karmas, G., Lakatos, P., Javitt, D. C., & Schroeder, C. E. (2007). Multisensory convergence in auditory cortex, I. Cortical connections of the caudal superior temporal plane in macaque monkeys. *Journal of Comparative Neurology*, 502(6), 894–923.
Sperry, R. W. (1950). Neural basis of the spontaneous optokinetic response produced by visual inversion. *Journal of Comparative and Physiological Psychology*, 43(6), 482–489.
Stricanne, B., Andersen, R. A., & Mazzoni, P. (1996). Eye-centered, head-centered, and intermediate coding of remembered sound locations in area LIP. *Journal of Neurophysiology*, 76(3), 2071–2076.
Suga, N., O'Neill, W. E., & Manabe, T. (1978). Cortical neurons sensitive to combinations of information-bearing elements of biosonar signals in the mustache bat. *Science*, 200, 778–781.
Tata, M. S., & Ward, L. M. (2005a). Early phase of spatial mismatch negativity is localized to a posterior "where" auditory pathway. *Experimental Brain Research*, 167(3), 481–486.
Tata, M. S., & Ward, L. M. (2005b). Spatial attention modulates activity in a posterior "where" auditory pathway. *Neuropsychologia*, 43(4), 509–516.
Tian, B., & Rauschecker, J. P. (2004). Processing of frequency-modulated sounds in the lateral auditory belt cortex of the rhesus monkey. *Journal of Neurophysiology*, 92(5), 2993–3013.

Tian, B., Reser, D., Durham, A., Kustov, A., & Rauschecker, J. P. (2001). Functional specialization in rhesus monkey auditory cortex. *Science*, 292(5515), 290–293.
Tourville, J. A., Reilly, K. J., & Guenther, F. H. (2008). Neural mechanisms underlying auditory feedback control of speech. *NeuroImage*, 39(3), 1429–1443.
Von Holst, E., & Mittelstaedt, H. (1950). Das Reafferenzprinzip (Wechselwirkungen zwischen Zentralnervensystem und Peripherie). *Die Naturwissenschaften*, 37, 464–476.
Wang, X. (2000). On cortical coding of vocal communication sounds in primates. *Proceedings of the National Academy of Sciences of the USA*, 97, 11843–11849.
Warren, J. D., Zielinski, B. A., Green, G. G. R., Rauschecker, J. P., & Griffiths, T. D. (2002). Analysis of sound source motion by the human brain. *Neuron*, 34, 1–20.
Warren, J. E., Wise, R. J., & Warren, J. D. (2005). Sounds do-able: Auditory-motor transformations and the posterior temporal plane. *Trends in Neurosciences*, 28(12), 636–643.
Weeks, R. A., Aziz-Sultan, A., Bushara, K. O., Tian, B., Wessinger, C. M., Dang, N., et al. (1999). A PET study of human auditory spatial processing. *Neuroscience Letters*, 262(3), 155–158.
Wernicke, C. (1874). *Der aphasische Symptomencomplex: Eine psychologische Studie auf anatomischer Basis*. Breslau: Cohn & Weigert.
Wernicke, C. (1881). *Lehrbuch der Gehirnkrankheiten für Aerzte und Studirende*. Kassel, Berlin: Verlag Theodor Fischer.
Wessinger, C. M., VanMeter, J., Tian, B., Van Lare, J., Pekar, J., & Rauschecker, J. P. (2001). Hierarchical organization of the human auditory cortex revealed by functional magnetic resonance imaging. *Journal of Cognitive Neuroscience*, 13(1), 1–7.
Wilson, S. M., Saygin, A. P., Sereno, M. I., & Iacoboni, M. (2004). Listening to speech activates motor areas involved in speech production. *Nature Neuroscience*, 7(7), 701–702.
Wise, R. J., Scott, S. K., Blank, S. C., Mummery, C. J., Murphy, K., & Warburton, E. A. (2001). Separate neural subsystems within 'Wernicke's area'. *Brain*, 124(Pt 1), 83–95.
Wolpert, D. M., Ghahramani, Z., & Jordan, M. I. (1995). An internal model for sensorimotor integration. *Science*, 269(5232), 1880–1882.
Wolpert, D. M., Doya, K., & Kawato, M. (2003). A unifying computational framework for motor control and social interaction. *Philosophical Transactions of the Royal Society of London B: Biological Sciences*, 358(1431), 593–602.
Yu, J. J., & Young, E. D. (2000). Linear and nonlinear pathways of spectral information transmission in the cochlear nucleus. *Proceedings of the National Academy of Sciences of the USA*, 97(22), 11780–11786.
Zatorre, R. J., & Belin, P. (2001). Spectral and temporal processing in human auditory cortex. *Cerebral Cortex*, 11(10), 946–953.
Zatorre, R. J., Bouffard, M., Ahad, P., & Belin, P. (2002). Where is 'where' in the human auditory cortex? *Nature Neuroscience*, 5(9), 905–909.
Zatorre, R. J., Bouffard, M., & Belin, P. (2004). Sensitivity to auditory object features in human temporal neocortex. *Journal of Neuroscience*, 24(14), 3637–3642.
Zatorre, R. J., Chen, J. L., & Penhune, V. B. (2007). When the brain plays music: auditory-motor interactions in music perception and production. *Nature Reviews Neuroscience*, 8(7), 547–558.
Zimmer, U., & Macaluso, E. (2005). High binaural coherence determines successful sound localization and increased activity in posterior auditory areas. *Neuron*, 47(6), 893–905.

Chapter 3
Task-Related Activation of Auditory Cortex

Henning Scheich and Michael Brosch

Keywords Associative learning • Behavioral meaning • Cognitive • Contingent negative variation • Cross-modal • Memory • Motivation • Multisensory • Operant conditioning • Pavlovian conditioning • Prediction error

Abbreviations

BF	best frequency
CR	conditioned response
CS	conditioned stimulus
STRF	spectrotemporal receptive field
UR	unconditioned response
US	unconditioned stimulus

3.1 Introduction

In everyday language, a task is a set of actions that accomplish a job. Examples of auditory tasks are a doorman opening the door after the door bell has been rung, a musician playing a piece of music, and a physician listening to the heartbeat of a patient. For the purpose of this chapter, an auditory task is defined as an operation requiring the production of an intentional connection between sounds, other context-relevant environmental stimuli, and behavioral actions, the latter of which are responses executed

H. Scheich (✉) • M. Brosch
Leibniz Institut für Neurobiologie and Center for Behavioral Brain Sciences,
Brenneckestraße 6, 39118 Magdeburg, Germany
e-mail: SCHEICH@LIN-MAGDEBURG.DE; BROSCH@LIN-MAGDEBURG.DE

Y.E. Cohen et al. (eds.), *Neural Correlates of Auditory Cognition*, Springer Handbook of Auditory Research 45, DOI 10.1007/978-1-4614-2350-8_3,
© Springer Science+Business Media New York 2013

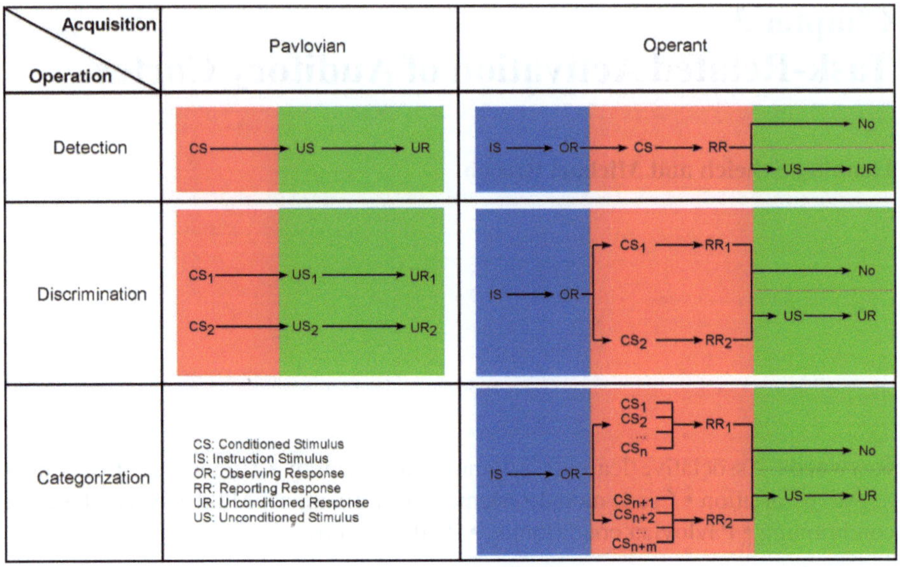

Fig. 3.1 Elements of tasks for different logical operations to be performed on stimuli and for different task acquisition schemes. Red shading indicates conditioned stimuli (CS) and, in the right tier, reporting responses (RR). Green shading indicates unconditioned stimuli (US) and unconditioned responses (UR). In the right column, blue shading indicates the instruction stimulus (IS) and the observing response (OR)

because subjects are motivated to do so. In the last example, for instance, the context is examination of a patient lying in her bed, a stethoscope is used for this examination, and the auditory analysis is performed for the purpose of a diagnosis. A prerequisite for being able to perform a task is learning the task. This distinguishes a task from innate, reflex-like behavior and automatic sensory–motor mappings.

Elements of different tasks performed under controlled laboratory situations are schematized in Fig. 3.1. These are instructions, conditioned stimuli (CS) and unconditioned stimuli (US), as well as observing and reporting (conditioned) and unconditioned behavioral responses (CR and UR). Thus most tasks used in the laboratory may be considered an orderly sequence of different sensory, behavioral, and motivational elements that must be associated for appropriate task performance. The rationale for learning these associations in experiments with humans is usually provided by verbal instructions whereas in experiments on animals, learning the task associations, in principle, occurs by trial and error. This requires a strong primary incentive and reinforcement feedback to own actions as motivations to engage in the task.

The specification "auditory task" renders the aforementioned definition somewhat ambiguous. It could mean either, in a narrow sense, a problem that can be solved by listening alone, such as a purely perceptual task directed at property differences of sounds or, in a wider sense, a problem that also encompasses association with other senses, behavioral actions, and reinforcers. In this chapter this distinction of perceptual and behavioral learning is maintained but it will become clear that most often in auditory tasks the two aspects grade into one another. Further, because the topic does not deal with machine processing of acoustics but with the brain, and especially with auditory cortex, a task carried out by an individual acquires a personal relevance. Solving a task is embedded in a meaningful context and at least requires a personal motivation. Consequently the sounds involved, whatever their previous meaning, undergo a task-related transformation of meaning. As discussed in Section 3.2, the meaning problem of sounds does not arise only with tasks but is fundamental for any audition already due to the way in which sounds naturally arise and propagate as occasional signals from material objects. For any subject, novel sounds are initially neutral and lack a specific meaning unless one observes their source. But they could surprise the observer and cause her to orient to the sound source. So in that sense novel sounds are treated as having a source that, however, needs to be defined for a reasonable attitude toward the sounds. In principle, this information that determines meaning of the novel sound cannot be deduced by directly addressing the sound through behavioral interactions (touching or saccading) but only indirectly by identifying its source and context of occurrence, that is, by extra-auditory information. Addressing the sound itself will only allow identification of its auditory perceptual properties. This is unlike the situation for vision or somatosensation of novel material objects that are comprehensively defined in a given spatial framework, that is, not only by location but also by size, shape, surface, and color at this location. Consequently vision and somatosensation of novel objects allow direct behavioral interaction with these objects in the same modality to know more about them. Therefore, it is no surprise that the phenomenology of task-related neuronal activities in auditory cortex to be described below is remarkably rich and multimodal and not restricted to the representation of the sounds alone. It is argued here that this phenomenology can be better understood by focusing on the meaning problem of sounds (Scheich & Ohl, 2010; Scheich et al., 2011).

Traditionally, in auditory system research on animals and humans the neuronal and network analysis at all levels has focused on mechanisms of sound representation and coding. This has been studied by passively exposing (anesthetized) subjects to sounds and parametrically varying sound properties. The large body of knowledge and concepts of functional organization and stimulus processing developed from this exposure approach also forms the basis of interpretation for much of the rapidly developing human auditory cortex research made possible with the advent of functional imaging methods, especially functional magnetic resonance imaging (fMRI). It should be noted, however, that the exposure approach, which does not control for what the subject finally does with the stimuli, implicitly follows a very special concept of functional specialization and hierarchical brain organization: It assumes that the sole role of auditory cortex is to analyze sound properties whereas other parts of the brain use this information to organize behavior and cognition (Fuster, 2005). This is saying that the same experimental approach can be used for auditory cortex as for the ascending auditory system, that is, that it is sufficient to measure sensory transfer functions. But it has recently been found that auditory cortex, and even primary auditory cortex, receives only a fraction of its inputs from the ascending auditory system and that the bulk of the inputs is from extremely diverse other subcortical and especially cortical sources (Budinger & Scheich, 2009). Thus, already primary auditory cortex should be considered an interface for combining ascending and descending auditory and nonauditory information from numerous sources. This does not yet predict that auditory task structure is reflected in auditory cortex but makes it implausible that the stimulus approach can reveal all functions of auditory cortex.

It should be acknowledged that stimulus exposure in its simplest form with tones has led to the distinction of multiple tonotopically organized from nontonotopically organized auditory cortical fields, and this still forms the main physiological property for distinguishing different fields. But the extension of this exposure approach with parametrically varied spectrally and temporally more complex stimulus dimensions has not led to a clearer division of functional specification of auditory cortex into fields devoted to such complex dimensions. In spite of some complex sensitivities of individual neurons that may be more often found in some fields than in others, the distinguishing properties of fields are statistically less pronounced than the common properties (Eggermont, 1998; Scheich & Ohl, 2010). This poor stimulus-related delineation of fields is all the more remarkable as different anatomical input connections in some spatial correspondence to the tonotopy-based divisions provide evidence of separate functional fields in animals (Kaas & Hackett, 2000).

In summary, the existence of multiple fields in auditory cortex is undisputed but not much is known of their functional roles from stimulus exposure experiments. Therefore, a number of groups studying auditory cortex have changed their research strategy toward monitoring auditory cortex while animals and humans perform auditory tasks. A number of methods have been used, including lesions, reversible pharmacological blocking, and cooling of cortex to visual imaging methods and multiple electrode recordings. The strategy has also been pursued with functional imaging in human subjects and has led to remarkable demonstrations of task-related changes of sound representation (Brechmann & Scheich, 2005; Brechmann et al., 2007). However, owing to the still weak evidence of human auditory cortex divisions and the nature of the imaging method, the changing sound representations are difficult to interpret (Ohl & Scheich, 2010; Scheich et al., 2011). First, fMRI is based on reactively increased flow of oxygenated blood around activated neuron clusters, with a spatial resolution in the millimeter range. Thus, this blood oxygen level–dependent (BOLD) response represents no direct neuronal signals, and it has a temporal resolution in the range of seconds. Second, as a difference measurement, it requires a suitable control state of auditory cortex to determine activation, which for tasks is a problem in itself. Third, lacking neuronal resolution, the imaged signal is incapable of separating the simultaneous excitatory and inhibitory processes and the contributions of the numerous cortical neuron types that are fundamental to neuronal computations in a cortical network. It only shows a spatial and temporal average of such processes. Therefore macroscopic imaging results are only occasionally be referred to in this chapter.

3.2 Meaning of Natural Sounds in Environments and in Tasks

Most environments are crowed with different sounds to which we easily adapt when they are familiar and soft. That we still hear them and make use of them subconsciously is obvious as our brain readily detects any change in an ongoing familiar sequence of sounds (mismatch response, e.g., Näätänen et al., 2011). We also use subconsciously many self-generated sounds as auditory feedback to control behaviors, such as when we walk on uneven grounds, handle objects, or shift gears in a car because the engine produces a higher pitch. Only when someone hears a novel sound or a known sound but unexpected in a given context might he pay special attention and wonder what this could mean. By contrast, already familiar sounds are instantly associated with an object, such as a croak close to a pond with a frog or a characteristic screech with a breaking car.

This sketch sets the scene for a reconsideration of one of the basic mysteries of the auditory modality, beyond pattern recognition and sound localization, namely, assigning meaning to sounds. Usually, this semantic question is brought up first in connection with high-level cognitive capabilities such as the use of language or in musical compositions. In languages, sequences of sounds, that is, of phonemes, symbolize a specific meaning. In music, sound sequences usually do not acquire such meaning but rather an "emotional" meaning, aesthetic and Gestalt principles that have a meaning for listeners, that is, can be interpreted.

It is argued here that the need for sound interpretation by learning is intrinsic to the behavioral use of audition from the start and does not arise only with specific auditory task. As described in Section 3.1, the need for interpretation is more fundamental and demanding than in the visual or somatosensory modality, in which the material objects of sight or touch can be directly described with properties in space and therefore directly behaviorally addressed. This is not the case for sounds that are occasional signals that occur only when forces act upon or within material objects and make them vibrate. Because the resulting sound waves, unlike light beams reflected by objects, do not in any specific way describe or image the objects and motions involved, a novel sound by itself is initially neutral to a listener.

Environmental sounds obtain their source meaning, that is, by "what" respectively "where and when" they are emitted from associating information on their source objects and/or the context of their occurrence, thus chiefly by nonauditory information. Note that these "what and where" concepts of sound meaning are fundamentally different from the purely auditory concepts of "what and where" processing streams, as initially formulated by Rauschecker and colleagues (see Rauschecker, Chapter 2). The latter refer to the auditory properties of a sound pattern (what) and to the localization of the sound source in space (where). Thus, they refer to the perceptual domain of audition and make no assumptions on the source meaning that is provided by the nonauditory information on the objects and contexts of sound generation.

This associative principle can now be applied to the meaning generation in auditory tasks.In section 3.1 it was noted that for task performance, other sensory including context, behavioral, and motivational elements all must be in some way associated. This shifts the emphasis from merely recognizing the source of a sound to developing a personally meaningful (motivated) behavior in response to or with the help of the sound. For most natural auditory tasks the source meaning of sounds is previously known and then does not even require vision to orchestrate a reasonable behavior, for example, when a cat in the dark makes a precise jump to catch a mouse after hearing rustling noises on the ground. But this reference of a familiar sound to a known source is not necessary for learning an auditory task. This is easily shown in the laboratory when animals may learn to jump a hurdle upon presentation of whatever sound from a loudspeaker at whatever position to avoid a subsequent footshock.

Thus, the emphasis of the meaning generation is on the association of a sound with a behavior that satisfies a motivation. The associated behavior after this learning may be sound-conditioned reflexes, conditioned intentional responses, and, in human subjects, also verbal responses depending on the types of tasks. But the connections with a specific behavior is merely a formal requirement for an experimental evidence of task performance, that is, serves as a quantifiable read-out. Otherwise one would not know whether and when the task is correctly performed. In natural practice of tasks, the association of a sound with behaviors is often in terms of several behavioral options that, dependent on the circumstances, become available to reach the same goal (Liddell, 1942). To catch a singing bird, a cat may hide on the ground, jump a fence, or climb a tree.

This goal aspect of a task leads to motivations. Motivation is a core element and driving force for any task performance, either as a primary incentive or as a reinforcement of chosen behavior. The concept of motivation covers multiple and usually very concrete items or situations such as food, water, social company, and sex for rewards, or adversities such as injury, pain, current, cold, heat, and stress to be avoided. Thus, these motivational elements have an innate meaning for an individual in terms of advantage or disadvantage determining basically behavioral approach or avoidance strategies. Notably, such concrete items or situations are perceived as nonauditory information similar to the behavioral elements of the task and must be associated with the sounds involved in the task.

Taking into account these various aspects of associating information in an auditory task, it is now possible to define generally the behavioral meaning of sounds in such tasks.

This behavioral meaning can be conceived as the associative information content of a sound, including context information, that allows an individual to solve a task, or more generally, to choose among sound-related behavioral alternatives (Scheich et al., 2011). This is an operational definition that allows one to search for the required associations and their mechanisms in the brain, for example, in auditory cortex. It should also be noted that it is fundamentally different from definitions of the source meaning of sounds (see earlier). This meaning covers the association of what causes the sound or, at least, where and when the sound reliably occurs (context), and not what an individual might do in consequence of the sound occurrence. But both have in common that the meanings of novel sounds can be derived only by associating them with nonauditory information.

If one assumes that auditory cortex is a site of the necessary associations to establish the task-related behavioral meaning of a sound, concepts of neuronal representation of auditory as well of nonauditory information become salient. For example, are all these components of a task represented by separate neuronal activity in auditory cortex and influence each other or do the nonauditory influences merely modify the representation of the sounds involved? Further, since learning of a task occurs, how do the dynamics of this process lead to a stable meaning of a sound represented? As the experimental evidence from auditory cortex described in this chapter points to multiple nonauditory representations as well as to modifications of the auditory responses, they can be tentatively classified into (1) nonauditory sensory motor representations, (2) motivational representations, and (3) changes of the representation of sound that become relevant in the task. There is another important type of influence on the sound representation, here called deductive "logical operations." These hierarchically related operations that will become crucial in the following are detection, discrimination, and categorization learning. They determine the same–different relationships of the acquired meaning of sounds with respect to other sounds (Fig. 3.2).

Detection, discrimination, and categorization are fundamental distinctions applicable to the meaning problem of any sensory item (Cohen & Nagel, 1993). As will be described, they implicitly result from the task structure, namely, whether only one or several sensory items acquire meaning during the learning of the same task.

In summary, there is good evidence that numerous details of the structure of a task, including auditory and nonauditory stimulus events and motivational aspects, are represented with high temporal fidelity even in primary auditory cortex. As a conceptual outlook on the following sections, it is our tenet that these co-representations of task contingencies can be understood only if one assumes that auditory cortex is not only involved in analysis of sounds per se but also tries to derive the correct behavioral conclusions on sounds from these contingencies to solve tasks. Even though the co-representation of nonauditory contingencies in auditory cortex can be established only in cooperation with other brain areas, this semantic concept (Scheich et al., 2011), nevertheless, attributes a much higher degree of task-related autonomy to auditory cortex than hitherto assumed (for state of the art see texts on auditory cortex by König et al., 2005 and Winer and Schreiner, 2011).

3.3 Auditory Task Structure for Animals

In spite of the artificial settings in laboratories, many auditory laboratory tasks closely mimic essential aspects of the ways animals and humans naturally make experiences with sounds and learn to adapt their behaviors. Because no one instructs animals in nature how to deal with a novel sound, they instruct themselves by trial and error while forming various associations with other information and making inferences about sounds. Historically, from the perspective of the associations formed, two types of learning processes are distinguished in animals, so-called Pavlovian conditioning (Pavlov, 1960) and operant (instrumental) conditioning (Thorndike, 1933). The term conditioning merely points to the fact that the learning process usually requires multiple cumulative experiences to separate correct associations from false associations (trial and error) and to consolidate the result by applications. This is irrespective of the fact that strong negative experiences may lead to lasting associations and subsequent avoidance in a one-trial learning process.

Contrary to common schemes of learning and memory principles (Squire & Zola-Morgan, 1991), conditioning and particularly operant conditioning, unless they train perceptual or motor skills, are not simple primordial forms of learning (Fig. 3.3). They allow acquisition of highly cognitive concepts and abstractions

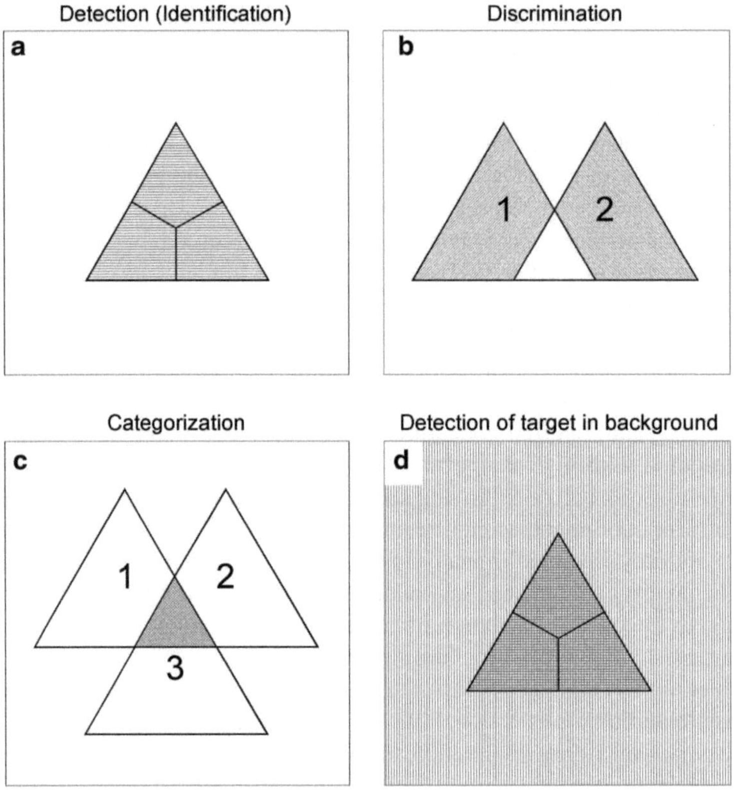

Fig. 3.2 Basic logical operations that can be performed on sounds, or any other sensory stimuli, during task learning. Illustrations by Venn diagrams in multidimensional property space of sounds (Cohen and Nagel, 1993). The logical operations are detection, discrimination, and categorization; they determine the deductive logical same–different relationships of a given sound to any other sounds. They are inherent requirements for attributing behavioral meaning (see text) to a given sound. (**a**) Detection learning of a given sound is based on potentially all of its spectrotemporal properties (horizontal shading in the triangle), even though some are most salient. Detection, also called identification, allows recognition of a given sound as the same upon reoccurrence (familiar sound as opposed to novel sounds). When a single sound has acquired a meaning by associating it to a behavioral response and reinforcement in a task, this meaning, in a test phase, may be spontaneously extended to other similar sounds. Consequently, behavioral responses to sounds after detection learning usually show a shallow generalization gradient toward similar sounds (Ohl et al., 2001). (**b**) Discrimination learning specifies the distinguishing properties of a given sound in relation to other sounds (here shown for one other). When sounds have partially the same properties this leads to the distinction of common properties (small white triangle) from exclusive properties (left or right shading). The latter become the basis of the discrimination. In discrimination learning, always several sounds, at least two, must be involved that either acquire different meanings or one acquires one meaning and the others explicitly not that meaning. For true discrimination learning; it is not sufficient to learn the meaning of one sound and to compare other sounds in a testing phase as control stimuli. (**c**) Categorization learning is a process by which a multitude of different sounds during training acquire the same meaning (here shown for three sounds). This is based on relevant common properties (crosshatched triangle) that become specified by experiences that the corresponding sounds have the same behavioral meaning. Thus, categorization defines sounds relationally like discrimination but, opposite to discrimination, with respect to common or similar properties. Nevertheless, categorization learning requires discrimination

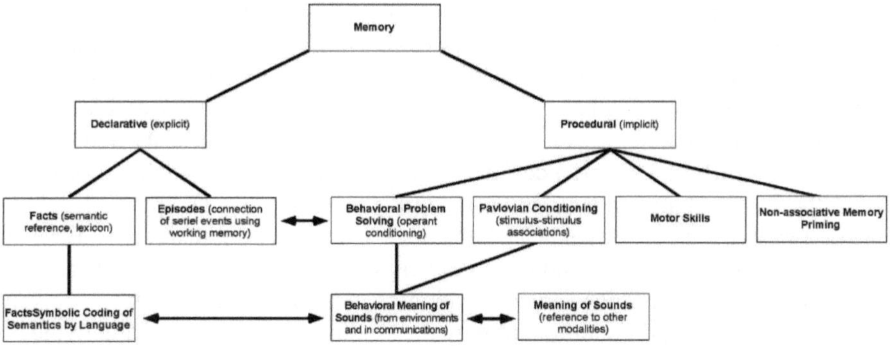

Fig. 3.3 Taxonomy of memory types from the perspective of memory contents that are established by different types of learning. Modification of the original scheme proposed by Squire and Zola-Morgan (1991). In essence, the separation of declarative and procedural memory is retained from this earlier proposal in the light of the human cognitive specializations. However, their independence is given up for episodic learning and operant conditioning because both are facets of experiences with personal involvement (self-learning) and, in the case of episodic experiences, frequently also incorporate decisions or deliberations on own actions. Episodic learning and memory and operant conditioning are therefore conceived to represent parallel and interactive processes in humans. With respect to the special nature of auditory memories, there seems to be a connection between Pavlovian conditioning as well as operant conditioning of sound meaning and acoustic coding of meaning in languages. Spoken languages refer to items and events by symbolic codes of sounds that are culturally established and not available to animals. However, the underlying principle that sounds always refer to something else that may not be visible or in reach but is behaviorally relevant is already used in animal auditory learning and memory

Fig. 3.2 (continued) as a first step to determine which properties of the sounds are different and which are comparable to specify the relevant common and to exclude the irrelevant accidental properties. A category typically covers a range of variability of the common properties of category members. Although this within-category variability is neglected leading to a unified categorical meaning, category boundaries are often extremely sharp. In particular, if a common boundary with a neighboring category is formed (e.g., for speech phonemes), percepts and behavioral responses may show step-like changes when a sound property change transcends the category boundary. Most relevant for cognition, a meaning category, once established, allows prediction of the meaning of novel sounds if they share the common relevant properties. (**d**) Target sound on background detection learning. This process, as a first step, requires one to discriminate the target sound (horizontal shading) from the simultaneous background sound (vertical shading) by discrimination before a target acquires a meaning. But different from usual discrimination learning (**b**), in which the difference is determined by sequential comparisons of sounds, this discrimination needs a simultaneous comparison of all sound properties that belong to either the target or the background. This is achieved mainly by a coherence analysis of sound properties in the temporal domain using mechanisms to determine co-modulations of elements that stem from one sound source and by onset-asynchrony analysis of elements that stem from different sound sources. Note that the mechanisms of this discrimination within an actual sound wave where the components sum up are fundamentally different from figure–ground separation in vision where a foreground object by concealing a background can be spatially defined by its boundaries to the background

such as meaning categories of items and sequential concepts of events (Selezneva et al., 2006; Scheich & Ohl, 2010; Scheich et al., 2011). The two types of conditioning, in principle, make no special assumptions on the motivation for performing a task and the corresponding behavioral meaning of sounds. The value assignment of novel sounds is theoretically ambivalent, which often elicits curiosity and exploratory behavior. But by experiences and the associations formed in nature or in the laboratory, a sound may turn out to be appetitive (ensuring regular approach behavior) or aversive (ensuring avoidance behavior). This, however, disregards different species-specific dispositions and biases of behavior, as those of so-called flight animals and of predators. This valence question of sound interpretation has to be kept separate from the required motor strategy in response to the sounds, namely the go or no-go strategy distinguished in tasks, which has been claimed to affect performance (Guitart-Masip et al., 2011). But avoiding punishment, for example, in the shuttlebox during discrimination learning, may require a go response for one sound and a no-go response for a different sound. Even though the prospect of rewards in most animals causes approach, suppressing a go-response, such as in an ambush for a prey, is also part of the natural behavioral repertoire of many animals.

3.3.1 Pavlovian Conditioning

Before describing the various and complex phenomenologies of task learning correlates in the auditory cortex, it is mandatory to understand thoroughly the task structures underlying the different types of conditioning because their elements are represented in auditory cortex (Ohl & Scheich, 2010; Brosch et al., 2011b; Scheich et al., 2011). In Pavlovian conditioning, as shown in the seminal dog experiments (Pavlov, 1960), a previously neutral sound, by frequently preceding the sight of food that promises feeding, acquires the meaning of a food announcer. Through these experiences the sound eventually stimulates salivation, which is usually triggered by the sight of food. The sight of food is the primarily meaningful unconditioned stimulus (US) and the sound becomes the conditioned stimulus (CS) that elicits a conditioned response (CR), that is, the same response as originally elicited by the US as an unconditioned response (UR; see Fig. 3.1). The behavioral output measure for the learning in this task is the increment of producing reliable CRs (salivations) to the CS sound across trials. Such learning data show a typical asymptotic time course. Similarly, in so-called fear conditioning of rodents, a sound frequently preceding a short footshock (that in some rodents elicits reflex-like initial freezing behavior), on reoccurrence will also elicit freezing. An example of a US-elicited protective reflex is the nictitating membrane response that can be triggered by an air puff to an eye in animals and humans (Thompson, 2005). The eyelid closure is easily conditioned by a preceding sound.

In essence, the structure of Pavlovian conditioning, as its core element, contains a stimulus–stimulus association between the CS and US. Thereby, the US already must have a behavioral meaning so that the preceding stimulus as an announcer can acquire the same meaning. In Pavlovian conditioning no development of a new behavior is involved that goes beyond the UR already connected to the meaningful US. Although most Pavlovian paradigms use responses that are either driven by the autonomic nervous system or are reflex-like, that is, cannot be voluntarily suppressed or expressed, in others there may be more active components of the response. For example, in the Pavlovian paradigm intensively used to study conditioning of midbrain dopaminergic neurons in monkeys (Schultz, 2006), subjects were trained to exhibit anticipatory licking after a visual or auditory stimulus but before some liquid was delivered from a spout just in front of the mouth. This is an active consumatory behavior that, unlike automatic behavior (salivation), can be voluntarily controlled. Nevertheless, the association between the sight of the spout and the delivery of water is so close in trained animals that the licking is almost automatic. The US with its behavioral meaning (such as reward or punishment) also provides the motivational aspect of the associative learning in the task. As demonstrated by the preceding examples using food sight, footshocks, and air puffs, the USs to which the sounds become associated are typically nonauditory. This is an important point for understanding how meaning is established for novel sounds and for understanding the phenomenology of training-induced and task-related activations in auditory cortex.

A key aspect of task structure is also the temporal contiguity of conditioned and unconditioned stimuli, a temporal relationship that is intimately connected to time domain neuronal mechanisms by which the association is formed (Schultz, 2006). This is not only clear by the fact that a change of a previously learned fixed temporal relationship between CS and US requires relearning. Even more critical is the distinction of a US (the sound) that starts earlier but overlaps with the onset of the US (so-called delay conditioning) from a CS that precedes the CS with some interval between the two (so-called trace conditioning). Whereas delay conditioning is a relatively simple learning task, corresponding trace conditioning using the same cues is not only more time consuming, or in lower mammals even impossible, but also seems to involve brain structures typically involved in temporally complex cognitive learning processes such as hippocampus and prefrontal cortex (Woodruff-Pak & Disterhoft, 2008).

The conceptualization of the characteristic asymptotic time course of Pavlovian conditioning has led to the idea of a prediction mechanism that can explain the asymptote and other properties (Rescorla & Wagner, 1972). In essence, the strength of the association between the CS and the US is controlled by a prediction error, that is, a sort of surprise of experiencing that an initially neutral sound is reliably followed by a meaningful US. The prediction error is large in initial trials, which increases the likelihood that already CSs are followed by CRs. Thus learning progress is fast. When the reliability of the CS–US sequence accumulates the prediction error in subsequent trials diminishes, and even though more frequent CRs occur the progress becomes asymptotic to a final level of performance. The Rescorla–Wagner model to a first approximation describes the behavioral data that have been obtained in auditory cortex research in Pavlovian conditioned as well as in operantly conditioned animals. Further, a neuronal correlate of a prediction error mechanism has been localized in auditory cortex of operantly conditioned monkeys (see 3.8.2; Brosch et al., 2011b) which, however, is different from the mechanism derived for midbrain dopaminergic neurons in Pavlovian conditioned monkeys (Schultz, 2006).

3.3.2 Operant (Instrumental) Conditioning

In auditory Pavlovian conditioning, by the association of an auditory CS with a US, the behavioral meaning inherent to the US and consequently also the behavior originally elicited by the US is transferred to the preceding CS. In essence, Pavlovian conditioning assigns a value and meaning to sensory stimuli that come from the external world or modify their meaning (Behrens et al., 2007) and in this sense is focused on what has been called "world learning" (Colomb & Brembs, 2010). But Pavlovian conditioning has a limited operating range and versatility, at least in the laboratory, because the choice of meaningful USs each automatically generating a different URs is rather limited. This is compatible with detection learning schemes but limits to perform stimulus discrimination with Pavlovian conditioning schemes. One of the most important applications of Pavlovian conditioning in modern life are advertisement strategies for selling products. They use combinations of product images with unrelated but desirable objects or with emotionally positive stimuli, including music, to make products attractive.

By contrast, operant conditioning (Skinner, 1938) is focused on assigning value and meaning to ones own actions that lead to personal goals (sometimes called self-learning, Colomb & Brembs, 2010). The core elements of the task structure are new behaviors, also termed conditioned responses, that try to cope with a situation. (Note that the term conditioned response has different meanings in the context of Pavlovian and operant conditioning, i.e., referring either to automatic or to controllable behavioral responses). These behaviors are either accidentally executed by animals or specially modified in a given situation and allow the gain of an advantage (reward) or avoidance of a disadvantage (punishment). For example,, the sight of food out of immediate reach might require climbing a tree or circumventing an obstacle toward catching a prey or opening a food dispenser. Conversely, the sight of a predator may also motivate climbing a tree for evasion or heading for a hole, and the experience of footshocks in a shuttlebox may lead to their avoidance by jumping across a hurdle. The consequences of the behavior in these cases are used as a feedback to drive the behavioral learning. These consequences in terms of stimuli can be of the same nonauditory nature as the US in Pavlovian conditioning, such as food or footshocks. They also represent the same motivational elements for the learning. But their role in the task structure is unlike that of an US as they control by feedback the adequate choices or modifications of behaviors.

What is then the role of a preceding sound, as a CS, and how is it integrated into the operant task structure? In some sense the role is similar to the CS in Pavlovian conditioning in that it announces the relevant subsequent events. However, the acquired meaning of the CS does not replace that of an US, for example, it does not indicate that food will be automatically available after the CS replaces the sight of food but that after the CS a specific action (a CR) is required; for example, a dispenser can be opened by lever pressing or by nose poking to obtain the food. In this case the action plan, including its aim, is transferred as a new meaning to the preceding sound. Thus, a novel stimulus–response association may be formed by operant conditioning in which neither the stimulus nor the response has a preconceived meaning in this context. Learning can be tested by measuring the actions as CRs (such as lever presses) after the CS. In contrast to Pavlovian conditioning with its stimulus–stimulus association, the behavioral outcome plays the central role by feedback (reinforcement) to the action. This can be characterized as an action-outcome association with the CS stimulus playing the role of the relevant context.

A simple realization of operant conditioning strategies occurs in the Skinner box. Here animals accidentally discover, for example, that performing certain actions (like pressing a lever) after a specific auditory stimulus is followed by food or other reward and subsequently systemize this discovery. This auditory detection experiment may be easily extended to a discrimination task in which a particular auditory stimulus announces that food can be obtained by lever pressing whereas no other auditory stimulus does. For task-related auditory cortex processing such discrimination experiments open mechanistic insights into the associations formed, especially by analyzing neuronal activities related to different types of mistakes that can occur (Selezneva et al., 2006). In the simplest case of discrimination experiments, correct behavioral response to one stimulus (hit) and a failure to respond (miss) can occur as well as generation of this response after the other stimulus (false alarm) or suppression of the response to this stimulus (correct rejection).

Owing to the relatively free choice of both CSs and CRs in comparison to Pavlovian conditioning, operant conditioning is highly versatile and allows one to address very complex questions of task-specific stimulus processing, procedural solutions, cognitive decisions, and motivational aspects in auditory cortex.

Often operant conditioning schemes contain Pavlovian elements (Fig. 3.1). One of these more complex examples of task-learning studied in auditory cortex is active footshock avoidance in a shuttlebox, in which a natural behavioral response becomes transferred during operant conditioning (Stark et al., 2004, 2007; Wetzel et al., 2008). A heuristic scheme encompassing auditory, nonauditory, procedural, and motivational aspects of the task is illustrated in Fig. 3.4. The conditioning starts with a CS, repetitive sounds, which after some delay are accompanied by the US, multiple mild footshocks, which elicit an escape responses (UR, Fig. 3.4a). If the animals cross the hurdle, from either side, they experience that the other compartment is safe, that is, they do not experience a US. This component of the training is essentially like in Pavlovian delay conditioning. But unlike in Pavlovian conditioning, the foot shocks are not eventually omitted by the experimenter to test whether the escape responses can be elicited by the sounds, that is, as CRs. Instead the training is continued. This leads to an orientation response toward and eventually to an approach of the hurdle already after the sound onset, even though animals still jump only after footshock onset, that is, animals reduce the exposure to the footshock. These behaviors together show that the animals have learned to expect the footshock after the CS onset.

Eventually, all animals in a shuttlebox will probe earlier jumping, that is, after the sound onset but already before the expected onset of the footshock, and consequently take the action of complete footshock avoidance (Fig. 3.4b). After very few experiences of successful footshock avoidance, the frequency of CRs rapidly

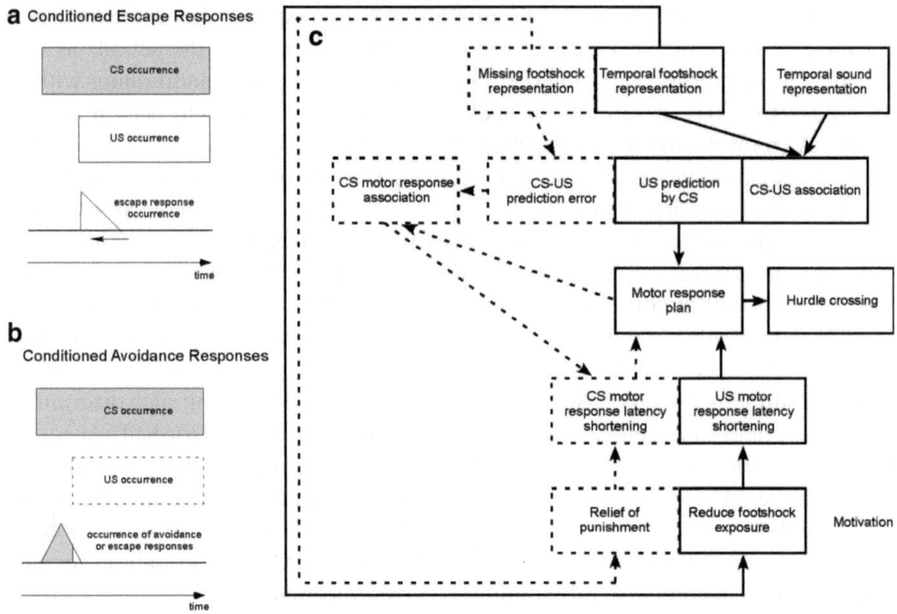

Fig. 3.4 Heuristic flow diagram of information processing in tone-conditioned escape and avoidance learning. (**a**) Conditioned escape responses. Temporal configuration of the tone CS, of the footshock US after a delay, and of the escape responses across the hurdle. Note that during learning escape occurs at successively shorter latencies after US onsets. A key element is the CS–US association that allows to predict the US and thereby provides behavioral response latency shortening to the US (small arrows), such that the hurdle jumps occur earlier in successive trials. This in turn results in a reduction of footshock exposure and serves as a motivational drive. The direct influence of the CS on the escape response is minor at this stage. (**b**) Conditioned avoidance responses. Temporal configuration of the tone CS the occasionally occurring footshock US (dashed), and of tone-conditioned avoidance responses and residual escape responses to the US. (**c**) Internal representations of the tone CS, of the footshock US, and the motor control together with some hypothetical neuronal functions and the interconnections of these elements, which can explain the transition from the initial escape strategy (solid arrows) to the avoidance strategy (dashed arrows). A key element is the CS–US association that allows to predict the US and thereby results in behavioral response latency shortening to the US. The hurdle jumps occur successively earlier. This in turn serves to reduce footshock exposure as a motivational drive. In contrast to the US influence, the direct influence of the CS on the behavioral response is minor at this stage. When at this stage of fast responses the US prediction by the CS leads to jumps that result in avoidance of the footshock, this generates a US prediction error due to predicted but lacking footshocks. This in turn promotes generation of a direct association between the CS and the actual motor planning and enables the CS to take control over the latency shortening mechanism, which results in further shortening the latency. The motivational drive is the relief of punishment. For the occasionally experienced footshocks the escape scheme is as in **a**

increases. Thus, the discovery that jumping sooner after CS onset entails complete US avoidance behavior leads to a change of role of the footshock US, namely from a driver of the escape response to a feedback information for successful or unsuccessful behavioral changes (Fig. 3.4c). The motivation to pursue this behavioral avenue is clearly relief from punishment, which seems to be equivalent to a direct reward (Konorski, 1967). Active footshock avoidance training, even though it starts with a Pavlovian punishment step, is very different from Pavlovian fear conditioning as described in the preceding text. Already the initial modification of escape behavior is distinct from a situation in which animals cannot avoid the footshock and freeze. The subsequent successful avoidance strategy, even though occasional footshocks leading to escape are still received, seems to solve the punishment problem for the animals altogether (at least for mild footshocks). Experienced animals continue various activities in the intertrial intervals without any sign of distress. They know, in principle, how to solve the problem.

Using an auditory CS to learn to behaviorally cope with a potentially rewarding or aversive situation is what auditory operant conditioning is all about. In the examples described a particular sound as a CS acquires the meaning of a command for a specific action, that is, "jump as fast as possible whenever the sound occurs." This can be conceptualized as *detection learning* of a sound. This operant conditioning can be extended to *discrimination learning*, which involves two different sounds that become a CS_1 and a CS_2 respectively, instigating alternative behaviors. For instance, in a shuttlebox the two sounds may acquire a go and no-go meaning, respectively (Ohl et al., 2001; Wetzel et al., 2008). Both detection and discrimination learning in the shuttlebox can be approximated by the Rescorla–Wagner model (1972). Shuttlebox learning is similar to an acoustic version of a traffic light system at street crossings installed for blind people. Even though the initial understanding of the "go" and "no-go" signal is by instruction to people, the behavioral framework to avoid casualties corresponds to the shuttlebox. From discrimination learning by operant conditioning it is also possible to develop *category learning* of different sounds based on similar sound properties that selectively acquire the same meaning (Ohl et al., 2001). As shown later, categorization into two meaning classes of sounds occurs spontaneously after a number of different samples of such sounds have acquired different meanings, respectively, by conditioning, and is then maintained. Thus, the categorization learning does not follow the Rescorla–Wagner model of gradual learning. Animals then realize on the basis of reoccurring similar sound properties that the categorical meaning of new sounds can be predicted.

Because in categorization commonly only some characteristic properties of items are selectively used while others are neglected as accidental, this process is typical of rule-based abstraction (Fig. 3.2c). The hallmark of rule-based abstractions, possible with operant conditioning, appears to be categorization of sequential changes of sounds (Fig. 3.5). In speech, different meanings are coded essentially by sequence permutations of a limited set of meaningless sounds, that is, phonemes. In terms of an auditory task structure this represents a step beyond attributing meaning to a single sound as a CS because the sequential change rather than the individual elements carries the information. This requires special associative mechanisms that can be studied with a simple model of sequential tones in an operant conditioning task in monkeys (Brosch et al., 2005). The results of this learning, described later, show that operant conditioning has a much wider cognitive scope and function during learning processes than hitherto believed and may even be a key element of speech recognition and language acquisition (Fig. 3.3).

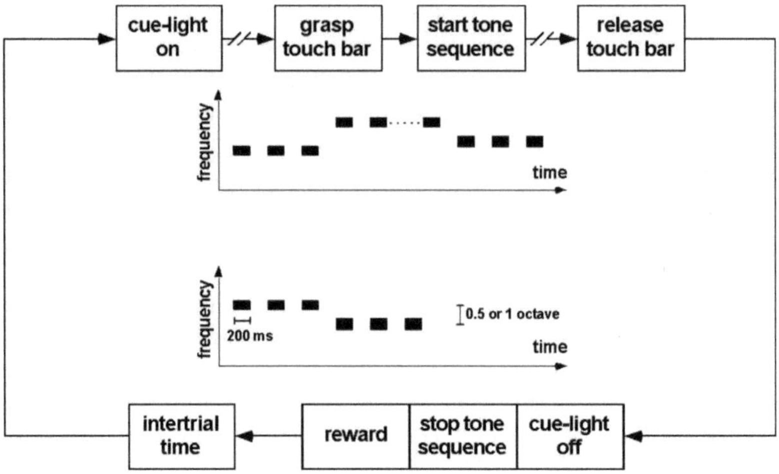

Fig. 3.5 Task structure used to study neuronal correlates of categorization of the direction of frequency steps in tone sequences (modified after Brosch et al., 2005)

3.4 Task-Related Modulations of Neuronal Responses in Auditory Cortex

A number of studies using a variety of paradigms has addressed the question of whether responses of auditory cortex neurons to sounds vary as different tasks are performed. The results are not yet fully conclusive which might be due to methodological differences between the studies, such as the use of different stimuli, different relationships between task-relevant stimuli and neuronal stimulus preferences, different deductive logical operations to be performed on auditory stimuli, different audiomotor mappings, and different reinforcements, and to ill defined nonperforming conditions. Generally, it appears that the responses of quite a large number of neurons are affected as animals become engaged in an auditory task. The percentage of neurons that do show significant changes ranges between 29% and 100% (average 58.4%) in different reports when performing and nonperforming conditions are compared. There is no consistency among the reports on how the response changes. The largest number of studies has observed that auditory responses are stronger in the performing condition (Miller et al., 1972; Ryan et al., 1984; Gottlieb et al., 1989; Scott et al., 2007); a few studies have observed the opposite relationship (Benson and Hienz, 1978; Otazu et al., 2009) while others observed no clear differences between the two conditions (Hocherman et al., 1976; Gilat & Perlman, 1984). The discrepancy also applies to evoked field potentials (Miller et al., 1972; Otazu et al., 2009). Although the auditory responses of many neurons change as animals become engaged in an auditory task, its influence appears to be moderate. Average changes were reported to be about 20%, both for discharges and local field potentials. In contrast to auditory responses hardly any study reported systematic and consistent changes of "spontaneous" activity. In fact, Scott et al. (2007) reported increases in about 70% of the neurons; this may be confounded, however, by activity related to consumption of the liquid reward during the "spontaneous" period. Reports on "spontaneous" activity must be considered cautiously because there may be no period during an experiment in which neuronal activity in auditory cortex is not affected by the task (Brosch et al., 2011a,b).

Task effects appear to be weaker when performing an auditory task is compared to performing some other nonauditory task involving, for example, visual or olfactory stimuli. This type of approach is able to control for effects of general attention, arousal, or motivational factors. The percentage of neurons that show significant response changes ranges between 8% and 66% (Hocherman et al., 1976; Miller et al., 1980; Benson et al., 1981; Otazu et al., 2009), and the changes in response strength range between ≈ 0% and 94%. It is also possible that sustained response components are more prone to vary with task engagement than phasic response components do. This is suggested by the few studies that have used tasks in which subjects required longer observation periods for proper task performance (Gottlieb et al., 1989; Brosch et al., 2011a), and in which changes were on average 250%.

Auditory responses may also depend on the type of association between sound and reward. Beaton and Miller (1975) trained two monkeys to detect tones. While response to high-frequency tones always yielded a reward, low-frequency tones yielded a reward only when a cue light was lit during the trial. In recordings from koniocortical auditory cortex and immediately associated regions, the majority of 15 neurons showed weaker phasic responses to the low-frequency tones when the monkeys could earn a reward compared when they could not earn any reward. This reward-associated inhibition was most prominent at frequencies remote from the best frequency (BF) of a neuron.

Anticipation also seems to affect phasic responses to sounds in auditory cortex. Hocherman and Yirmiya (1990) had one monkey discriminate a tone burst from a broad-band noise burst by shifting a lever in different directions. Shortly before auditory stimulation, a light was flashed, either on the left or on the right side. This occurred in approximately 75% of the trials on the same side toward which the next lever movement was required (consistent pairing). In the remaining trials the flash occurred on the opposite side (inconsistent pairing). The behavioral changes resulting from the inconsistent pairing were accompanied by increases or decreases of the acoustic responses in about one quarter of 94 neurons studied in auditory cortex.

More anticipatory effects were shown by Jaramillo and Zador (2011). Rats had to listen to a sequence of unmodulated and modulated tones and had to move in one of two directions depending on whether the modulated tone had a low or a high frequency. Rats could and did predict whether the modulated tone occurred early or late in the sequence because trials were arranged in blocks in which the modulated tones occurred either early or late during the sequence. In the vast majority of 44 neurons, anticipation resulted in an increase of the response to the unmodulated tones and this anticipatory effect was the stronger the closer was the frequency to a neuron's BF. It increased over time the closer the unmodulated tones were to the anticipated time of the modulated tone, however, by sparing the modulated tone.

Task-related changes in response strength seem to be paralleled by changes in receptive field properties. This has been demonstrated for spectral aspects of receptive fields but not yet for auditory spatial receptive fields.

3.5 Task-Related Changes of Receptive Fields

In a series of experiments Fritz et al. (2003, 2005) trained ferrets with a shock avoidance procedure to detect a fixed target tone amid a background sequence of noise bursts, which in some experiments also consisted of tones of different frequency. The noise bursts were also used to determine spectrotemporal receptive fields (STRFs). This analysis revealed, as animals switched from being passively exposed to the sounds to performing the task, that the frequency selectivity of neurons in primary auditory cortex changed. Depending on the initial receptive field structure, excitatory subfields were enhanced and inhibitory subfields were weakened, without any net changes of the total responsiveness, neither to the background nor to the target sounds. These receptive fields changes took place in the about three quarters of the 191 neurons studied. When after initial conditioning with a specific frequency a new frequency was conditioned, the receptive field of the neuron changed at the new frequency. When the background sounds also consisted of tones with a frequency different from that of the target, an additional weakening was observed at this frequency. The degree of receptive field change correlated to the task mastery of the animals. The changes could persist for several minutes after the animals stopped performing the task. Receptive field changes were inverted when purely appetitive reinforcement was used in which animals received water only when they licked during the target tone (David et al., 2012). Thus, responses at the target frequency were weakened and responses at the reference frequency were enhanced. This might reflect that neurons increase their response to sounds that are related to reinforcement or require motor behavior. In the first reinforcement condition, the animals could safely lick throughout the entire experiment, except for the time the target tones were present, which required the animals to interrupt licking to avoid a mild electric shock. Thus occurrence of the target tone prompted a motor command to temporally interrupt licking. In the second reinforcement condition, the animals had to resist from licking throughout the background sound to avoid a timeout from the experiment (equivalent to the interruption of licking in the first reinforcement condition) and licked only during the target tone.

When the tone detection task was repeated in the presence of concurrent noise maskers, the authors found that STRF gain decreased during task performance in most cells (Atiani et al., 2009). By contrast, STRF shape changes were excitatory and specific, and were strongest in cells with BFs near the target tone. The net effect of these adaptations was to accentuate the representation of the target tone relative to the noise by enhancing responses of near-target cells to the tone during high-signal-to-noise ratio (SNR) tasks while suppressing responses of far-from-target cells to the masking noise in low-SNR tasks. These adaptive STRF changes were largest in high-performance sessions, confirming a close correlation with behavior.

In contrast to STRF changes, no or only small task-related changes have been found for auditory spatial receptive fields when animals switch between different tasks. A possible reason for these negative results could be details of the behavioral procedures used such that more substantial changes of spatial receptive fields may be observed when, for example, a specific sound location is reinforced. This is suggested by the study of Benson and Hienz (1978), who found that 14 of 15 neurons responded more strongly to a sound when it was presented on the side the animals attended to.

Two studies found that the azimuth tuning of auditory cortex neurons changed nonsystematically with eye position, both when animals performed a visual fixation task and when the eyes were only spontaneously moved without reinforcement (Werner-Reiss et al., 2003; Fu et al., 2004). The type of changes were unrelated to the relationship between the center of gaze and the location of the sound source. Scott and colleagues (2007) failed to see changes in interaural level or interaural temporal difference tuning in auditory cortex neurons when animals were switched between performing a lateralization task or not being able to so.

A mild shrinkage of the size of spatial receptive fields was reported by Lee and Middlebrooks (2011) when animals were switched between performing one of two auditory detection tasks and being idle. In the first task, animals had to release a hold key when a buzz occurred in a sequence of background noise bursts played at different azimuths. In the second sound elevation task, animals listened to the same sequence of background sounds but now had to detect when a sound was played from an elevation 40°–80° above the standard elevation. The effect of engaging in the tasks was that spatial receptive fields, expressed as equivalent rectangular receptive fields, changed from 185° for the idle condition to 176° and 165° for the two task conditions, with no significant difference between the latter two. A reason for the marginal changes could be that in none of the two tasks receptive fields properties were assessed that were task relevant, that is, neither spectral or temporal tuning nor elevation tuning but only a task-irrelevant feature, namely azimuth tuning.

3.6 Learning-Induced Changes of Auditory Responses

3.6.1 Tone Detection and Discrimination Tasks

Two types of learning induced, that is, long-term, changes of spectral receptive field tuning to tones have been described in primary auditory cortex neurons. One type is a learning-induced shift of the BF of auditory cortex neurons (BF shift plasticity) toward a frequency that has gained special significance in single-tone classic or operant conditioning tasks (for review see Weinberger, 2010). According to our definition (Fig. 3.2) that associating the properties of a single sound pattern to a behavioral meaning without contrasting this to a different meaning of other sounds during training, this is a detection paradigm. Neurons that initially had BFs above the conditioned frequency lowered their best frequencies after conditioning and closely matched the conditioned frequency. Correspondingly, neurons with initially BFs below the conditioned frequency raised their BFs. As a consequence, the population of neurons tuned to the conditioned frequency increased.

The second type is a learning-induced slope sharpening of receptive fields around the conditioned frequency (Ohl & Scheich, 1996, 2001; Witte & Kipke, 2005). This type was found in a Pavlovian conditioning paradigm involving training with multiple nonconditioned frequencies in addition to a frequency that was paired with an electrodermal stimulus (see Ohl & Scheich, 2005, for discussion). Consequently, this was a discrimination paradigm according to our definition (Fig. 3.2).

By this type of retuning the frequency conditioning generated a steepening or even a notch on low-frequency or high-frequency slopes of receptive fields of those neurons that were conditioned with a frequency different from their BF (Fig. 3.6). The slope sharpening developed fast, within minutes of exposure to the multiple frequencies, while the spectral receptive field was continuously measured. The effect was extinguished as fast when the conditioned frequency was no longer paired with the electrodermal stimulus, that is, lost its special significance. However, the slope sharpening was maintained when the auditory and electrodermal stimulation was stopped for the same time as needed for the extinction. Even though response changes at other frequencies including some change of BF could occur in parallel to slope sharpening in some neurons, these were not systematic phenomena. Thus, the population of neurons having a BF at the conditioned frequency will not increase by this slope sharpening type of retuning. Neurons conditioned at their BF will not change BF but may generally increase their responses.

A population view of neighboring neurons in a map reveals several differences as a consequence of these two types of retuning (Scheich et al., 2011). In short, for the BF-shift plasticity, salience of the conditioned frequency representation is increased by recruiting maximal excitatory responses of neighboring neurons to the tonal conditioned frequency at the expense of a lower than normal excitatory activity of these neurons in response to nonconditioned frequencies. Therefore, in a usual tonotopic map established with BF measurements of neurons plotted on location of these neurons, the spatial representation of the conditioned frequency by BFs would expand in the map at the expense of the representation of neighboring BFs. This has indeed been observed in auditory cortex for the representation of a trained frequency in perceptual learning (Recanzone et al. 1993; cf. Brown et al., 2004). It should be noted that these perceptual learning tasks, even though they used same–different comparisons of tones, were not discrimination tasks in the sense defined in the preceding text: For correct performance of the task, it was not necessary for the animal to associate a detected frequency difference with a difference in meaning.

For the slope sharpening type of retuning, the salience of the conditioned frequency representation is increased by selectively lowering the excitatory responses of neighboring neurons to the conditioned frequency without changing their BF. In a population view of neurons, a sharp contrast is formed between the conditioned frequency and neighboring frequencies because the representation of the conditioned frequency is spatially narrowed in contrast to an expansion by BF retuning. The slope sharpening leads to more sensitive discrimination of the conditioned

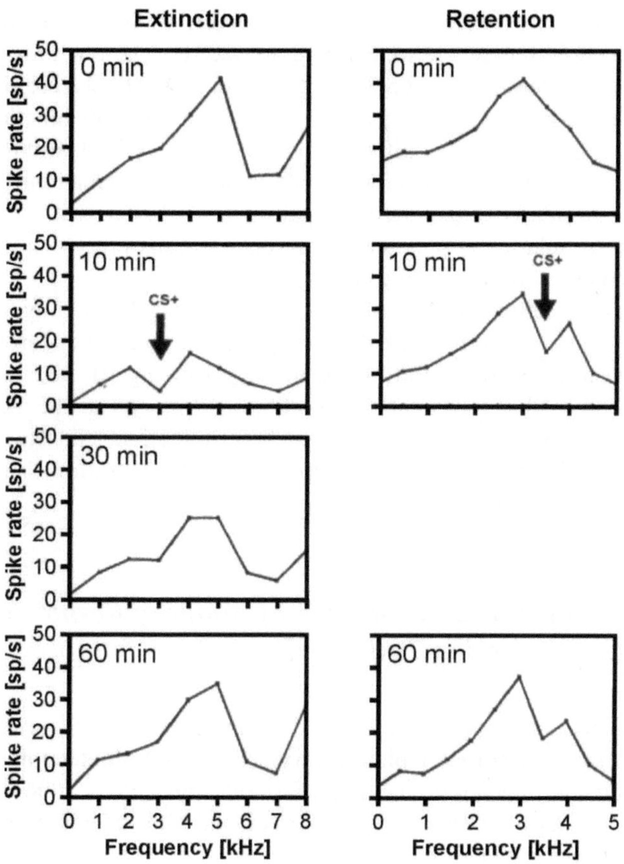

Fig. 3.6 Discrimination learning-induced changes of spectral receptive fields of primary auditory cortex neurons in the gerbil and their extinction and retention over time. Receptive fields and their changes in the two neurons (left and right column) were obtained by continuously stimulating the animals and measuring responses to tone pips of randomized frequencies (abscissa). At some point the 3-kHz (left column) or the 3.5-kHz tone (right column), but no other tones, were paired with an electrodermal body stimulus in a Pavlovian paradigm. Plots shows the receptive fields before conditioning (0 minutes) and 10, 30, and 60 minutes after conditioning. The pairing with a non-best frequency tone in the context of also presented other irrelevant frequencies (discrimination paradigm) created a notch in the receptive field (arrows) with steepening of the slopes around the conditioned frequency (increase of contrast for the conditioned frequency). In the left column, the conditioned frequency was subsequently no longer paired with shocks (extinction training), leading to a full recovery of the original receptive field at 60 minutes. In the right column, all stimulation was stopped until re-measuring the receptive field at 60 minutes. In this case the training-induced receptive field notch was retained at least until this time. (Modified from Ohl and Scheich, 1996)

frequency from neighboring frequencies. The relevance of local slopes in neuronal tuning for discrimination has also been demonstrated for sound localization processing in the midbrain (Harper & McAlpine, 2004).

Expansion of the neuronal representation of a specific tone frequency in primary auditory cortex of the rat has also been observed after pairing a tone with electrical stimulation of nucleus basalis (Bakin & Weinberger, 1996; Kilgard & Merzenich, 1998) or the ventral tegmental area (Bao et al., 2001). These source nuclei of the cholinergic and dopaminergic modulatory system respectively are thought to be instrumental for cortical learning plasticity yet serve different functions (Schultz, 2006; Thiele, 2009). Thus, BF-shifts as a type of plasticity for learnt tone representation seem to occur in detection tasks and increase saliency of a tone by overrepresentation. The similarity to the effects of pairing a single tone with cholinergic and dopaminergic stimulation argues for a representation that reflects simple detection of a stimulus that has become relevant.

In summary, detection and discrimination tasks, albeit shown only for tone learning, appear to lead to different types of retuning of auditory cortex neurons. Detection increases the saliency of a conditioned tone by shifting BF and increasing the number of neurons that represent the conditioned tone yet at the expense of the representation of neighboring frequencies. Discrimination leads to an increased response contrast between the neurons representing the conditioned frequency and the neurons representing neighboring frequencies.

3.6.2 Learning-Induced Changes of Auditory Responses in Categorization Tasks

In the two electrophysiological studies that are available on categorical sound representation in auditory cortex, a picture very different from detection and discrimination learning has emerged (Ohl et al., 2001; Selezneva et al., 2006). One study in gerbils (Ohl et al., 2001) investigated the directional categorization of rising versus falling frequency-modulated tones (FM sweeps) irrespective of frequency range. The emergence of behavioral categorization was observed by training gerbils in a shuttlebox avoidance paradigm as illustrated in Fig. 3.4 for detection learning only. In contrast to detection, this involved a discrimination learning of mirror-imaged rising FM with a go meaning and falling FM with a no-go meaning here called pairs. Thus, all other parameters of the two FM sweeps in a pair were the same except the direction of the frequency change. The successive training with different pairs aimed at making this directional discrimination independent from frequency range of the sweeps, that is, at formation of two categories. The learning of the two meanings for each novel pair required relearning, as shown by typical asymptotic learning curves for every pair. Eventually with this procedure, immediately correct responses to novel pairs occurred. This categorization stage was spontaneously reached after an individually different number of training blocks.

The whole process took weeks to develop with daily sessions of 60 trials with randomized rising and falling sweeps of a pair. It was electrophysiologically monitored by continuously recording local epidural field potentials with an implanted electrode array in primary auditory cortex.

Single trial analysis in the β- and γ-band with the state-vector method revealed that before learning, the presentation of an FM sweep in naïve animals resulted in a

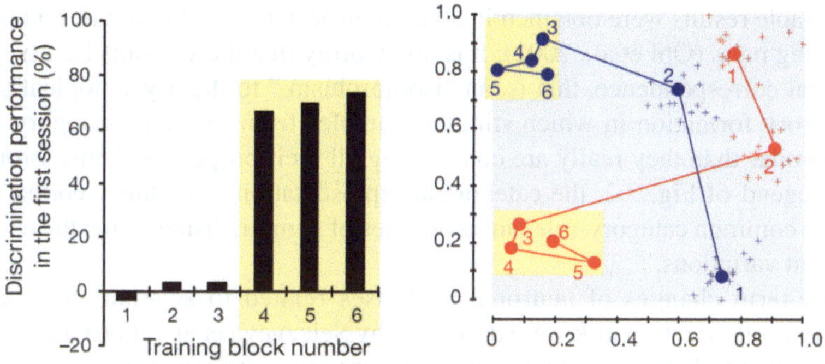

Fig. 3.7 Correspondence between the behavioral transition from the discrimination phase to the categorization phase and the emergence of category-representing spatial patterns of local field potentials over primary auditory cortex of a gerbil. On the left, bars represent discrimination performance for each novel pair of rising and falling FM tone at the beginning of their training block. Serial numbers on the abscissa refer to blocks with novel pairs. The categorization phase is indicated by the abruptly increased discrimination performance (emphasized by the yellow rectangles). The right panel shows similarity–dissimilarity relations between activity patterns (states) found for rising and falling FM tones (red and blue dots, respectively) in the sequence of training blocks (numbers). The relative dissimilarity between any pair of states and among states is represented by the distance of the corresponding state points in this two-dimensional display within arbitrary coordinates. Note the emergence of dot clustering (indicating high state similarity) within categories and largest dot separation across categories when transfer from discrimination phase to categorization phase occurred, that is, after the third pair. (Modified after Ohl et al., 2001)

short latency spatiotemporal activity pattern in primary auditory cortex (Fig. 3.7). In this way, a short latency deviation of the state vector from baseline that was observed in each trial provided a single-trial equivalent of an FM-evoked potential. This also occurred in trained animals, but an additional deviation from baseline occurred yet with a longer and variable latency across trials (70–2000 ms). These initially rare events were observed more frequently in trials when the animals had experience with different FM pairs and were observed in every trial after behavioral categorization was achieved. The knowledge of the time bins in which the secondary state change occurred in each trial allowed comparison of the corresponding spatiotemporal patterns in auditory cortex across different FM pairs and separately for rising and falling FM. It was found that these secondary spatiotemporal patterns of activity reflect the formation of categories in each animal (Fig. 3.7, right side). In essence, a formal mathematical description of these patterns according to their similarity showed that the representation of rising and falling FM was initially very dissimilar for each pair and across pairs. However, when behavioral categorization was achieved (with the fourth pair in this case, Fig. 3.7, left side), representations of all rising FM became very similar as well as representations of all falling FM whereas representations in a comparison across categories became maximally dissimilar. Thus the representational patterns reflected categorical similarities even though the novel FM samples 4–6 with respect to their frequency ranges were not similar to the initially trained samples 1–3 and covered as much variations.

Comparable results were obtained in all animals tested, yet after a different number of training pairs (Ohl et al., 2001). It is noteworthy that these results bear a striking structural correspondence, that is, an "isomorphism," to the psychological process of category formation in which stimulus samples from the same category appear more similar than they really are considering all their properties. Thus, as outlined in the legend of Fig. 3.2, the categorical representation in auditory cortex indeed may use common category-relevant properties of stimuli, disregarding the category-irrelevant variations.

Long-term changes of neuronal responses related to acquisition of another auditory categorization task were described by Selezneva et al. (2006). In this study, monkeys had to listen to various sequences of tones with tone repeats and up-and-down frequency steps. They were required to report, by bar release, the occurrence of a down-step, irrespective of the frequency of the tones and the step size (Fig. 3.5). Thus, different from the FM categorization, the categorization of tone steps covered the sequence aspect of different sounds, a problem typical for processing of speech signals and music. The authors found that, compared to an untrained monkey, neurons in primary and posterior auditory cortex responded, on average, more strongly (approximately 20%) to reward-associated down-steps than to other tone steps. The categorical nature of the response to the step was even more clearly shown by comparing how the responses from the first to the second tone of a step changed. This analysis revealed that the response increased for down-steps only but not for up-steps or tone repeats. Interestingly, this response preference prevailed in error trials, that is, when the animals released the bar after an up-step or failed to do so after a down-step. The learning-induced response preference was also preserved after performance of the task, that is, when the animals were exposed to the same stimuli but were not able to perform the task.

3.7 Neuronal Responses to Self-Generated Sounds

Studies into self-triggered sounds, speech production, and vocalizations suggest mostly suppressive effects onto neuronal responses to auditory stimuli. This was first demonstrated in an electroencephalographic (EEG) study on human subjects by Schafer and Marcus (1973) in which the authors compared click-evoked potentials at the vertex (Cz-A1) that were either self-triggered or externally triggered. They found that self-triggering resulted in an approximately 50% reduction of different components of the evoked potential compared to the situation in which the same self-generated click sequence was replayed to the subject. Interestingly, a much smaller motor-induced suppression was reported when potentials evoked by self-triggered sounds were compared to potentials evoked by an externally triggered isochronous click sequence with an interclick interval corresponding to the average interval of self-stimulation (approximately 5 s). A possible interpretation of this finding is that the suppression of the evoked potential resulted from the depressing influence of a motor efference copy (or by a forward model generated from the efference copy that predicts the sensory consequences of one's own actions [Wolpert, 1997]; see also Winkler et al., 2009), which renders auditory cortex more sensitive to other sounds. Alternatively, the suppression could be related to more general anticipatory mechanisms that predict the time of occurrence of auditory stimuli. Independent support for motor-induced suppression was found in magnetoencephalographic (MEG) studies with similar tasks (Martikainen et al., 2005; Aliu et al., 2009. The latter study also showed that motor-induced suppression even occurs when a sound is delayed by several hundred milliseconds relative to the motor act; in this case motor-induced suppression is no longer stimulus specific and specific to details of the motor act. This suggests that different neuronal mechanisms are involved for self-triggered sounds that occur immediately after an action, or are delayed.

Motor-induced suppression of auditory evoked potentials has also been observed in studies using self-generated speech signals (e.g., Curio et al., 2000; Houde et al., 2002) and animal vocalizations (Müller-Preuss & Ploog, 1981; Eliades & Wang, 2003, 2005, 2008). A problem of these studies, however, is that there is limited control over what subjects hear (sounds reach the inner ear also through bone conduction) and that motor actions and sound production typically overlap in time. These studies, but also those in which sounds are generated by means other than one's own larynx, bear the potential confound of the acoustic reflex, which reduces sound transmission by approximately 20 dB.

More compelling demonstrations of motor-induced changes of auditory evoked potentials are those that report enhanced responses to self-triggered sounds. Liu and colleagues (2010) performed an EEG study on subjects who listened to a steady-state vowel for several seconds through earphones that was either prerecorded or actively produced by them. During sound exposure the pitch of the vowel decreased by 1/6 of an octave either for a short period after the subject had pressed a button or at several random times. The authors found that the self-triggered pitch perturbations elicited larger auditory evoked potentials than the externally triggered perturbations, both when the vowel was actively produced by the subjects and when a prerecorded version was replayed to them.

Motor-related modulations of responses to self-generated vocalizations have also been observed in the activity of many single cells in various auditory cortical fields (AI, C, and L) and mostly from upper layers during spontaneously or electrically evoked vocalizations (Müller-Preuss & Ploog, 1981; Eliades & Wang, 2003, 2005, 2008). Relative to a 4-s period before vocalizing, the firing decreased to about 23% of baseline in 82% of the instances the animals vocalized. The suppression started several hundreds of milliseconds before and ended with the vocalization, suggesting the suppression could not be explained solely by the acoustic reflex. The suppression also affected responses to other auditory stimuli that occurred unrelated to the vocalization. Suppressive effects were significantly reduced, however, when, through a realtime headphone system, the animal's own vocalization was shifted in pitch. These findings were interpreted to suggest a role of vocalization-induced suppression for auditory self-monitoring. It is possible that such a mechanism is also effective for other motor-induced sound productions, such as during monkey drumming (Remedios et al., 2009). Vocalization-induced response enhancement, which was observed during the remaining 18% of the vocalizations, was reflected by an approximate doubling of the firing rate and typically started concomitantly with the vocalization. Neurons responded nearly normally to other external auditory stimuli that occurred unrelated to the vocalization but had little, if any, sensitivity for pitch perturbations of vocalization feedback. Thus vocalization-induced enhancement did not affect the neurons' sensitivity for other external auditory stimuli, independent of whether or the animal vocalized.

3.8 Neuronal Firing Related to Nonauditory Events

Although most studies considered thus far have focused on how task engagement affects the responses of auditory cortex neurons to sounds, a few studies have unequivocally revealed that auditory cortex also represents nonauditory events that are associated with the task. These findings extend substantially the interpretation of a larger number of observations of multimodal response properties of neurons in various auditory cortical fields (reviewed in Brosch & Scheich, 2005; Ghazanfar & Schröder, 2006), namely that it serves the merging of the senses (Stein & Meredith, 1993) or simply makes us hear better (Ghazanfar & Schröder, 2006). Brosch et al. (2011b) have conceptualized findings to suggest that "auditory cortex consists of two representational systems that are operating on different time scales (Fig. 3.8). The first system utilizes phasic firing changes in the 10- to 100-ms range and represents the timing and type of auditory and auditory-related events. The second system utilizes slow firing changes in the 1- to 10-s range and represents the association between consecutive behaviorally relevant events." Thus auditory cortex is not considered to be solely a highly evolved and sophisticated sound analyzer (e.g., Griffiths & Warren, 2002; Nelken & Calford, 2010) but handles also other information that supports and

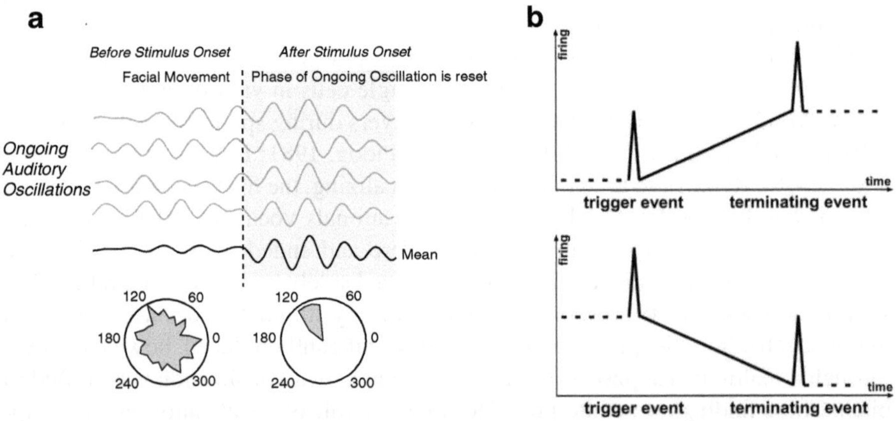

Fig. 3.8 Scheme of the two proposed representational systems for task performance in auditory cortex phasic responses to events and slow firing increases (left) or decreases between events (right). (Modified after Brosch et al., 2011a)

accelerates performance of auditory tasks (Scheich et al., 2011), that is, facilitates the mapping of sounds onto appropriate behavior and controls behavior such as to optimize future sound processing. This view poses auditory cortex to function as a "semantic processor" deducing the task-specific meaning of sounds by learning.

3.8.1 Neuronal Responses to Nonauditory Stimuli

In the categorization task for tone steps (Section 3.6.2; Fig. 3.5), Brosch and colleagues (2005) observed that up to 10% of auditory cortex neurons can exhibit phasic responses to a cue light (the instruction stimulus) with a median latency of approximately 100 ms. In many cases, such light responses were seen only when the animals were engaged in the task and disappeared, or were much weaker when the animals did not perform. In the nonperforming state some neurons also exhibited a weak response to the offset of the cue (Brosch et al., 2011a). Responses of neurons in auditory cortex and inferior colliculus to visual stimuli have also been seen with other experimental designs. In these experiments monkeys were required to fixate a small light and, after it was extinguished to make a saccade toward one of nine different speakers from which currently a sound was presented (Metzger et al., 2006); or monkeys were required to release a bar after the termination of a tone, which could be preceded by a visual stimulus and that thus contained important information for performing the auditory task (Brosch et al., 2011c).

After an instruction stimulus many neurons exhibit firing that is related to movements, that is, to the grasping of the response bar (Brosch et al., 2005). Such firing can commence several hundreds of milliseconds before the hand made contact with the bar, was most frequently seen immediately after the contact, and ended within less than 700 ms. The firing before contact was interpreted to be related to the execution of the limb movement toward the bar. The firing after contact could be related to the limb movement as well, but also be due to the tactile stimulation of the hand after it was laid on the bar (Fu et al. [2003] have described neurons in auditory cortex that respond to passive somatosensory stimulation). Further evidence for motor-related activity in auditory cortex was obtained by Brosch et al. (2005) when the firing was analyzed relative to bar release, that is, the reporting response. Observation of this firing did not implicate an exclusive relationship between auditory cortex activity and motor behavior, because the bar release was also tightly related to the auditory categorization process and to reward (expectation). An important property of the firing related to the observing or to the reporting response is that it is task-dependent; when a monkey was switched from categorizing the tone sequence to detecting the onset of a visual flickering while all other task contingencies were maintained, grasping and release-related firing disappeared.

There is also evidence that the firing of neurons in auditory cortex is related to specific aspects of sensory–motor mappings. Vaadia and colleagues (1982) trained one rhesus monkey to perform a two-alternative forced choice auditory discrimination reversal task. Immediately after pressing a lever, either a tone or a noise burst was presented, which the monkey had to discriminate by shifting the lever to the left or to the right side. After approximately 100 trials the sound–motor contingencies were reversed. Among 146 neurons recorded from primary and lateral auditory cortex, 23 neurons responded differently to the auditory stimuli, depending on whether the stimulus was associated with a right or a left lever shift. Thus a neuron fired most strongly when the noise was presented and a leftward shift was required. Although not explicitly stated for all neurons, the differential firing always seemed to be expressed as a sustained change in firing rate lasting for a few hundred milliseconds until the lever shift was executed. None of the neurons exhibited firing that was related to the execution of lever movements only.

More and independent support for motor-related activity in auditory cortex has been obtained in other studies. Yin and colleagues (2008) found that 12.5% of neurons in auditory cortical fields AI and R exhibited firing that was related to bar release in a task in which monkeys had to detect a specific tone sequence. In most neurons, it could be distinguished from reward expectation by observing that the firing did not change when blocks of trials with different reward delays were tested. A small number of neurons exhibited such firing related to reward timing. Release-related firing also started about a few hundred milliseconds before bar release and initially was expressed as suppression that became maximal shortly before bar release and then rapidly became excitatory near bar release. Similar firing was reported in trials in which the monkeys made false alarms to nontarget sound sequences, although it was weaker than for hits. For an amplitude modulation discrimination task, Niwa et al. (2012) confirmed that firing in primary auditory cortex was related either to motor execution/decision or simply to different reward timing.

More recently, Brosch et al. (2011b) clearly demonstrated that firing in auditory cortex can also be related to the reinforcement given at conclusion of an auditory task, corroborating previous evidence (Brosch et al., 2005; Yin et al., 2008; see Metzger et al., 2006 for inferior colliculus). In this study monkeys performed the auditory categorization task (Fig. 3.5) with a performance-dependent reward schedule; the size of a liquid reward that could be earned for a correct response depended on the correctness of the behavioral response in the preceding trial. The reward was large if the preceding trial was correctly performed; it was small if the preceding trial was incorrect and unrewarded. By comparing trials with different reward sizes the authors found neurons that discriminated reward size by the strength of their firing. That means these reward responses were tuned to a specific nonauditory feature, namely the amount of water. Three hundred twenty-four of 626 multiunits showed reward-size responses shortly after reward delivery (or at the expected time of reward after incorrect behavioral responses). This firing was strongest about 400 ms after reward delivery and disappeared within the following 2000 ms. When an extra large reward was occasionally administered in place of the standard large reward even stronger reward-size responses were observed. When reward delivery was delayed reward-size responses emerged at a later period after bar release commensurate to the reward delay. This parallels the observation of reward anticipating responses described by Yin et al. (2008).

3.8.2 Neuronal Activity Between Elements of an Auditory Task

Although these phasic responses are considered to represent the timing and some features of auditory and nonauditory events associated with auditory task performance, also slow changes of firing or sustained decreases or increases of firing were observed to occur between task-relevant events (Fig. 3.8; Brosch et al., 2011a; see also Walter et al., 1964; Birbaumer et al., 1990; Abolafia et al., 2011). Such events can be auditory or visual stimuli, motor actions and reinforcers. When monkeys performed the categorization task shown in Fig. 3.5, a group of neurons in auditory cortex was found that started to slowly decrease their firing after the monkeys had made the observing response (grasp the touch bar) after the onset of the visual instruction stimulus. The firing of some of these neurons continued to decrease until the beginning of the tone sequence approximately 2 s later. For other neurons the decrease continued after the beginning of the tone sequence, and it ended only when the monkeys released the bar. Another class of neurons exhibited an opposite activity pattern, that is, they slowly increased their firing after bar grasp until the beginning of the tone sequence. No relationship between the types of events, task requirements, and the two flavors of slow firing changes (increases or decreases) has yet been established. The level of sustained firing, however, was found to be related to the cognitive demands of the task. After increasing firing until the beginning of the tone sequence the firing level remained constant during the initial four tones of the sequence and then rapidly decreased until the bar was eventually released. The forth tone was always the event when the monkeys had to categorize the direction of the tone step. The slope of the decrease was related to the time when the monkey would exhibit the reporting response. The decrease was fast when the monkeys reported the occurrence of the down-step by exhibiting the reporting response. The decrease was slow when the monkeys reported the occurrence of the up-step by withholding a response. In this case, the forth tone was repeated a few times until the tone frequency stepped down, requiring monkeys to exhibit a reporting response. Because different cognitive operations were required at the first frequency change (categorization of the direction of a frequency step), and at the second frequency change (detection of this change, which was always a down-step) the level of slowly changing firing was interpreted to represent an endogenous brain state during which the categorical sensory information contained in the phasic responses to the tones was transformed into a decision (Selezneva et al., 2006).

Slow firing changes have been observed in auditory cortex with other behavioral procedures and other sounds. Shinba et al. (1995) observed slow firing increases in the auditory cortex of rats while they performed a visual detection task in which the animals could receive a water reward for pressing a lever while a light stimulus was on, which was preceded by a brief warning tone. While the rats performed this paradigm, 9 of a total of 36 neurons were found that started to increase their firing shortly after the tone and then their firing remained elevated at least until the lever press. When the warning tone was presented unrelated to the light stimulus no slow firing changes were observed.

Slow firing changes have also been observed in auditory cortex of a monkey trained to perform an auditory working memory task (Gottlieb et al., 1989) in which two sequential tones had to be compared with each other. The firing of more than half of 120 neurons, mostly from posterior auditory cortex, either slowly increased after the first tone, slowly decreased, or remained at a fairly constant and elevated rate. A small fraction of the neurons exhibited sustained firing that was frequency specific during. Thus, the slow firing changes seemed to reflect information about a preceding auditory event. Working memory-related activity has also been reported by Sakurai (1994).

By contrast, Lemus et al. (2009a) failed to find evidence for activity related to auditory working memory. In that study monkeys had to compare two 500-ms pulse trains with different pulse rates, separated by a delay of 1–3 s, and to indicate, by button press, whether the pulse rate had increased or decreased. They found that the responses of neurons in left primary auditory cortex reliably differentiated between the pulse rates, but only during the trains and not during the delay period. Such differential firing, however, was observed in ventral premotor cortex (Lemus et al., 2009b). This was interpreted to indicate that primary auditory cortex is exclusively associated with the sensory but not with the cognitive components of this task.

Slow firing changes have also been seen with Pavlovian conditioning. This type of associative learning does not seem to require auditory cortex but still modifies auditory cortex (Ohl & Scheich, 1996; Letzkus et al., 2012). Particularly, some neurons have been described in this brain region that exhibited a slow firing increase in delay fear conditioning in which rats received a mild foot shock after a sound (Quirk et al., 1997; Armony et al., 1998). Such firing increases were not present before conditioning and disappeared within 10 or fewer extinction trials. Neither was such firing observed in rats that had received an amygdala lesion before recordings, which was taken as evidence that slow firing changes in auditory cortex were enabled by the amygdala. That Pavlovian conditioning produces slow firing changes (increases as well as decreases) in auditory cortex has also been observed in paralyzed but unanesthetized cats, which received a tail shock several seconds after a brief noise burst (Kitzes et al., 1978).

Other potential properties of slow firing changes in auditory cortex are suggested by studies that have been conducted in subcortical stages of the auditory system, which might receive these properties from auditory cortex or convey them to auditory cortex (Komura et al., 2001, 2005; Metzger et al., 2006). In nonprimary auditory thalamus, Komura and colleagues (2001, 2005) found slow firing increases that emerged after the beginning of a sound lasting for 2 s after which required animals were required to lick to receive reward. When different types of rewards were compared it was found that the slope of the firing increase varied with reward value. An experimental manipulation in which the delay period was suddenly extended from one to the next trial indicated that the time of expected reward was the important event at which the firing increase peaked. During the first trials after this change the tonic firing continued to peak at the time when the reward had arrived before the change was introduced. Within a few trials, however, the time course of firing changed such that it increased more slowly and peaked at the end of the new delay period. When the reward was withheld, the neurons initially continued to show slow firing changes after the stimulus. However, this firing completely disappeared within about 10 trials, in parallel with the licking. This suggests that the crucial factor controlling the emergence of slow firing changes is the presence of a reinforcer, reflecting the ultimate reason why a subject performs a task. This has also been shown for auditory cortex, in which Brosch et al. (2011a) found that the main condition for their emergence was that animals are engaged in performing an auditory task and that slow firing changes typically do not emerge when animals do not participate in such task, or when animals are only passively exposed to the same sounds. By contrast, slow firing changes occurred in false alarm trials, that is, in trials in which the correct reporting response was made, however, at the wrong moment.

The association between behaviorally significant events provided by slow firing changes might even be directed in some cases, that is, this type of firing might provide some sort of either prospective coding of an upcoming event or retrospective coding about a preceding event. It is possible that the firing increases after an observing response provide a neuronal correlate for the anticipation of the upcoming sounds or a neuronal correlate of preparation or of priming. The anticipatory function has most clearly been demonstrated in auditory thalamus (Komura et al., 2001). When the delay period after which a reward was administered was extended to a value longer than the standard delay, the firing increase was unchanged during the initial trials with the extended delay such that the firing continued to peak at the end of the standard delay. An example that slow firing changes could represent retrospective coding is provided by its frequency specificity, which was seen in an animal while it was performing a working memory task (Gottlieb et al., 1989). The slow firing changes after the first tone seemed to maintain information about this tone until the moment at which the animal had to compare its frequency with that of the subsequent tone. Another example of retrospective coding may be the firing in auditory cortex observed by Brosch et al., 2011b (see Section 3.8.1). In that study a performance-dependent reward schedule with a variable reward size was used to train monkeys to performed an auditory categorization task (Fig. 3.5). In addition to the described coding of the reward size, there was also neuronal firing lasting and starting a few seconds after reward delivery or its expected time that distinguished rewarded from unrewarded trials and represented the reward prediction error, that is, the mismatch between the expected and delivered reward. This was followed by a period in which the firing was related to the size of the reward that could be expected for the upcoming trial, thus providing another example of prospective coding.

Associations between task elements may also be represented in specific spatiotemporal firing patterns in auditory cortex (Villa et al., 1999). In rats trained on a self-paced auditory go/no-go discrimination task with frequency sweeps, the authors found that different spatiotemporal firing patterns emerged in trials in which animals would move and in trials in which animals would not move, independent of the correctness of the upcoming behavioral responses. As only go trials could result in reward, it is not clear whether the differential patterns reflected different behavioral responses or different reward expectations. Similar observations were made later with a similar behavioral procedure in which artificial vowels with different formants needed to be categorized (Villa, 2005).

3.9 Summary

Theoretical concepts of audition in conjunction with modified auditory and heterogeneous nonauditory activities found in auditory cortex during task performance point to its role as a semantic processor. Notably, these activities during each task trial reflect not only identification of auditory target features but also in many details the associations formed with other information for behavioral execution of the task. In this way the behavioral meaning of the sounds seems to be determined locally, namely what to do with a sound in a task-specific fashion. Even though many details of activation changes and activation states during a task trial need clarification, the available evidence suggests that it might be possible to recognize from these activities which basic types of logical operations were involved, for example, detection, discrimination, or categorization of sounds. Also motivational aspects related to approach or avoidance, prediction of events, and reinforcements as well as prediction errors seem to be discriminable from the activities. It is obvious that these facets of a task cannot be deduced by auditory cortex alone but only in cooperation with numerous other cortical and subcortical brain areas. Recent evidence suggests that the necessary anatomical connections are available even for primary auditory cortex but become functional only during engagement in auditory tasks. This new view on auditory cortex implies that hierarchical concepts of brain organization reserving cognitive functions to "higher order" cortices must be modified. The information flow from sensory cortex to such cortical areas is undebatable but they seem to feed information back to sensory cortex for local cognitive processing.

Acknowledgments This work was supported by the Deutsche Forschungsgemeinschaft (SFB 779, SFB TR 31, SFB TR 62) and the Europäischer Fond für regionale Entwicklung (EFRE 2007-2013).

References

Abolafia, J. M., Martinez-Garcia, M., Deco, G., & Sanchez-Vives, M. V. (2011). Slow modulation of ongoing discharge in the auditory cortex during an interval-discrimination task. *Frontiers in Integrative Neuroscience*, 5(5), doi: 10.3389/fnint.2011.00060.

Aliu, S. O., Houde, J. F., & Nagarajan, S. S. (2009). Motor-induced suppression of the auditory cortex. *Journal of Cognitive Neuroscience*, 21, 791–802.

Armony, J. L., Quirk, G. J., & LeDoux, J. E. (1998). Differential effects of amygdala lesions of early and late plastic components of auditory cortex spike trains during fear conditioning. *Journal of Neuroscience*, 18, 2592–2601.

Atiani, S., Elhilali, M., David, S. V., Fritz, J. B., & Shamma, S. A. (2009). Task difficulty and performance induce diverse adaptive patterns in gain and shape of primary auditory cortical receptive fields. *Neuron*, 61, 467–480.

Bakin, J. S., & Weinberger, N. M. (1996). Induction of a physiological memory in the cerebral cortex by stimulation of the nucleus basalis. *Proceedings of the National Academy of Sciences of the USA*, 93, 11219–11224.

Bao, S., Chan, V. T., & Merzenich, M. M. (2001). Cortical remodelling induced by activity of ventral tegmental dopamine neurons. *Nature*, 412, 79–83.

Beaton, R., & Miller, J. M. (1975). Single cell activity in the auditory cortex of the unanesthetized, behaving monkey: Correlation with stimulus controlled behavior. *Brain Research*, 100, 543–562.

Behrens, T. E., Woolrich, M. W., Walton, M. E., & Rushworth, M. F. (2007). Learning the value of information in an uncertain world. *Nature Neuroscience*, 10, 1214–1221

Benson, D. A., & Hienz, R. D. (1978). Single-unit activity in the auditory cortex of monkeys selectively attending left vs. right ear stimuli. *Brain Research*, 159, 307–320.

Benson, D. A., Hienz, R. D., & Goldstein, M. H., Jr. (1981). Single-unit activity in the auditory cortex of monkeys actively localizing sound sources: Spatial tuning and behavioral dependency. *Brain Research*, 219, 249–267.

Birbaumer, N., Elbert, T., Canavan, A. G., & Rockstroh, B. (1990). Slow potentials of the cerebral cortex and behavior. *Physiological Reviews*, 70, 1–41.

Brechmann, A., & Scheich, H. (2005). Hemispheric shifts of sound representation in auditory cortex with conceptual listening. *Cerebral Cortex*, 15, 578–587.

Brechmann, A., Gaschler-Markefski, B., Sohr, M., Yoneda, K., Kaulisch, T., & Scheich, H. (2007). Working memory specific activity in auditory cortex: Potential correlates of sequential processing and maintenance. *Cerebral Cortex*, 17, 2544–2552.

Brosch, M., & Scheich, H. (2005). Non-acoustic influence on neural activity in auditory cortex. In König, R., Heil, P., Budinger, E., & Scheich, H. (Eds.), *Auditory cortex: Towards a synthesis of human and animal research* (pp. 127–143), Mahwah, NJ: Lawrence Erlbaum.

Brosch, M., Selezneva, E., & Scheich, H. (2005). Nonauditory events of a behavioral procedure activate auditory cortex of highly trained monkeys. *Journal of Neuroscience*, 25, 6797–6806.

Brosch, M., Selezneva, E., & Scheich, H. (2011a). Formation of associations in auditory cortex by slow changes of tonic firing. *Hearing Research*, 271, 66–73.

Brosch, M., Selezneva, E., & Scheich, H. (2011b). Representation of reward feedback in primate auditory cortex. *Frontiers in Systems Neuroscience*, 5(5), doi: 10.3389/fnsys.2011.00005.

Brosch, M., Babanin, M., Selezneva, E., Huang, Y., & Scheich H. (2011c). Task demands and motivation affect neuronal activity in the auditory cortex of nonhuman primates. *Society for Neuroscience Abstracts*, 556, 2.

Brown, M., Irvine, D. R., & Park, V. N. (2004). Perceptual learning on an auditory frequency discrimination task by cats: Association with changes in primary auditory cortex. *Cerebral Cortex*, 14, 952–965.

Budinger, E., & Scheich, H. (2009). Anatomical connections suitable for the direct processing of neuronal information of different modalities via the rodent primary auditory cortex. *Hearing Research*, 258, 16–27.

Cohen, M. R., & Nagel, E. (1993). *An introduction to logic*. Indianapolis, IN: Hackett.

Colomb, J., & Brembs, B. (2010). The biology of psychology 'simple' conditioning? *Communicative and Integrative Biology*, 3, 142–145.

Curio, G., Neuloh, G., Numminen, J., Jousmaki, V., & Hari, R. (2000). Speaking modifies voice-evoked activity in the human auditory cortex. *Human Brain Mapping*, 9, 183–191.

David, S. V., Fritz, J. B., & Shamma, S. A. (2012). Task reward structure shapes rapid receptive field plasticity in auditory cortex. *Proceedings of the National Academy of Sciences of the USA*, 109(6), 2144-2149, doi: 10.1073/pnas.1117717109.

Eggermont, J. J. (1998). Representation of spectral and temporal sound features in three cortical fields of the cat: Similarities outweigh differences, *Journal of Neurophysiology*, 80, 2743–2764.

Eliades, S. J., & Wang, X. (2003). Sensory-motor interaction in the primate auditory cortex during self-initiated vocalizations. *Journal of Neurophysiology*, 89, 2194–2207.

Eliades, S. J., & Wang, X. (2005). Dynamics of auditory-vocal interaction in monkey auditory cortex. *Cerebral Cortex*, 15, 1510–1523.

Eliades, S. J., & Wang, X. (2008). Neural substrates of vocalization feedback monitoring in primate auditory cortex. *Nature*, 453, 1102–1106.

Fritz, J., Shamma, S., Elhilali, M., & Klein, D. (2003). Rapid task-related plasticity of spectrotemporal receptive fields in primary auditory cortex. *Nature Neuroscience*, 6, 1216–1223.

Fritz, J. B., Elhilali, M., & Shamma, S. A. (2005). Differential dynamic plasticity of A1 receptive fields during multiple spectral tasks. *Journal of Neuroscience*, 25, 7623–7635.

Fu, K. M., Johnston, T. A., Shah, A. S., Arnold, L., Smiley, J., Hackett, T.A., et al. (2003). Auditory cortical neurons respond to somatosensory stimulation. *Journal of Neuroscience*, 23, 7510–7515.

Fu, K. M., Shah, A. S., O'Connell, M. N., McGinnis, T., Eckholdt, H., Lakatos, P., et al. (2004). Timing and laminar profile of eye-position effects on auditory responses in primate auditory cortex. *Journal of Neurophysiology*, 92, 3522–3531.

Fuster, J. M. (2005). *Cortex and mind: Unifying cognition*. New York: Oxford University Press..

Ghazanfar, A. A., & Schröder, C. E. (2006). Is neocortex essentially multisensory? *Trends in Cognitive Science*, 10, 278–285.

Gilat, E., & Perlman, I. (1984). Single unit activity in the auditory cortex and the medial geniculate body of the rhesus monkey: Behavioral modulation. *Brain Research*, 324, 323–333.

Gottlieb, Y., Vaadia, E., & Abeles, M. (1989). Single unit activity in the auditory cortex of a monkey performing a short term memory task. *Experimental Brain Research*, 74, 139–148.

Griffiths, T. D., & Warren, J. D. (2002). The planum temporale as a computational hub. *Trends in Neuroscience*, 25, 348–353.

Guitart-Masip, M., Fuentemilla, L., Bach, D. R., Huys, Q. J., Dayan, P., Dolan, R. J., & Duzel, E. (2011). Action dominates valence in anticipatory representations in the human striatum and dopaminergic midbrain. *Journal of Neuroscience*, 31, 7867–7875.

Harper, N. S., & McAlpine, D. (2004). Optimal neural population coding of an auditory spatial cue. *Nature*, 430, 682–686.

Hocherman, S., & Yirmiya, R. (1990). Neuronal activity in the medial geniculate nucleus and in the auditory cortex of the rhesus monkey reflects signal anticipation. *Brain*, 113, 1707–1720.

Hocherman, S., Benson, D. A., Goldstein, M. H., Jr., Heffner, H. E., & Hienz, R. D. (1976). Evoked unit activity in auditory cortex of monkeys performing a selective attention task. *Brain Research*, 117, 51–68.

Houde, J. F., Nagarajan, S. S., Sekihara, K., & Merzenich, M. M. (2002). Modulation of the auditory cortex during speech: An MEG study. *Journal of Cognitive Neuroscience*, 14, 1125–1138.

Jaramillo, S., & Zador, A. M. (2011). The auditory cortex mediates the perceptual effects of acoustic temporal expectation. *Nature Neuroscience*, 14, 246–251.

Kaas, J. H., & Hackett, T. A. (2000). Subdivisions of auditory cortex and processing streams in primates. *Proceedings of the National Academy of Sciences of the USA*, 97, 11793–11799.

Kilgard, M. P., & Merzenich, M. M. (1998). Cortical map reorganization enabled by nucleus basalis activity. *Science*, 279, 1714–1718.

Kitzes, L. M., Farley, G. R., & Starr, A. (1978). Modulation of auditory cortex unit activity during the performance of a conditioned response. *Experimental Neurology*, 62, 678–697.

Komura, Y., Tamura, R., Uwano, T., Nishijo, H., Kaga, K., & Ono, T. (2001). Retrospective and prospective coding for predicted reward in the sensory thalamus. *Nature*, 412, 546–549.

Komura, Y., Tamura, R., Uwano, T., Nishijo, H., Kaga, K., & Ono, T. (2005). Auditory thalamus integrates visual inputs into behavioral gains. *Nature Neuroscience*, 8, 1203–1209.

König, R., Heil, P., Budinger, E., & Scheich, H. (2005). *Auditory cortex: Towards a synthesis of human and animal research*, Mahwah, NJ: Lawrence Erlbaum.

Konorski, J. (1967) *Integrative activity of the brain*. Chicago: University of Chicago Press

Lee, C. C., & Middlebrooks, J. C. (2011). Auditory cortex spatial sensitivity sharpens during task performance. *Nature Neuroscience*, 14, 108–114.

Lemus, L., Hernández, A., & Romo, R. (2009a). Neural codes for perceptual discrimination of acoustic flutter in the primate auditory cortex. *Proceedings of the National Academy of Sciences of the USA*, 106, 9471–9476.

Lemus, L., Hernández, A., & Romo, R. (2009b). Neural encoding of auditory discrimination in ventral premotor cortex. *Proceedings of the National Academy of Sciences of the USA*, 106, 14640–14645.

Letzkus, J. J., Wolff, S. B., Meyer, E. M., Tovote, P., Courtin, J., Herry, C., & Lüthi, A. (2012). A disinhibitory microcircuit for associative fear learning in the auditory cortex. *Nature*, 480, 331–335.

Liddell, H. S. (1942). The conditioned reflex. In F. A. Moss (Ed.), *Comparative psychology*. Upper Saddle River, NJ: Prentice-Hall.

Liu, H., Behroozmand, R., & Larson, C. R. (2010). Enhanced neural responses to self-triggered voice pitch feedback perturbations. *NeuroReport*, 21, 527–531.

Martikainen, M. H., Kaneko, K., & Hari, R. (2005). Suppressed responses to self-triggered sounds in the human auditory cortex. *Cerebral Cortex*, 15, 299–302.

Metzger, R. R., Greene, N. T., Porter, K. K., & Groh, J. M. (2006). Effects of reward and behavioural context on neural activity in the primate inferior colliculus. *Journal of Neuroscience*, 26, 7468–7476.

Miller, J. M., Sutton, D., Pfingst, B., Ryan, A., Beaton, R., & Gourevitch, G. (1972). Single cell activity in the auditory cortex of Rhesus monkeys: Behavioral dependency. *Science*, 177, 449–451.

Miller, J. M., Dobie, R. A., Pfingst, B. E., & Hienz, R. D. (1980). Electrophysiologic studies of the auditory cortex in the awake monkey. *American Journal of Otolaryngology*, 1, 119–130.

Müller-Preuss, P., & Ploog, D. (1981). Inhibition of auditory cortical neurons during phonation. *Brain Research*, 215, 61–76.

Näätänen, R., Kujala, T., & Winkler, I. (2011). Auditory processing that leads to conscious perception: A unique window to central auditory processing opened by the mismatch negativity and related responses. *Psychophysiology*, 48, 4–22.

Nelken, I., & Mike, B., & Calford, M. B. (2010). Processing strategies in auditory cortex: Comparison with other sensory modalities. In J. Winer & C. E. Schreiner (Eds.), *The auditory cortex* (pp. 643–656). New York: Springer.

Niwa, M., Johnson, J. S., O'Connor, K. N., & Sutter, M. L. (2012). Activity related to perceptual judgment and action in primary auditory cortex (A1). *Journal of Neuroscience*, 32(9), 3193–3210.

Ohl, F. W., & Scheich, H. (1996). Differential frequency conditioning enhances spectral contrast sensitivity of units in auditory cortex (field Al) of the alert Mongolian gerbil. *European Journal of Neuroscience*, 8, 1001–1017.

Ohl, F. W., & Scheich, H. (2005). Learning-induced plasticity in animal and human auditory cortex. *Current Opinion in Neurobiology*, 15, 470–477.

Ohl, F. W., Scheich, H., & Freeman, W. J. (2001). Change in pattern of ongoing cortical activity with auditory category learning. *Nature*, 412, 733–736.

Otazu, G. H., Tai, L. H., Yang, Y., & Zador, A. M. (2009). Engaging in an auditory task suppresses responses in auditory cortex. *Nature Neuroscience*, 12, 646–654.

Pavlov, I. P. (1960). *Conditioned reflexes*. New York: Dover.

Quirk, G. J., Armony, J. L., & LeDoux, J. E. (1997). Fear conditioning enhances different temporal components of tone-evoked spike trains in auditory cortex and lateral amygdale. *Neuron*, 19, 613–624.

Recanzone, G. H., Schreiner, C. E., & Merzenich, M. M. (1993). Plasticity in the frequency representation of primary auditory cortex following discrimination training in adult owl monkeys. *Journal of Neuroscience*, 13, 87–103.

Remedios, R., Logothetis, N. K., & Kayser, C. (2009). Monkey drumming reveals common networks for perceiving vocal and nonvocal communication sounds. *Proceedings of the National Academy of Sciences of the USA*, 106, 18010–18015.

Rescorla, R. A., & Wagner, A. R. (1972). In H. Black & W.F. Prokasy (Eds.), *A theory of Pavlovian conditioning: Variations in the effectiveness of reinforcement and nonreinforcement, Classical Conditioning II* (pp. 64–99). New York: Appleton-Century-Crofts.

Ryan, A. F., Miller, J. M., Pfingst, B. E., & Martin, G. K. (1984). Effects of reaction time performance on single-unit activity in the central auditory pathway of the rhesus macaque. *Journal of Neuroscience*, 4, 298–308.

Sakurai, Y. (1994). Involvement of auditory cortical and hippocampal neurons in auditory working memory and reference memory in the rat. *Journal of Neuroscience*, 4, 2606–2623.

Schafer, E. W., & Marcus, M. M. (1973). Self-stimulation alters human sensory brain responses. *Science*, 181, 175–177.

Scheich, H., & Ohl, F. W. (2010). A semantic concept of auditory cortex function and learning. In J. Winer & C. E. Schreiner (Eds.), *The auditory cortex* (pp. 369–386). New York: Springer.

Scheich, H., Brechmann, A., Brosch, M., Budinger, E., Ohl, F.W., Selezneva, E., et al. (2011). Behavioral semantics of learning and crossmodal processing in auditory cortex: The semantic processor concept. *Hearing Research*, 271, 3–15.

Schultz, W. (2006). Behavioral theories and the neurophysiology of reward. *Annual Reviews in Psychology*, 57, 87–115.

Scott, B. H., Malone, B. J., & Semple, M. N. (2007). Effect of behavioral context on representation of a spatial cue in core auditory cortex of awake macaques. *Journal of Neuroscience*, 27, 6489–6499.

Selezneva, E., Scheich, H., & Brosch, M. (2006). Dual time scales for categorical decision making in auditory cortex. *Current Biology*, 16, 2428–2433.

Shinba, T., Sumi, M., Iwanami, A., Ozawa, N., & Yamamoto, K. (1995). Increased neuronal firing in the rat auditory cortex associated with preparatory set. *Brain Research Bulletin*, 37, 199–204.

Skinner, B. F. (1938). *The behavior of organisms: An experimental analysis*. New York: Appleton-Century-Crofts.

Squire, L. R., & Zola-Morgan, S. (1991). The medial temporal lobe memory system. *Science*, 253, 1380–1386.

Stark, H., Rothe, T., Wagner, T., & Scheich, H. (2004). Learning a new behavioral strategy in the shuttle-box increases prefrontal dopamine. *Neuroscience*, 126, 21–29.

Stark, H., Rothe, T., Deliano, M., & Scheich, H. (2007). Theta activity attenuation correlates with avoidance learning progress in gerbils. *NeuroReport*, 18, 549–552.

Stein, B. E., & Meredith, M. A. (1993). *The merging of the senses*. Cambridge, MA: MIT Press.

Thiele, A. (2009). Optimizing brain processing. *Nature Neuroscience*, 12, 1359–1360.

Thompson, R. F. (2005). In search of memory traces. *Annual Reviews in Psychology*, 56, 1–23.

Thorndike, E. L. (1933). A prove of the law of effect. *Science*, 77, 173–175.

Vaadia, E., Gottlieb, Y., & Abeles, M. (1982). Single-unit activity related to sensorimotor association in auditory cortex of a monkey. *Journal of Neurophysiology*, 48, 1201–1213.

Villa, A. E. (2005). Spatio-temporal patterns of spike occurrences in feely-moving rats associated to perception of human vowels. In R. König, P. Heil, E. Budinger, & H. Scheich (Eds.), *Auditory cortex: Towards a synthesis of human and animal research* (pp. 275–294). Mahwah, NJ: Lawrence Erlbaum.

Villa, A. E., Tetko, I. V., Hyland, B., & Najem, A. (1999). Spatiotemporal activity patterns of rat cortical neurons predict responses in a conditioned task. *Proceedings of the National Academy of Sciences of the USA*, 96, 1106–1111.

Walter, W. G., Cooper, R., Aldridge, V. J., McCallum, W. C., & Winter, A. L. (1964). Contingent negative variation: An electric sign of sensorimotor association and expectancy in the human brain. *Nature*, 203, 380–383.

Weinberger, N. M. (2010). Reconceptualizing the primary auditory cortex: Learning, memory and specific plasticity. In J. Winer & C. E. Schreiner (Eds.), *The auditory cortex* (pp. 465–491). New York: Springer.

Werner-Reiss, U., Kelly, K. A., Trause, A. S., Underhill, A. M., & Groh, J. M. (2003). Eye position affects activity in primary auditory cortex of primates. *Current Biology*, 13, 554–562.

Wetzel, W., Ohl, F. W., & Scheich, H. (2008). Global versus local processing of frequency-modulated tones in gerbils: An animal model of lateralized auditory cortex functions. *Proceedings of the National Academy of Sciences of the USA*, 105, 6753–6758.

Winer, J., & Schreiner, C. E. (2011). *The auditory cortex*. New York: Springer.

Winkler, I., Denham, S. L., & Nelken, I. (2009). Modeling the auditory scene: Predictive regularity representations and perceptual objects. *Trends in Cognitive Science*, 13(12), 532–540.

Witte, R. S., & Kipke, D. R. (2005). Enhanced contrast sensitivity in auditory cortex as cats learn to discriminate sound frequencies. *Cognitive Brain Research*, 23, 171–184.
Wolpert, D. M. (1997). Computational approaches to motor control. *Trends in Cognitive Science*, 1, 209–216.
Woodruff-Pak, D. S., & Disterhoft, J. F. (2008). Where is the trace in trace conditioning? *Trends in Neuroscience*, 31, 105–112.
Yin, P., Mishkin, M., Sutter, M., & Fritz, J. B. (2008). Early stages of melody processing: stimulus-sequence and task-dependent neuronal activity in monkey auditory cortical fields A1 and R. *Journal of Neurophysiology*, 100, 3009–3029.

Chapter 4
The Role of Auditory Cortex in Spatial Processing

Gregg H. Recanzone

Keywords Aging • Area A1 • Area CL • Macaque monkeys • Maximum likelihood • Neural computation • Perception • Plasticity • Population encoding • Population vector • Sound localization

4.1 Introduction

One of the primary functions of the sensory nervous system is to determine and represent where stimuli originate from. This is critical for identifying predators and prey, as well as mate selection and attending objects of interest. Audition is one of the three primary senses that can provide some information about extrapersonal space, together with vision and olfaction. The latter two sensory systems define the range of spatial localization abilities, with olfactory spatial perception being extremely poor under most naturalistic conditions (Doty, 2001; Kauer & White, 2001), and visual spatial perception being very accurate (Westheimer, 1984; Shapley & Lennie, 1985). Auditory spatial perception is somewhere in between (Blauert, 1997), but holds a significant advantage over vision in that it is sensitive to all regions of extrapersonal space, not just the visual field, which is restricted to frontal space in most primates.

As well as being ethologically vital in terrestrial mammals, a second consideration of auditory spatial perception is that it is necessarily computed by the nervous system. This is because, unlike the visual system, the sensory epithelium cannot encode acoustic spatial location. Neural computations are not restricted to the perception of acoustic space, of course, as any but the simplest percepts are dependent on the nervous system computing and extracting particular stimulus features (see Schnupp, Honey, and Willmore, Chapter 5; Poremba and Bigelow, Chapter 6;

G.H. Recanzone (✉)
Center for Neuroscience and Department of Neurobiology, Physiology and Behavior, University of California at Davis, 1544 Newton Ct., Davis, CA 95618, USA
e-mail: ghrecanzone@ucdavis.edu

Y.E. Cohen et al. (eds.), *Neural Correlates of Auditory Cognition*, Springer Handbook of Auditory Research 45, DOI 10.1007/978-1-4614-2350-8_4,
© Springer Science+Business Media New York 2013

Steinschneider, Chapter 7; and Ghazanfar and Chandrasekaran, Chapter 8). In the visual domain, this would include the recognition of different shapes and patterns beyond those that can be encoded by single retinal ganglion cells, epitomized perhaps by the recognition of faces and other complex shapes (Tsao & Livingston, 2008). A key difference between attempting to understand the representation of acoustic space and attempting to understand the perception of visual shapes is that we know what the nervous system is trying to encode with acoustic space: azimuth and elevation. In contrast, visual shapes are represented in some way, but we do not know the cardinal axes by which this representation is demarcated, or even the number of axes. Thus, similar to visual motion processing in which there are essentially two dimensions to be encoded (direction and velocity; Born & Bradley, 2005), the representation of auditory space can be broken down into two orthogonal dimensions. These more easily definable dimensions make understanding the neural mechanisms underlying auditory spatial perception tractable, although there has been relatively less effort in this endeavor when compared to other sensory systems, particularly vision with regard to visual spatial and visual motion perceptions. This chapter defines the information available to the nervous system that can be used to compute acoustic space, the necessary neural structures involved in acoustic space perception, and a handful of lesion and physiological studies in animal models that probe potential neural codes that plausibly underlie auditory spatial perception.

4.2 Auditory Spatial Localization Cues

The perception of auditory space is dependent on the processing of several different spatial cues that begins at the brain stem and midbrain regions. In terrestrial mammals in which the sounds arrive through the air, there are three physical cues that are used to compute the location in space where a sound originated from. The primary physical cues in azimuth are the interaural timing (and phase) difference (ITD) cues and the interaural level (intensity) difference (ILD) cues (see Blauert, 1997; Recanzone & Sutter, 2008). These cues arise due to the distance between the two ears and the shadowing of the stimulus by the head and torso. This head geometry results in the stimuli being louder and arriving earlier at the near ear versus the far ear. These difference cues are initially processed by different regions within the superior olivary complex of the brain stem (Masterton et al., 1967; Casseday & Neff, 1975; Grothe et al., 2010) and transmitted up through the nucleus of the lateral lemniscus to the inferior colliculus in the midbrain.

The third cue can also provide azimuth information but is primarily involved in processing cues in elevation under normal conditions. This cue is the spectral peaks and primarily notches that occur due to a transformation of the acoustic signal by the head, torso, and pinna, and the resulting sound that vibrates the tympanic membrane is termed the head-related transfer function (HRTF; Wightman & Kistler, 1989a,b). These are termed "spectral cues" and help explain why elevation localization is not as impaired as azimuth localization when stimuli are presented monaurally, for example, by ear plugging (Blauert, 1997; Van Wanrooij & Van Opstal, 2007).

4.2.1 *Plasticity of Spatial Cues During Development*

One interesting aspect of these cues is that they change over time with changes in the geometry of the head and body and the shape of the pinna. This is most dramatic during development, as the newborn head progressively grows to that of an adult, and further changes, although more gradually, with age. There is a difference in the spatial tuning of auditory cortical neurons as a function of age, with younger animals showing broader spatial tuning versus adult animals. This is true across different stimulus intensities (see Mrsic-Flogel et al., 2003) and slowly refines from early postnatal time to adulthood. Similarly, the spatial receptive fields in the superior colliculus are also much broader in infant and very young ferrets compared to older ferrets (Campbell et al., 2008). One question that arises from these findings is whether this refinement of spatial receptive fields with age is due to central or peripheral factors. It can be imagined that the exuberant projections of early axonal processes, which are ultimately pruned to the adult form, would give rise to larger spatial receptive fields. Also of note is that there are considerable differences between the pinnae of neonatal and adult ferrets. To discriminate between these two possibilities, HRTFs were recorded in adult ferret ear canals and played back to the infant animals; thus the infants were "hearing" through adult ears. As a result, the spatial receptive field sharpened considerably, and became much more similar to those seen in the adults. This was true both for the auditory cortical neurons as well as the superior collicular neurons (Mrsic-Flogel et al., 2003; Campbell et al., 2008). These findings can be interpreted to indicate that peripheral structures are in large part responsible for the poorer localization ability of younger animals.

4.3 Sound Localization Ability as Measured by Psychophysical Performance

Decades of psychophysical studies in humans, cats, ferrets, and monkeys as well as several other species have converged on a few fundamental properties of sound localization ability. These properties are also consistent with the known processing of the different cues as described earlier and all indicate that populations of auditory cortical neurons are necessary for the perception of where sounds are in space. The two main parameters of an acoustic stimulus that influences sound localization ability are the stimulus spectrum and the stimulus intensity. The spectrum of the stimulus influences the localization ability differently for azimuth than for elevation, but in general the wider the spectral bandwidth, the more accurate the listeners are at localizing the sound. The intensity of the stimulus also influences localization ability. Sounds that are near detection threshold are very difficult to localize, whereas localization ability increases as the intensity increases, and then remains relatively constant across a wide intensity range. In each case, the localization ability is consistent with a neuronal population encoding scheme, in which increasing numbers

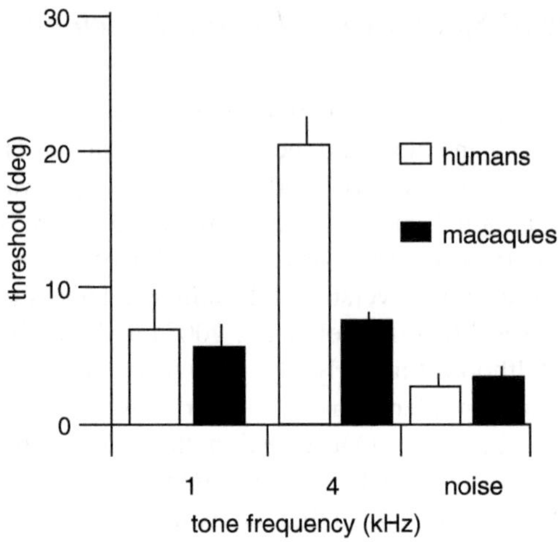

Fig. 4.1 Sound localization thresholds in humans and macaques. Each bar represents the mean and standard deviation of thresholds measured in humans (open) and rhesus macaques (filled) to tone frequencies of 1 kHz, 4 kHz, and broadband noise (noise). Thresholds were determined using a go/no-go procedure in which subjects listened to 200-ms duration acoustic stimulus at either 30 dB above sensation level (humans) or 65 dB SPL (monkeys). Stimuli were presented from directly in front of the subject three to seven times before changing location in azimuth. Threshold was defined as the amount of change in location necessary to be detected on 50% of the trials. Humans and macaques have similar psychophysical thresholds, except for 4-kHz tones, in which humans have considerably higher thresholds. This is likely due to the poor interaural cues at this frequency owing to the size of the human head. (Data derived from published values found in Recanzone et al., 1988 for humans and Recanzone et al., 2000b for monkeys)

of neurons are recruited to processes the spatial location with increases in spectral bandwidth and/or stimulus intensity.

4.3.1 Effects of Stimulus Spectrum

The influence of stimulus spectral bandwidth has been studied extensively and reveals that certain frequencies are particularly poorly localized. These frequencies are species-dependent and are thought to be due to regions in which the two interaural cues are in a range that is not particularly effective given the size and geometry of the head. This is around 4 kHz in humans, as the interaural time/phase cues are too fast to be accurately encoded by the auditory nerve, and the head is too small to shadow lower frequencies effectively compared to higher frequencies (Middlebrooks & Green, 1991; Blauert, 1997). Fig. 4.1 shows the performance of both humans and monkeys localizing tonal stimuli at two different frequencies (1 and 4 kHz) as well as broadband noise (Recanzone et al., 1998, 2000b). The thresholds are similar between the two species for the 1-kHz tone stimulus measured using a standard go/no-go paradigm as shown, but this similarity is also seen across a broad number of different paradigms (Recanzone, unpublished data). The thresholds are also the same for broadband noise

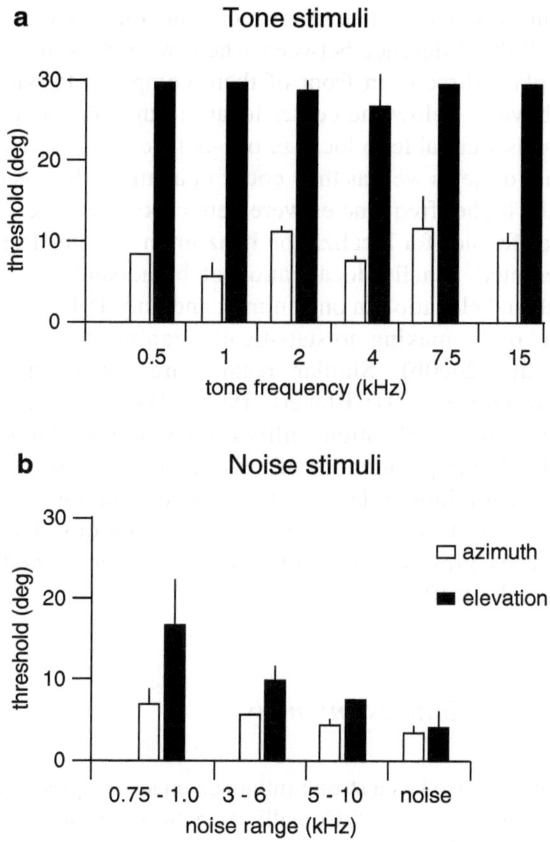

Fig. 4.2 Psychophysical thresholds in azimuth and elevation for macaque monkeys. Thresholds are shown for tonal stimuli (**a**) and either one-octave band-passed noise or broadband noise (**b**) for localization in both azimuth (open bars) and elevation (filled bars). Two monkey subjects performed the same task as described in Fig. 4.1. Tones were virtually impossible to localize in elevation, whereas localization in azimuth was within a few degrees across a broad frequency range. Localization of one-octave band-passed noise improved with higher frequency components, and localization of broadband noise stimuli was equivalent for azimuth and elevation. (Data taken from values published in Recanzone et al., 2000b)

stimuli between the two species, which is the most easily localized stimuli of those tested. Indeed, depending on the task and the duration and intensity of the stimulus, humans can discriminate sound localization to within about 1° (Stevens & Newman, 1936). As mentioned previously, there is a clear difference between the species for 4 kHz tone stimuli, but this difference is restricted to this particular frequency range, almost certainly due to the differences in the head size between the two species.

The bandwidth of the stimulus spectrum differentially influences localization in azimuth and elevation, as a broad spectrum is necessary for the spectral peaks and notches to be evident. Fig. 4.2 shows the results from macaque monkeys localizing different sounds in both azimuth (open bars) and elevation (filled bars) for tones (Fig. 4.2a) and one-octave band-passed or broadband noise (Fig. 4.2b).

Monkeys have an extremely difficult time localizing tones in elevation, and generally could not tell the difference between when two tone stimuli were presented from the same place directly in front of them compared to when one was presented at 30° above or below the center location, the largest spatial separations tested. The animals were able to localize one-octave band-passed noise in elevation, although not quite as well as they could in azimuth. In this case, the stimuli with energy at the higher frequencies were better localized in elevation, although there was little difference for localization in azimuth as a function of the stimulus frequencies presented. Finally, localization of broadband noise was equivalent between azimuth and elevation in one animal, and only 1° larger in elevation than in azimuth in the other, making no statistically significant difference in this case (Recanzone et al., 2000b). Similar results are found in human subjects (Middlebrooks & Green, 1991; Blauert, 1997). The fact that broader stimulus spectrum leads to better localization ability is consistent with a population encoding scheme in that larger populations of neurons would be recruited, particularly at the brain stem and midbrain levels where spectral tuning is sharper compared to the cerebral cortex. Thus, there would be more neurons able to calculate the different spatial cues, presumably resulting in a better representation of the sound source (Grothe et al., 2010).

4.3.2 Effects of Stimulus Intensity

The stimulus intensity also has a strong influence on spatial processing, but in a very nonmonotonic manner (Altshuler & Comalli, 1975; Su & Recanzone, 2001; Recanzone & Beckerman, 2004). Fig. 4.3 shows the localization performance of humans and macaque monkeys localizing broadband noise stimuli at four different absolute intensity levels using two different behavioral paradigms (see legend). For the loudest stimuli (solid bars), localization performance is quite good at 90°, directly opposite each ear. As the stimulus intensity decreases, there is only a slight decrease in localization ability until the stimulus becomes quite low (25 dB SPL), which was 10–15 dB above detection threshold as reported by the human subjects, and based on behavioral audiograms of the monkey subjects (Juarez-Salinas et al., 2010). These data are similar to those of other studies in human localization (e.g., Comalli & Altshuler, 1976; Sabin et al., 2005), again supporting the notion that populations of neurons represent acoustic space, as increases in stimulus intensity will initially increase the population size of activated neurons given the variance in threshold of primate auditory cortical neurons (e.g., Recanzone et al., 1999, 2000a). A second consideration is that the macaque and human thresholds are quite similar when compared across the same acoustic stimuli and the same tasks. This finding, coupled with the similarity in the functional organization of auditory cortex (Hackett et al., 2001; Rauschecker & Scott, 2009; Rauschecker, Chapter 2), demonstrate that the macaque monkey is a good animal model for studying basic acoustic processing and perception, such as the location of a sound in space.

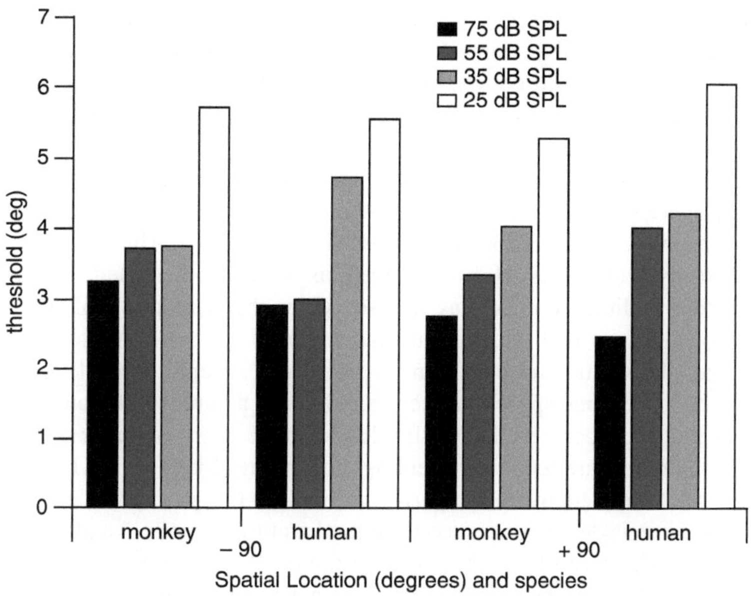

Fig. 4.3 Psychophysical thresholds as a function of stimulus intensity. Results from human and rhesus macaque monkey subjects localizing 200-ms duration broadband noise stimuli presented at four different intensity levels. Humans were performing a task in which a single stimulus was presented, and they were asked to identify from which of 16 possible locations spanning 360° in azimuth the stimulus originated. Accuracy at this task for stimuli presented opposite the left ear (–112.5 to –67.5°) is shown to the left, and for stimuli presented opposite the right ear (67.5 to 112.5°) on the right. Thresholds were derived using the formulation described in Recanzone et al. (1998) correlating absolute and relative threshold measurements. Macaque monkeys were performing a go/no-go task as described in Fig. 4.1, except the stimuli started at either –90° (left) or +90° (right). Louder stimuli were easier to localize, as shown by the lower thresholds. Thresholds were largest for the quietest stimuli (solid bars). Humans and macaques showed similar discrimination thresholds and the same trends in localization ability as a function of stimulus intensity. (Data from humans were taken from those published in Miller & Recanzone, 2010 and data from monkeys were taken from those published in Recanzone & Beckerman, 2004)

4.4 Effect of Auditory Cortical Lesions on Sound Localization Performance

Multiple early studies in a variety of mammals have shown that lesions of auditory cortex result in deficits in the ability of the animal to make purposeful movements indicating the location of the sound stimulus. One of the more compelling examples comes from relatively small lesions in cats that were trained to approach the speaker that emitted a tone or noise stimulus in order to receive a reward (Jenkins & Merzenich, 1984). After training, the animals showed very few errors for both noise and tonal stimuli. The animals were then given unilateral lesions in A1 that were restricted to a single frequency representation based on electrophysiologically identifying the tonotopic organization for that particular animal. After recovery, animals showed a selective, contralesional deficit in localizing tones that had been represented in the lesioned cortex, but no deficits in ipsilesional space, or for tone frequencies outside of the lesioned area. These results indicate that primary auditory cortex is necessary for performance at this task.

A similar set of studies in primates provided the same result, although the frequency specificity was not determined as in the Jenkins and Merzenich study (Thompson and Cortez, 1983; Heffner & Heffner, 1990). The results from the macaque study (Heffner & Heffner, 1990), as well as others, led the authors to conclude that auditory cortex was necessary for the perception of the acoustic stimulus location, not that the stimulus had occurred. This is consistent with other studies that show that animals can instinctively orient to brief, unexpected sounds in contralesional space, but they cannot make a purposeful motor movement indicating that they perceived the stimulus location (e.g., Beitel & Kaas, 1993). Thus, although there are reflexive orienting movements, likely driven by subcortical structures and directed by the superior colliculus, the *perception* of where the sound came from is dependent on contralateral auditory cortex. An alternative way of considering these results is that without the cerebral cortex there is a memory deficit whereby the animal cannot continue to act on the location of the sound once it is over.

More recent studies in carnivores have explored which areas of auditory cortex are critical for sound location perception. Studies in the ferret have shown that the chemical inactivation largely restricted to A1 produce contralesional deficits, which are largest for brief stimuli (Smith et al., 2004). These studies were extended by comparing the effects of small lesions, restricted to primary auditory cortex and larger lesions affecting all of A1 and many other auditory cortical areas (Nodal et al., 2010). In this case, as suggested by earlier lesion work, the ability to localize brief sounds was compromised in both lesion groups. However, with longer stimuli the ferrets with lesions restricted to A1 had nearly normal performance at both approaching the speaker that emitted the sound as well as orienting toward it during the initial head movement. Ferrets with larger lesions still showed a deficit, indicating that although A1 is an important contributor to processing acoustic space cues, other cortical areas are similarly involved in auditory spatial perception. Similar results have been seen in the cat using reversible cooling of different cortical areas (Maholtra et al., 2004). In these studies, cooling A1 as well as two other cortical regions, the posterior auditory field and the cortical areas within the anterior ectosylvian sulcus, caused contralesional deficits in the ability of the cats to approach the active speaker. There were also other cortical areas, notably the anterior auditory field (AAF), in which cooling had no influence on this localization behavior. Notably, cooling of this region did show a deficit in a nonspatial auditory task (Lomber & Malhotra, 2009), indicating that the carnivore auditory cortex may also parse the processing of different acoustic signals into at least two main pathways, similar to the visual cortex and discussed in more detail later (see also Rauschecker and Tian, 2000; Rauschecker, Chapter 2).

4.5 Plasticity of Sound Localization Processing

The preceding sections described how auditory cortex is necessary for sound localization, and also that the localization cues are altered over time while the head changes shape during development. There are also instances in which the localization cues can be altered in adulthood, either over the short term by infections, injuries, or natural aging processes that will attenuate the sensitivity to sound differentially between the two ears. Auditory cortex has been shown to be plastic both developmentally and in adults (e.g., Recanzone et al., 1993; Kilgard et al., 2001; de Villers-Sidani et al., 2007) and that the adult plasticity is not passive but requires attention by the animal to the acoustic environment, or activation of the neuromodulatory systems. These plasticity studies raise the question whether there is similar plasticity in sound localization in adults, and what the role of auditory cortex is in this plasticity (also see Fritz, David, and Shamma, Chapter 9 and Kilgard and Liu, Chapter 10).

4.5.1 Plasticity in Localization in Elevation

One clear example of adult sound localization plasticity can be shown by plugging one ear with a specialized plug that differentially influences the spectral cues (Hofman et al., 1998). Plugging initially results in a severe inability to localize broadband sounds in elevation, and most subjects localize all sounds across different azimuths along a single elevation contour, although the particular elevation will vary between individuals. This localization deficit recovers over the course of about 5–6 weeks, and then is normal again immediately after unplugging. This indicates that there are two independent representations of acoustic space after the plasticity induced by the altered spectral cues: one that is functional given the altered cues and the original, unaltered representation.

The initial localization deficit after ear plugging could actually be due to either the alteration of the spectral cues themselves in the one ear or to the disruption of the comparison between the two sets of spectral cues between the two ears. The spectral cues in each ear are basically the same (or very similar) and provide essentially the same information about the elevation of a stimulus, which is why plugging one ear has little effect on localization in elevation (Middlebrooks & Green, 1991; Blauert, 1997). Therefore the deficit that occurs when changing the spectral cues in one ear could be due to the change itself, or it could be due to the mismatch in the information about the elevation location between the two ears. This was directly tested by comparing the adapted responses and altering the spectral cues between the two ears in a number of different variations. These results were consistent with the effect being the result of the altered cues in the plugged ear (Van Wanrooij & Van Opstal, 2005). This indicates that acoustic space is mapped independently in each ear and this spatial mapping is altered by the changed spectral cues before the information is integrated binaurally.

4.5.2 *Plasticity in Localization in Azimuth*

A second series of experiments investigated the role of A1 on this adult plasticity. A different style of monaural ear plug can change the intensity of the stimuli at one ear, resulting in an alteration of azimuth localization, but less so in elevation (e.g., Van Wanrooij & Van Opstal, 2007). This phenomenon has been studied more extensively in ferrets, in which a similar result is obtained. In this case, ferrets are trained to approach one of several speakers that span the entire range of azimuth at about ear level. After insertion of an ear plug there is an initial deficit (Kacelnik et al., 2006). In normal ferrets, monaural ear plugging results in deficits in localizing sounds on the same side as the plug. There is plasticity in this localization ability, however, as there is generally a return to normal localization within several days. As with the human studies described earlier, removal of the plug did not produce any immediate deficits; rather, the ferrets were equivalent in their performance as before the plug was introduced. An extension of these findings was comparing whether the animals readily adapted to these changes, or if training at the localization task was necessary. When the localization ability of animals that showed continuous training was compared to animals that were only sporadically tested, those animals that were well practiced showed the improvement, whereas those that did not, or received visual localization training, did not show the improvement. Thus, localization in both azimuth and elevation is plastic in adult animals, but requires that the animal is engaged in a localization task. This type of result is very reminiscent of findings of changes in cortical representations as a function of practice that are also dependent on whether the animal is engaged in the task (e.g., Recanzone et al., 1992, 1993).

4.5.3 *Role of Auditory Cortex in Sound Localization Plasticity*

4.5.3.1 A1 and Plasticity After Monaural Plugging

The influence of auditory cortex in this plasticity was investigated by making small lesions restricted to A1 in one group of ferrets and larger lesions including A1 (Nodal et al., 2010). This caused sound localization deficits in the large lesion group, but the effect was much less in the small lesion group, as described earlier. However, after monaural ear plugging, ferrets in both groups were unable to adapt to the new cues and the post-plugging deficits remained in spite of repeated training. This was not due to an overall deficit in localization ability, as animals were able to localize as well as during the pre-plug performance immediately after the plug was removed. These results indicate that A1 alone is not absolutely necessary for localization under some conditions, but is a vital participant in the adaptive plasticity mechanisms that can compensate for altered peripheral input. This idea is also consistent with studies in the barn owl, in which ear plugging results in neuronal changes in the thalamus and forebrain to compensate for the altered localization cues (Miller & Knudsen, 2001, 2003).

This raises the question of what part of auditory cortex is necessary for this adaptive plasticity. The cerebral cortex is a multi-layered structure, in which the neurons in layer V project to subcortical targets. In auditory cortex, a major subcortical target is the inferior colliculus. The inferior colliculus is sensitive to acoustic space (e.g., Groh et al., 2003) and is an obligatory relay nucleus of the ascending auditory system. In one set of experiments, chemically conjugated microbeads were injected into the inferior colliculus and were allowed to transport retrogradely to layer V of auditory cortex (Bajo et al., 2010). After transport, the neurons that contained the microbeads were selectively lesioned, sparing the remaining tissue of primary auditory cortex. This procedure did not affect sound localization behavior, indicating that A1 remained largely intact and that the lesion was confined to the neurons providing the descending input to the inferior colliculus. When tested after the placement of the monaural ear plug, however, there was no evidence of the adaptive plasticity seen in normal ferrets. This results as well as those in the barn owl (Miller & Knudsen, 2001, 2003) indicates that the descending projections from auditory cortex to the inferior colliculus is a key component of adaptive plasticity that maintains the cortical representation of acoustic space, and ultimately to the perception of acoustic space.

4.5.3.2 A1 and Plasticity After Environmental Enrichment

These previous studies indicate that auditory cortex is necessary for normal spatial localization abilities as well as after an acute but intense alteration of the localization cues. Presumably more subtle changes in the animal's environment can also alter sound localization abilities, as these are more generally experienced. One example is from rearing rats in enriched environments, compared to standard laboratory housing. "Environmental enrichment" has been used in several different contexts, and generally results in greater activity and a higher resolution representation of the sensory structures that are being investigated. With respect to sound localization, raising rats in an environment where there are several different speakers presenting tonal stimuli does result in sharper spatial tuning of auditory cortical neurons (Cai et al., 2009; Zhang et al., 2009). This sharper tuning also correlates with improved spatial discrimination by rats raised in enriched environments, both with respect to the accuracy in which the rats can localize sounds as well as the reaction time to respond to the sound (Cai et al., 2009). The degree of the correlation, and which aspects of the sharper spatial tuning could account for the improved performance, was not rigorously tested. Nonetheless, these studies provide further evidence that the representation of acoustic space is alterable in a number of different ways, both during development as well as in adults.

4.6 Functional Properties of Auditory Cortical Neurons

Given the importance of sound localization, the necessity of the cerebral cortex, and the tractable nature of understanding what the nervous system is attempting to represent, it should be no surprise that considerable effort has been extended toward understanding the neural representation of acoustic space. Early studies in the midbrain structure of both barn owls (tectum; Knudsen & Konishi, 1978) as well as in mammals (superior colliculus; King & Palmer, 1983; Middlebrooks & Knudsen, 1984) showed that there is a topographic map of acoustic space in this subcortical structure. However, as noted earlier, the cerebral cortex is necessary for the percept of acoustic space, and it is likely that these topographic maps in the superior colliculus are dealing more with the orientation toward the stimulus, although higher order functions are also likely (e.g., Lovejoy & Krauzlis, 2010; Nummela & Krauzlis, 2010).

Investigations of acoustic space representations in the cerebral cortex were initially conducted primarily in the anesthetized cat, and multiple studies showed that there were spatially selective receptive fields in primary and secondary auditory cortical areas (Imig et al., 1990; Rajan et al., 1990a; Brugge et al., 1996). These studies did not reveal, however, any type of topographic organization, although there were instances in which the centers of the receptive fields were clustered locally (Rajan et al., 1990b; Recanzone et al., 2000b). These studies strongly indicate that acoustic space is not coded as a relatively simple "place code" across a topographically organized map in auditory cortex.

The findings in cats have since been extended to other mammals, most notably the ferret and macaque monkey (Woods et al., 2006; King et al., 2007). In all three species, several generalities of auditory spatial receptive fields can be made. First, the receptive fields are generally quite large, with some spanning an entire 360° around the animal, whereas others are more restricted to on hemisphere or even a quadrant (90°). Receptive field size seems to be slightly smaller in the alert macaque compared to the cat preparations; however, a recent study has shown that spatial tuning in cat auditory cortex is sharpened when the animal attempts to localize the stimulus (Lee & Middlebrooks, 2011). Nonetheless, it is extremely rare to find receptive fields that approach the size of the estimates in absolute space (Makous & Middlebrooks, 1990; Recanzone et al., 1998). These large spatial receptive fields also constitute an argument against a "labeled line" coding mechanism and suggest that a population code is more likely in operation, described in more detail later. Alternatively, given the lesion results described in the preceding text, it may be that primary auditory cortex is functioning more to pre-process the information, and then this is relayed to one or more higher-order cortical areas where the percept is generated, which is then translated into a motor action as measured in most animal studies.

4.6.1 A Spatial Processing Stream in Auditory Cortex

The working hypothesis that multiple cortical areas are involved in spatial processing was most strongly proposed by Rauschecker (1998) where he likened the primate auditory cortex with that of the primate visual cortex, composed of a dorsal spatial processing stream and a ventral nonspatial processing stream (Ungerleider & Mishkin, 1982; Ungerlieder & Haxby, 1994; see Rauschecker and Tian, 2000; Rauschecker, Chapter 2). In the case of auditory cortex, the spatial processing stream would start in the core region, notably A1, and then information would be sequentially processed in the caudal belt regions, the caudal parabelt, and ultimately in parietal and frontal spatial processing regions (Fig. 4.4). In contrast, nonspatial

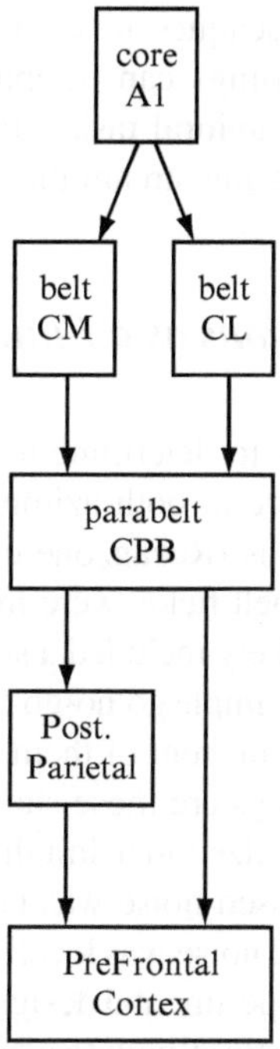

Fig. 4.4 Schematic representation of the hypothesized spatial processing components of primate auditory cortex. Stimuli are initially processed by primary auditory cortex (A1), which then projects to the caudal belt fields, the caudomedial field (CM) and caudolateral field (CL). These cortical areas project in turn to the caudal parabelt (CPB), which then projects to regions within the posterior parietal lobe and the prefrontal cortex. A more complete description of primate auditory cortex, including nonspatial processing areas, can be found in Kaas & Hackett (2000), Rauschecker et al. (2000), and Rauschecker, Chapter 2

information would be processed from the more rostral regions of auditory cortex, through the rostral belt and rostral parabelt, to be transmitted to nonspatial regions in the temporal and frontal lobes (Rauschecker, 1998). This hypothesis is supported by anatomical (Romanski et al., 1999; Kaas & Hackett, 2000; Romanski & Averback, 2009) and electrophysiological (Rauschecker et al., 1997; Tian et al., 2001; Woods et al., 2006) studies in macaques as well as imaging and psychophysical studies in humans (see Rauschecker & Scott, 2009). The next section explores several studies in detail that investigate *how* auditory cortical neurons could potential encode acoustic space along this caudal spatial processing pathway.

4.7 Auditory Cortical Representations of Acoustic Space

Recent studies have sought to take advantage of the differences in sound localization ability as a function of stimulus spectrum and intensity in the alert macaque monkey. As noted previously, the sound localization abilities and the basic organization of auditory cortex of humans and macaques are equivalent. The macaque is also a very tractable animal, and thus recordings can be made while the animal is not only awake but also performing a behavioral task. These studies have provided strong support of a caudal spatial processing stream that functions on a population level.

4.7.1 Cortical Representations as a Function of Stimulus Spectrum

The ability of macaque monkeys to determine a change in location of an auditory stimulus presented in frontal space in both azimuth and elevation was investigated using different spectral bandwidths (Recanzone et al., 2000). In this case, the core field A1 as well as more caudal belt fields were investigated. Those belt fields were originally identified as CM, but likely included a significant number of neurons in CL as well. Animals were trained on a simple go/no-go task to discriminate when a stimulus changed in location from directly in front of them to a more peripheral location. The behavioral data from these monkeys are those shown in Figs. 4.1 and 4.2 and reveal that tonal stimuli were poorly localized in azimuth and extremely difficult to localize in elevation, one-octave band-passed noise was better localized, particularly for the higher frequencies, and broadband noise was localized best in both azimuth and elevation. One key advantage of this experimental design is that the neuronal responses can be measured to two different acoustic stimuli to which they are responsive (a tone near their characteristic frequency and a band-passed or broadband noise) and be related to different levels of localization performance. Any general mechanisms of acoustic space encoding should be consistent with these two stimulus types.

Single-neuron recordings from auditory cortex in these animals showed that most neurons were sensitive to the spatial location of the stimulus (Fig. 4.5). Consistent with the behavioral performance, a greater percentage of neurons were spatially selective for the noise stimuli compared to the tone stimuli. In addition, there were proportionally more neurons spatially selective in azimuth compared to elevation, with very few showing any spatial selectivity in elevation for tonal stimuli. Neurons in the caudal belt showed a much sharper spatial tuning than neurons in A1, again predominately centered in contralateral space as with the core field neurons, often responding to all tested locations in frontal space. This is consistent with previous studies indicating that the caudal fields had sharper spatial tuning than the more rostral fields (Rauschecker et al., 1995). Some neurons were not spatially tuned in that quadrant, however. This is likely due to the technique of defining the spatial extent of

Fig. 4.5 Percentage of neurons sensitive to locations within frontal space. (**a**, **b**) Results for A1 neurons, where **a** shows the results for tone stimuli and **b** shows the results for one-octave and band-passed noise. (**c**) Results from caudal belt neurons, which were tested with a more limited stimulus set. Spatial sensitivity was defined by measuring the response across 17 locations positioned in concentric circles from straight ahead (1 location), 15° eccentric (8 locations), or 30° eccentric (8 locations). Regression analysis between firing rate and eccentricity broken down into azimuth (open bars) or elevation (filled bars) was then conducted and neurons with a statistically significant nonzero slope were considered spatially sensitive. Each neuron was tested with both a tone at a single frequency and either a band-passed or broadband noise on randomly interleaved trials.

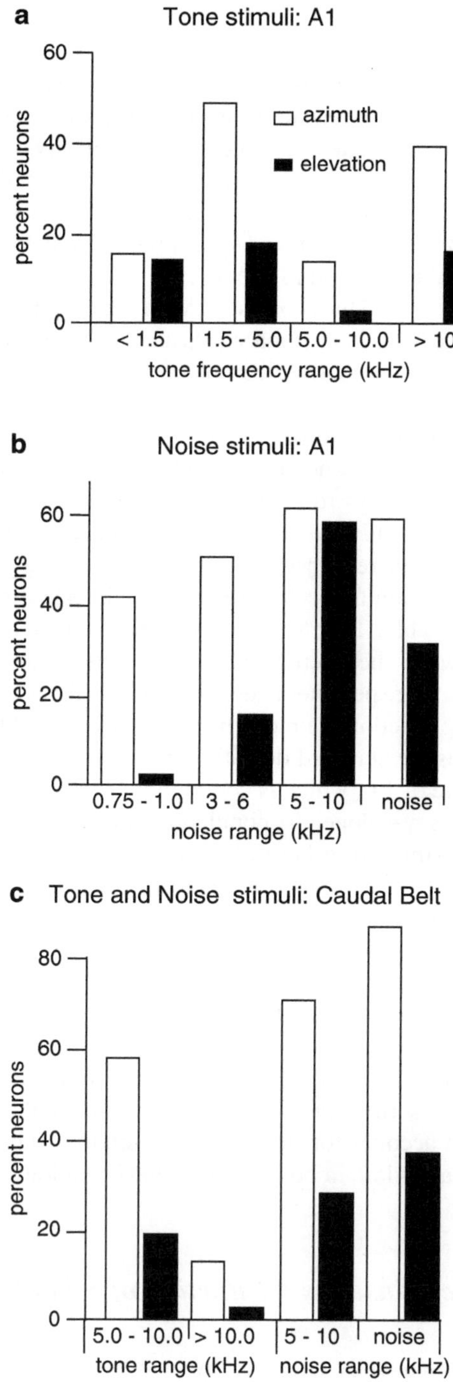

Fig. 4.5 (continued) A much greater percentage of neurons were spatially sensitive in azimuth than in elevation for the noise stimuli in A1 and for all stimuli in caudal belt neurons (comprising CM and CL). The percentage of A1 neurons spatially sensitive to noise stimuli was generally greater in azimuth than in elevation for A1 neurons except for the 5–10 kHz band-passed stimulus. (Data are taken from those published in Recanzone et al., 2000b)

the receptive field, as subsequent work showed that many neurons have considerably reduced response toward the far reaches of ipsilateral space (Woods et al., 2006).

Receptive field sizes were quite large, and the vast majority of neurons responded to all of the locations tested. When spatial receptive fields were defined as locations that elicited at least 75% of the maximum response, only about 5% of neurons in A1, and no neurons in the caudal belt, had spatial receptive fields constrained to 60° of frontal space. This value dropped to about 1.5% when the criterion of 50% of the maximum response was used. Thus, consistent with previous reports in anesthetized animals, spatial receptive fields in alert macaques are much larger than the localization ability would predict (Makous & Middlebrooks, 1990; Recanzone et al., 1998) and no sign of a topographic map was evident, similar to previous findings in the anesthetized cat (Rajan et al., 1990b).

The key issue to address is whether this degree of spatial tuning is consistent with the ability of the monkeys to localize these different stimuli. To investigate this, the difference in the population response for the location directly in front of the monkey was compared to the responses measured at increasingly eccentric locations in both azimuth and elevation for each of the tonal and noise stimuli for which both types of data were available. This was chosen as the behavioral task required the monkeys to indicate when they perceived the stimulus to have changed location from directly ahead to a more peripheral location. This analysis was done by defining where along the tuning function the mean firing rate was equal to that for the location directly ahead plus the standard deviation of that firing rate. This is similar in many ways to a d' value of 1.0, which is commonly considered threshold (Green & Swets, 1974). When this was done, the population of caudal belt neurons was much more accurate at discriminating between different locations (Fig. 4.6b). This discrimination ability across neurons could not fully account for the discrimination by the monkey, suggesting that a different sort of code may be in operation. However, if the sample were restricted to the most spatially selective neurons, the prediction by the neural population improved considerably. Using this restricted number of neurons, the population from A1 still did not correlate well with the psychophysical performance, but the neurons from the caudal belt did (Recanzone et al., 2000b). These correlations were noted only for locations in contralateral space, consistent with the lesion results described previously. These results indicate that a spike rate population code could account for differences in sound localization abilities as a function of stimulus bandwidth in both azimuth and elevation.

4.7.2 Cortical Responses as a Function of Stimulus Intensity

These results were subsequently extended by investigating the effects of stimulus intensity on both localization performance as well as the neural responses across more cortical fields (Woods et al., 2006). In this study, the rostral field, R, of the core was investigated as well as A1. Caudal belt fields were more extensively sampled and the middle medial (MM), caudomedial (CM), caudolateral (CL), and middle

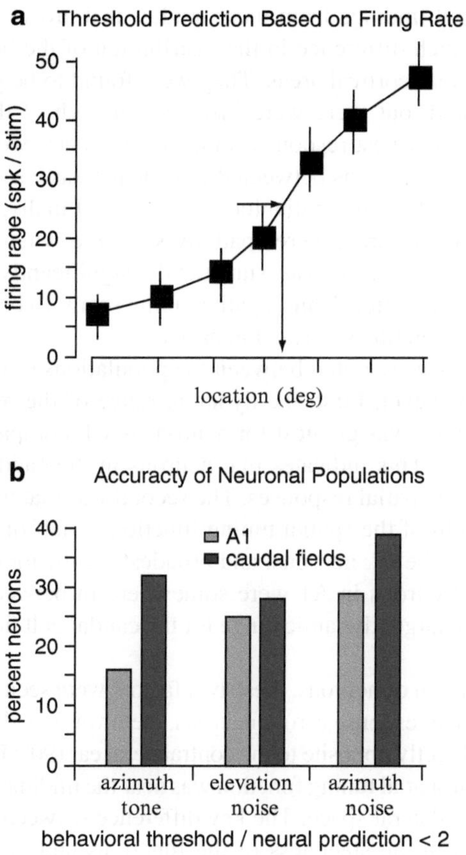

Fig. 4.6 Predicting behavioral thresholds from neural responses. (**a**) Schematic of a model neuron response as a function of stimulus location. Lines demark standard deviations of the response. Arrows show the change in location equivalent to a firing rate of the mean + 1 standard deviation of the response from straight ahead. This is similar to a *d*′ of 1.0, which is commonly taken as threshold. (**b**) Percentage of neurons that are within a factor of 2 in predicting the behaviorally measured thresholds. Neurons in the caudal fields consistently are more accurate at predicting thresholds compared to neurons in A1, particularly in azimuth. (Data taken from those published in Recanzone et al., 2000b)

lateral (ML) fields were all investigated. Finally, the entire 360° of azimuth space, at the level of the interaural axis (0° elevation) were tested. Stimulus intensity has a strong effect on the ability to localize noise stimuli (Fig. 4.3), but the influence of intensity on the neuronal responses of neurons in auditory cortex are commonly not monotonic (Recanzone et al., 1999, 2000a). Therefore it has remained a question as to why the neural activity can vary to a much greater extent than does the sound localization ability (e.g., Su & Recanzone, 2001 Recanzone & Beckerman, 2004), particularly if a population spike rate code is in operation. These experiments therefore addressed these questions.

There were several findings from these studies that are worth note. The first is that there was not much difference in the distribution of the best direction of neurons across the different cortical areas. They were found to be predominately in the contralateral hemifield, but there were instances in each of the six cortical areas where neurons with their best responses in ipsilateral space were encountered. The second commonality of neurons between the cortical areas was that most responded to the majority, if not all, of the spatial locations tested. Finally, most neurons, again regardless of the cortical area, were spatially selective when testing whether the response in the best direction was statistically significantly different from the response in the worst direction. Thus, spatial tuning seems to be a ubiquitous feature of many core and belt auditory cortical neurons.

Several differences were noted between the populations of neurons in the different cortical areas, however. First, the dynamic range of the response between the best and worst direction was greatest for neurons in CL compared to neurons in all other fields (Fig. 4.7). This indicates that neurons in the caudal belt fields have a greater fidelity of their spatial responses. The second was that the spatial bandwidth, measured as the width of the spatial tuning function at half of the maximum firing rate, was sharpest for the CL neurons, and broadest for the more rostral and medial fields, R and MM. Neurons in A1 were somewhere in between. Thus, the spatial tuning had not only a larger dynamic range for the caudal belt fields, but also sharper spatial tuning.

Across the population of neurons, these two factors were seen to interact. By taking the normalized average response across neurons, there was clearly a peak in activity in contralateral space directly opposite to the contralateral ear (90°; Fig. 4.8). The steepest part of the population spatial tuning function was near the midline, and the lowest firing rates were seen in ipsilateral space. The key difference between the different cortical areas, however, was in the steepness of the functions and the difference between stimulus intensities. The population of neurons in CL showed the sharpest overall tuning, with the steepest slope between ipsilateral and contralateral locations, and would therefore seem the most likely candidates to carry the most spatial information as a population compared to the neurons in the other cortical fields.

4.7.3 Potential Population Encoding Models

To test whether these differences in firing rate could plausibly account for sound localization performance, several different population coding models were tested. They key in this experiment was to determine which, if any, of the population models could accurately encode the sound localization performance across the different tested intensities. The second feature that had to be accounted for was that the neural responses were best for contralateral space, and would therefore be predicted to be relatively poor in ipsilateral space given the results of the lesion studies described in the preceding text. This may seem trivial, but low firing rates can also carry a great deal of information about where the stimulus is not. For example a low firing rate of the neuron shown in Fig. 4.6a would indicate that the stimulus is in ipsilateral space

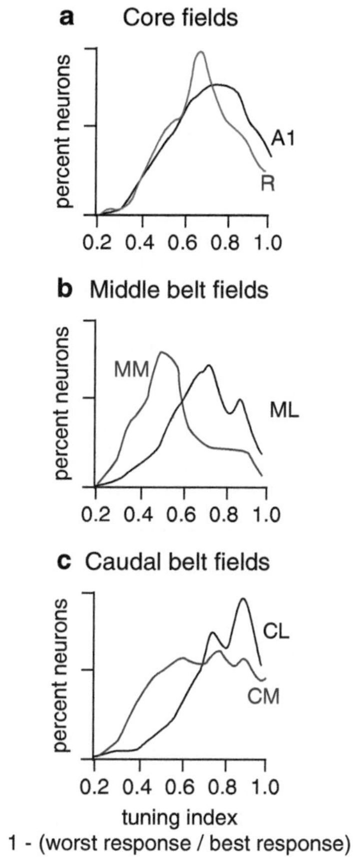

Fig. 4.7 Frequency distribution of tuning index across cortical areas. The tuning index is defined as 1 – (response in the best direction/response in the worst direction), where the worst direction is the direction with the lowest firing rate and not 180° opposite the best direction. Values near 1.0 indicate much greater response in the best direction compared to the worst direction, whereas values near 0 indicate the same response in the two directions. This index provides a measure of the dynamic range of the neuron as a function of spatial location. Data were taken from single neuron recordings to 200-ms duration broadband noise bursts presented from 1 of 16 locations located every 22.5° spanning 360° in azimuth at 0° in elevation in three alert rhesus macaque monkeys. The intensity of the stimulus was set at 25, 35, 55, or 75 dB SPL, randomly interleaved during each trial. (**a**) Results from neurons located in the core fields, A1 and the rostral field (R). (**b**) Results from the two belt fields located near the middle of the belt, the middle medial area (MM) and the middle lateral area (ML). (**c**) Results from the caudal belt fields CM and CL. Each line shows the mean across all four stimulus intensities, which were not statistically significantly different. Neurons in CL had the highest tuning index values across neurons, whereas neurons in MM had the smallest tuning index values. (Data taken from those published in Woods et al., 2006)

and not in contralateral space. Inspection of Fig. 4.8 shows that there are relatively flat portions of the spatial tuning functions in both contralateral and ipsilateral space, so any coding scheme would have to treat the higher firing rates differently than the lower firing rates.

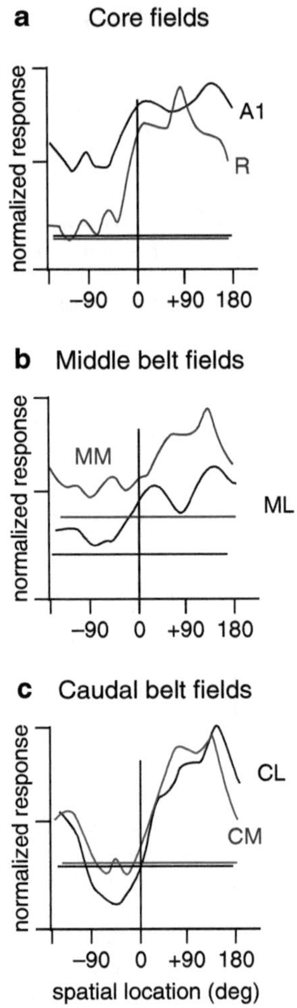

Fig. 4.8 Population response profiles across cortical areas. For each cortical area, the response to each stimulus was normalized by the response at the best location and then these responses were pooled across all neurons recorded and averaged. The horizontal line shows the normalized spontaneous activity. Panels show cortical areas following the conventions of Fig. 4.7. For neurons in all cortical areas, the population responds more to contralateral (rightward) locations compared to ipsilateral locations, with the steepest part of the tuning functions near the midline (vertical line). As with the tuning index, the neurons in MM have the shallowest tuning, whereas the population response in CL neurons show the most change between locations, and some ipsilateral locations actually showed inhibition. Data shown are from 55 dB SPL, but similar functions were seen at 75 dB SPL, but the tuning was degraded for the lower intensity stimuli. (Data are taken from those published in Woods et al., 2006)

4.7.3.1 Population Coding Based on Topography

Perhaps one of the theoretically simplest models is that topographically unique distributions of neurons are activated for different locations in space. As noted previously, there is no sign of a topographic representation of acoustic space in auditory cortex, so whatever the representation may be, it would have to be non-topographic. Nonetheless, it is not necessary for adjacent locations in space to be represented at adjacent locations in the cortex, only that unique locations could be distributed across a cortical area. For this to be the case, one would expect there to be both a continuous distribution of best locations as well as fairly sharp spatial tuning such that the overlap between adjacent locations that are clearly discriminable is very small, if it exists at all. This also does not seem to be the case, as spatial receptive fields are still very broad (median spatial bandwidth was >90° for CL neurons and well over 180° in many cortical areas) and the distribution of best locations was not uniform across contralateral acoustic space (Woods et al., 2006).

4.7.3.2 Population Coding Based on Neuronal Probabilities

An alternative method is to make no assumptions about the topography of the distributions, but to use each neuron as an element that provides a probability of where the stimulus may have originated from. Neuronal responses do have some variance (e.g., Fig. 4.6a), and therefore if one were to take the response of a neuron on a single trial to estimate the spatial location, there would inherently be some variance, or error, in the estimate. An example of how this could potentially work is shown for two hypothetical cells in Fig. 4.9a. For the cell represented by the black squares, on average the response at +90° is about 13 spikes/stimulus, although for that location the trial-by-trial variance ranged from about 9 to 17 spikes/trial. As the spatial tuning is somewhat broad (but average for the population of A1 neurons), 13 spikes/stimulus is also well within the range of the neuron's response to 67.5°, although not as common. Thus, if this neuron fires 13 spikes, the probability is high that it is from 90° but there is also a real possibility that it is from 67.5°. The probability that it is from –90°, given that it fired 13 spikes, is vanishingly small (nearly 0) and other locations provide other probabilities. One can then calculate the probability that the actual stimulus location was at each of the different 16 locations given 13 spikes on that particular trial, then normalize by the total and generate a probability map of where the stimulus may have come from (Fig. 4.9b). As can be seen, the amount of information that this single neuron provides is quite limited, and there is a reasonably large probability that the model will be incorrect, whereas the psychophysical data indicate that there should be few if any errors for a stimulus from this direction at this intensity.

A similar translation from a firing rate to a probability distribution is shown for another cell (gray circles in Fig. 4.9a) where the 90° stimulus is not at the best direction. In this case, the cell would signal that the highest probabilities are that the stimulus came from near 90° but also actually higher for another location (Fig. 4.9b). This is due to the firing rate being equivalent for two different locations given the spatial tuning functions and the variance in the response. Again, if only this cell were available to make the estimates, they would not match the psychophysical performance. In this case the neuron predictions are much worse than the psychophysical results. This can be remedied, of course, if one uses more than one

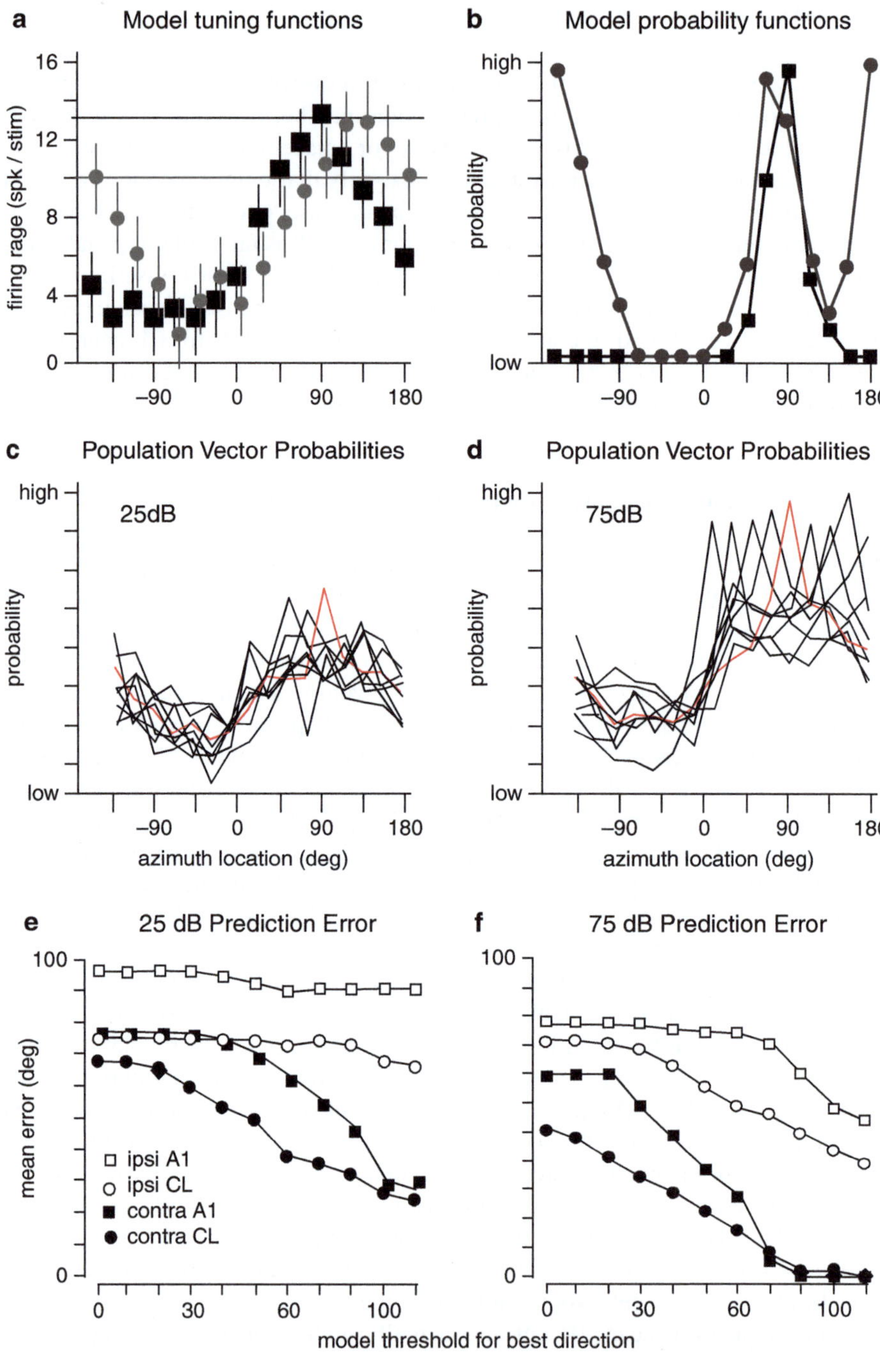

Fig. 4.9 Neuronal population modeling of acoustic space representations. (**a**) Two hypothetical neurons with similar firing rates and variances but best directions at either 90° (black) or 135° (gray). In the model, the firing rate on a svingle trial is used to generate the probability functions of where the model would predict the stimulus actually came from (**b**). In the example shown, if the stimulus came from 90° and the black cell fired 13 spikes (black horizontal line), it would correspond to high probabilities that the stimulus came from 90° (black squares of **b**), but also some

neuron. From the two neurons shown in Fig. 4.9, the two probability maps, if appropriately combined, could enhance the probability that the stimulus came from the location that it actually did, and the model estimate would be correct. One obvious drawback of this approach is that if any of the neurons generate a probability of 0 for a given location, multiplying this probability with all the others results in 0; thus in this model every neuron has "veto power" and can eliminate a particular location from further consideration. Although in the two-neuron example this would actually be a benefit, when comparing across large populations this is not the case. Given the variance in the responses, and the breadth of neurons that have best directions in ipsilateral space, there are few locations where the population is "unanimous" and that any location ends up with a nonzero possibility. This is clearly nonbiological, and there are several ways one can get around this mathematical problem. One is to set a lower limit on the minimum probability that is nonzero. This can be formulated in a number of ways, but in most cases would be quite small. This prevents elimination of any possibility, but it only eases the mathematical problem somewhat, and does really nothing with the biological implausibility of the assumption. The other is to use a different mathematical construct, or a different mathematical technique. These two alternatives are addressed in turn (Miller & Recanzone, 2010).

4.7.3.3 Acoustic Space Coding Based on Population Vectors

A different mathematical construct is to model each neuronal response as a vector and to then combine these vectors across the population. This strategy is similar to the population vector model that had been used successfully in modeling arm movements

Fig. 4.9 (continued) probability that it came from the adjacent locations. If the gray neuron fired 10 spikes for the same stimulus, there would actually be a higher probability that the model would predict that the stimulus came from 67.5° or –157.5° (gray circles in **b**). Neurons responding to stimuli outside of their best direction typically would have bilobed distributions. Comparing across neurons, as in **b**, would provide more information not only of where the stimulus could be (near 90°) but also where it likely was not (–67.5 to 0° in this example). See text for further explanations on the limitations of this approach. (**c**, **d**) The probabilities of estimating any particular location based on the population vector model. The red line shows the probability function for stimuli that are presented from 90°. Each black line represents the probability function for the other locations from 0 to 180°. At 25 dB (**c**) there are not always clear peaks for the individual functions, as in some cases the population vector peaks at the incorrect location. At 75 dB (**d**), each function has a sharp peak at the correct spatial location, giving rise to the clearly separated peaks across contralateral space. (**e**, **f**) Illustrations of how the model estimates vary depending on the threshold that the model considers. A "winner-take-all" strategy is at the far right of each panel, where the model threshold is 100%, whereas using the entire probability function is at the far left of each panel, where the model threshold is 0%. Ipsilateral locations are shown in open symbols and contralateral locations are shown in the closed symbols. Estimates based on A1 neurons are shown as squares and CL neurons as circles. The performance of the model is consistently poor for ipsilateral locations, consistent with the lesion results. The performance can vary widely for contralateral locations depending on the threshold that the model uses, but CL neurons are consistently more accurate than A1 neurons. (Data for **c–f** are taken from those published in the supplemental information of Miller & Recanzone, 2010)

in the motor cortex (Georgopoulos et al., 1986, 1988). In this case, each neuron was represented as a single vector, with the direction the direction of the best response and the magnitude normalized to that best direction. Thus, across the population there were many more vectors toward contralateral space compared to ipsilateral space. To estimate the location of an individual stimulus, a single trial for that direction was taken from each neuron at random, and the corresponding vector was calculated based on the vector in the best direction. For example, if the stimulus was at 90° to the right, a random trial from that direction was taken. The estimate by that neuron was then that particular response normalized by the average response for that location. As an example, if the average response was 20 spikes/trial and on this particularly trial the response was 18 spikes, that neuron's contribution was that there was a 90% probability that the stimulus came from 90° to the right. For a second example, a random trial for when a stimulus was presented at 90° was taken from a neuron with a best direction at 180°. If the best response averaged 33 spikes/ trial in that neuron, and the single trial elicited 11 spikes, then that neuron's contribution was that there was a 33% probability that the stimulus came from 180°. Thus, one could calculate the probability that the stimulus came from each of the 16 different directions given the input from a random sample of single trials, and repeat this process many times to obtain an average probability of a particular estimate. Such an example for one direction, again 90° to the right, is shown in Fig. 4.9c as the red line. This plot shows the probability (ordinate) that the stimulus actually came from the location on the abscissa when the stimulus in fact came from +90°. The peak of this function is at +90°, in agreement with the actual stimulus location. However, there is still a finite (and reasonably large) probability that the stimulus came from ± 22.5° from the actual location, but a very low probability that the stimulus came from ipsilateral space. Fig. 4.9d shows the probability functions for each of the locations at 75 dB sound pressure level (SPL), again with the red line showing the estimates for +90°. Across these locations, the probabilities that the model will choose the correct location are high for contralateral space, but less accurate for ipsilateral space, consistent with the lesion results. Comparisons between the 25 dB stimuli (Fig. 4.9c), where localization is poor, and the 75 dB stimuli (Fig. 4.9d), where the localization is very good, is also consistent with the psychophysical results. There were several inconsistencies, however, when taking the peaks of the functions as the estimates, as the model was poor at differentiating between the different cortical areas, and was much more accurate than the behavioral data predicted.

The finding that the model was better than predicted when taking only the peaks of the functions could be remedied if one did not simply use the peak, but instead imposed a threshold of how much of the probability function the model would consider. If this is done, then choosing the correct threshold is very subjective and it has a strong influence on how well the model performs. An example of how the threshold influences the errors is shown form A1 and CL populations of neurons for both 25 and 75 dB SPL in Figs. 4.9e and f. For both ipsilateral (open symbols) and contralateral (closed symbols) locations, the errors decrease from using the entire probability functions (left) to just the peak (right). Errors for ipsilateral locations remain high regardless of the threshold, consistent with the lesion studies, whereas errors become much smaller for contralateral locations. It is also clear that errors based on CL neurons are generally smaller than those based on A1 neurons until the highest thresholds are imposed. However, at no particular threshold value were the errors consistent with the behavioral performance across the different stimulus intensities. Thus, although biologically more plausible, this model does not fully account for sound localization performance.

4.7.3.4 Acoustic Space Coding Based on Maximum Likelihood Estimation

The third strategy, which uses the technique of maximum likelihood estimation, is to apply a model originally developed for visual motion discrimination (Jazayeri & Movshon, 2006). This model proved to be quite accurate at modeling the responses of neurons in extrastriate cortical areas on the perception of direction and velocity discriminations. In the case of sound localization, each neuron is assigned to a pool based on the best direction. Again, individual trials are taken at random from each pool of neurons, and the probability that the stimulus came from the direction represented by that pool is proportional to the firing rate at the best direction. Then, instead of multiplying these values, the logarithm is taken and these values are then added. The estimate is then based on the "winner-take-all" strategy for that particular trial.

The results of this model are shown in Fig. 4.10. In this case, for contralateral locations, there is a clear difference in the accuracy of the estimates between the different cortical areas, with neurons in area CL providing the estimates that are most closely matched with those observed psychophysically (dashed horizontal lines of Fig. 4.10). The other caudal belt area tested, CM, was also very accurate, but did not quite match that from the psychophysical results. Cortical areas in the core were next in accuracy, and surprisingly the accuracy of the model based on neurons in area R were quite similar to those in area A1, although area R was previously not considered to be part of the "where" processing pathway. The middle belt areas showed the poorest performance, again consistent with a caudal "where" processing stream. This model has three advantages over the other two discussed in the preceding text. The first is that there are no artificial probabilities of 0 and therefore no neuron has complete veto power. The second is that there is no threshold independently set by the investigator. Finally, single trials are used, and thus the natural variance in the neural response is accounted for. Thus, this model is biologically plausible and has been shown to be effective in other cortical systems (extrastriate cortex) and may be a general cortical mechanism of representing stimulus features in a population-based, nontopographic way.

4.7.3.5 Acoustic Space Coding Not Based on Firing Rate

The preceding results using the maximum likelihood estimator indicate that the overall firing rate of cortical neurons in the caudal belt in primate auditory cortex do contain enough information to account for sound localization ability. An issue that remains unresolved, of course, is whether this is the code that is actually used by the nervous system, and whether a code based only on the firing rate is appropriate.

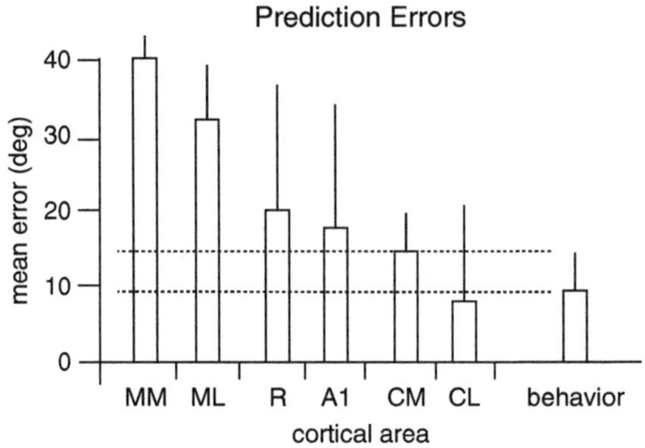

Fig. 4.10 Model performance using the maximum likelihood estimator. Each bar shows the mean unsigned error of the model localizing contralateral locations for all intensities combined based on the neuronal responses for a different cortical area. Cortical areas are sorted by decreasing error. Vertical lines show the standard deviation. At the far right are the errors from human subjects localizing these same stimuli. Dashed horizontal lines show the human average (lower line) and average +1 standard deviation (upper line) for comparison to the model data. Errors generated by this model based on the firing of CL neurons were not statistically significantly different from the behavioral performance, but the errors for the model based on all other cortical areas were. (Data taken from those published in Miller & Recanzone 2010)

There is additional information that the nervous system could potentially use to encode acoustic space, and the two most likely candidates are the latency of the response (Brugge et al., 1996; Reale et al., 2003) and the temporal structure of the response (Middlebrooks et al., 1994; Harrington et al., 2008). Both of these possibilities have been explored previously, most usually in the anesthetized cat, but some studies in alert animals have also been conducted (Mickey & Middlebrooks, 2003; Lee & Middlebrooks, 2011) and those results were largely consistent with what is seen in the anesthetized preparation. To address this, Woods et al. (2006) also considered the spatial tuning functions of the same neurons based on the latency of the response. In this case, there was little spatial structure in many neurons, although some did show similar and in some cases improved, spatial tuning based on the firing rate latency. Using this as the model for population encoding, however, did not reveal any improvement and there was much poorer correlation with the psychophysically measured results. The second possibility, that the temporal structure of the response could contribute, was not specifically tested in the monkeys, although extensive studies by Middlebrooks and colleagues have shown that the information contained in the pattern of activity is a more accurate neural code, and contains more information, than spike rate or spike latency alone. One potential drawback of such a neural code is that the neurons generally are as accurate at encoding sounds in ipsilateral space as in contralateral space, which is in contrast to the lesion results described earlier. Nonetheless, it remains an open question how much the temporal structure of the neural response contributes to the perception of acoustic space in primate auditory cortex.

4.7.4 *Effects of Natural Aging on Sound Localization and Cortical Spatial Processing*

A third and final feature that codes of auditory spatial perception have to account for in addition to the effects of stimulus spectrum and intensity is the deficits in sound localization performance as a consequence of natural aging. One of the hallmarks of natural aging is the reduced ability to localize sounds in space (Kubo et al., 1998; Abel et al., 2000; Marrone et al., 2008) as well as to encode the temporal structure of the sound (Snell & Frisnia, 2000; Snell et al., 2002). Both of these deficits likely contribute to the main complaint of the aged hearing impaired, the inability to understand speech in noisy environments (Gordon-Salant & Fitzgibbons, 1993; Phillips et al., 2000). If the neural coding of acoustic space is processed along the "where" pathway from the core to the caudal belt and beyond, one would then predict that the spatial tuning of neurons in these areas would be broader than in younger animals. This possibility was tested recently by Juarez-Salinas et al. (2010), who recorded the spatial tuning profiles of auditory cortical neurons in aged animals and compared them to those seen in the younger animals reported in Woods et al. (2006). These studies concentrated on the core region A1, as well as the caudal belt field CL. These two cortical areas were chosen because they represent the initial input to the auditory cortex (A1), and where spatial tuning is refined to the degree that can account for localization ability (CL; see Fig. 4.10). When the spatial tuning was assessed using the dynamic range index, there was no difference in the distributions of neurons in A1 between young and old monkeys. However, there was a significant difference for the distributions of neurons in area CL, with young monkeys having much greater dynamic range compared to aged monkeys. What was worthy of note was the magnitude of this difference. In young animals, there was an increase in dynamic range of spatial tuning of almost 20% between A1 and CL. In contrast, the dynamic range in aged animals actually decreased by about 1%, which was not statistically significant. This finding was also seen using other metrics of spatial tuning. Thus, a key consequence of aging with respect to sound localization is that there is no refinement of spatial tuning between the core and caudal belt fields as is seen in younger animals. This has important implications not only for acoustic space perception, but also in cortical processing in general. If it is in fact the case that natural aging leads to a diminished capacity to extract information from one cortical area to another, the effects would be most dramatic for the most complex perceptions. This is consistent with the deficits observed in general cognitive decline with aging.

4.8 Summary and Conclusions

This chapter has highlighted some of the recent studies exploring the neural correlates of auditory spatial perception. Psychophysical studies have shown how spectral integration improves localization ability, indicating that spatial perception is likely based on populations of neurons. Similar studies investigating localization as a function of stimulus intensity also support a population coding scheme. Lesions of auditory cortical areas indicate that auditory cortex is necessary for the perception of acoustic space, although more recent studies have indicated that primary auditory cortex, although important, is not absolutely critical. Primary auditory cortex is, however, necessary for adaptive plasticity as a consequence of ear plugging, primarily based on the cortico-collicular pathway. Electrophysiological studies in both carnivores and primates indicate that auditory cortical neurons are not organized in a spatially topographic fashion, and have receptive fields that are much larger than would be predicted by psychophysical studies. However, studies in alert monkeys provide strong support for a caudal spatial processing stream, where the firing rate across populations of caudal belt neurons contains enough information to account for sound localization ability. This localization ability is degraded as a consequence of natural aging, as is the spatial processing along this caudal pathway, providing further support of this model.

One hallmark of auditory cortical research on spatial localization is how the field has effectively integrated several different experimental approaches and model systems. Psychophysical, anatomical and electrophysiological studies have provided a clearer understanding of the physical cues present to compute the location of a sound in space, as well as the neural machinery that performs these computations and ultimately gives rise to the percept. The rodent, carnivore, monkey, and human model systems all have their strengths and weaknesses, but taken together they have provided a strong understanding of potential neural codes of acoustic spatial perception as well as how these representations, and perceptions, can be modified both during development and in adults. Future studies can build on these successes to test these hypotheses further to not only find the correlates but also directly manipulate the neural activity to generate predictable localization deficits and enhancements.

References

Abel, S. M., Giguère, C., Consoli, A., & Papsin, B. C. (2000). The effect of aging on horizontal plane sound localization. *The Journal of the Acoustical Society of America*, 108, 743–752.

Altshuler, M. W., & Comalli, P. E. (1975). Effect of stimulus intensity and frequency on median horizontal plane sound localization. *The Journal of Auditory Research*, 15, 262–265.

Bajo, V. M., Nodal, F. R., Moore, D. R., & King, A. J. (2010). The descending corticocollicular pathway mediates learning-induced auditory plasticity. *Nature Neuroscience*, 13, 253–260.

Beitel, R. E., & Kaas, J. H. (1993). Effects of bilateral and unilateral ablation of auditory cortex in cats on the unconditioned head orienting response to acoustic stimuli. *Journal of Neurophysiology*, 70, 351–369.

Blauert, J. (1997). *Spatial hearing*. Cambridge, MA: MIT Press.

Born, R. T., & Bradley, D. C. (2005). Structure and function of visual area MT. *Annual Review of Neuroscience*, 28, 157–189.

Brugge, J. F., Reale, R. A., & Hind, J. E. (1996). The structure of spatial receptive fields of neurons in primary auditory cortex of the cat. *The Journal of Neuroscience*, 16, 4420–4437.

Cai, R., Guo, F., Zhang, J., Xu, J., Cui, Y., & Sun, X. (2009). Environmental enrichment improves behavioral performance and auditory spatial representations of primary auditory cortical neurons in rat. *Neurobiology of Learning and Memory*, 91, 366–376.

Campbell, R. A., King, A. J., Nodal, F. R., Schnupp, J. W., Carlile, S., & Doubell, T. P. (2008). Virtual adult ears reveal the roles of acoustical factors and experience in auditory space map development. *The Journal of Neuroscience*, 28, 11557–11570.

Casseday, J. H., & Neff, W. D. (1975). Auditory localization: Role of auditory pathways in brain stem of the cat. *Journal of Neurophysiology*, 38, 842–858.

Comalli, P. E., & Altshuler, M. W. (1976). Effect of stimulus intensity, frequency and unilateral hearing loss on sound localization. *The Journal of Auditory Research*, 16, 275–279.

de Villers-Sidani, E., Chang, E. F., Bao, S., & Merzenich, M. M. (2007). Critical period window for spectral tuning defined in the primary auditory cortex (A1) in the rat. *The Journal of Neuroscience*, 27, 180–189.

Doty, R. L. (2001). Olfaction. *Annual Review of Psychology*, 52, 423–452.

Georgopoulos, A. P., Schwartz, A. B., & Kettner, R. E. (1986). Neuronal population coding of movement direction. *Science*, 233, 1416–1419.

Georgopoulos, A. P., Kettner, R. E., & Schwartz, A. B. (1988). Primate motor cortex and free arm movements to visual targets in three-dimensional space. II. Coding of the direction of movement by a neuronal population. *The Journal of Neuroscience*, 8, 2928–2937.

Gordon-Salant, S., & Fitzgibbons, P. J. (1993). Temporal factors and speech recognition performance in young and elderly listeners. *Journal of Speech and Hearing Research*, 36, 1276–1285.

Green, D. M., & Swets, J. A. (1974). *Signal detection theory and psychophysics*. Huntington, NY: Kreiger.

Groh, J. M., Kelly, K. A., & Underhill, A. M. (2003). A monotonic code for sound azimuth in primate inferior colliculus. *Journal of Cognitive Neuroscience*, 15, 1217–1231.

Grothe, B., Pecka, M., & McAlpine, D. (2010). Mechanisms of sound localization in mammals. *Physiological Reviews*, 90, 983–1012.

Hackett, T. A., Preuss, T. M., & Kaas, J. H. (2001). Architectonic identification of the core region in auditory cortex of macaques, chimpanzees and humans. *The Journal of Comparative Neurology*, 441, 197–22.

Harrington, I. A., Stecker, G. C., Macpherson, E. A., & Middlebrooks, J. C. (2008). Spatial sensitivity of neurons in the anterior, posterior and primary fields of cat auditory cortex. *Hearing Research*, 240, 22–41.

Heffner, H. E., & Heffner, R. S. (1990). Effect of bilateral auditory cortex lesions on sound localization in Japanese macaques. *Journal of Neurophysiology*, 64, 915–931.

Hofman, P. M., Van Riswick, J. G. A., & Van Opstal, A. J. (1998). Relearning sound localization with new ears. *Nature Neuroscience*, 1, 417–421.

Imig, T. J., Irons, W. A., & Samson, F. R. (1990). Single-unit selectivity to azimuthal direction and sound pressure level of noise bursts in cat high-frequency primary auditory cortex. *Journal of Neurophysiology*, 63, 1448–1466.

Jazayeri, M., & Movshon, J. A. (2006). Optimal representation of sensory information by neuronal populations. *Nature Neuroscience*, 5, 690–696.

Jenkins, W. M., & Merzenich, M. M. (1984). Role of cat primary auditory cortex for sound-localization behavior. *Journal of Neurophysiology*, 52, 819–847.

Juarez-Salinas, D. L, Engle, J. R., Navarro, X. O., & Recanzone, G. H. (2010). Hierarchical and serial processing in the spatial auditory cortical pathway is degraded by natural aging. *The Journal of Neuroscience*, 30, 14795–14804.

Kaas, J. H., & Hackett, T. A. (2000). Subdivisions of auditory cortex and processing streams in primates. *Proceedings of the National Academy of Sciences of the USA*, 97, 11793–11799.

Kacelnik, O., Nodal, F. R., Parsons, C. H., & King, A. J. (2006). Training-induced plasticity of auditory localization in adult mammals. *PLOS Biology*, 4, e71.

Kauer, J. S., & White, J. (2001). Imaging and coding in the olfactory system. *Annual Review of Neuroscience*, 24, 963–979.

Kilgard, M. P., Pandya, P. K., Vazquez, J., Gehi, A., Schreiner, C. E., & Merzenich, M. M. (2001). Sensory input directs spatial and temporal plasticity in primary auditory cortex. *Journal of Neurophysiology*, 86, 326–338.

King, A. J., & Palmer, A. R. (1983). Cells responsive to free-field auditory stimuli in guinea-pig superior colliculus: Distribution and response properties. *The Journal of Physiology*, 342, 361–381.

King, A. J., Bajor, V. M., Bizley, J. K., Campbell, R. A. A., Nodal, F. R., Schultz, A. L., … Schnupp, J. W. H. (2007). Physiological and behavioral studies of spatial coding in the auditory cortex. *Hearing Research*, 229, 106–115.

Knudsen, E. I., & Konishi, M. (1978). A neural map of auditory space in the owl. *Science*, 200, 795–797.

Kubo, T., Sakashita, T., Kusuki, M., Kyunai, K., Uneo, K., Hikawa, C., et al. (1998). Sound lateralization and speech discrimination in patients with sensorineural hearing loss. *Acta Oto-Layngologica Supplementum*, 538, 63–69.

Lee, C. C., & Middlebrooks, J. C. (2011). Auditory cortex spatial sensitivity sharpens during task performance. *Nature Neuroscience*, 14, 108–114.

Lomber, S. G., & Malhotra, S. (2009). Double dissociation of 'what' and 'where' processing in auditory cortex. *Nature Neuroscience*, 11, 609–616.

Lovejoy, L. P., & Krauzlis, R. J. (2010). Inactivation of primate superior colliculus impairs covert selection of signals for perceptual judgements. *Nature Neuroscience*, 13, 261–266.

Malhotra, S., Hall, A. J., & Lomber, S. G. (2004). Cortical control of sound localization in the cat: unilateral cooling deactivation of 19 cerebral areas. *Journal of Neurophysiology*, 92, 1625–1643.

Makous, J. C., & Middlebrooks, J. C. (1990). Two-dimensional sound localization by human listeners. *The Journal of the Acoustical Society of America*, 87, 2188–2200.

Marrone, N., Mason, C. R., & Kidd, G., Jr. (2008). The effect of hearing loss and age of the benefit of spatial separation between multiple talkers in reverberant rooms. *The Journal of the Acoustical Society of America*, 124, 3064.

Masterton, B., Jane, J. A., & Diamond, I. T. (1967). Role of brainstem auditory structures in sound localization. I. Trapezoid body, superior olive, and lateral lemniscus. *Journal of Neurophysiology*, 30, 341–359.

Mickey, B. J., & Middlebrooks, J. C. (2003). Representation of auditory space by cortical neurons in awake cats. *The Journal of Neuroscience*, 23, 8649–8663.

Middlebrooks, J. C., & Green, D. M. (1991). Sound localization by human listeners. *Annual Review of Psychology*, 42, 135–159.

Middlebrooks, J. C., & Knudsen, E. I. (1984). A neural code for auditory space in the cat's superior colliculus. *The Journal of Neuroscience*, 4, 2621–2634.

Middlebrooks, J. C., Clock, A. E., Xu, L., & Green, D. M. (1994). A panoramic code for sound location by cortical neurons. *Science*, 264, 842–844.

Miller, G. L., & Knudsen, E. I. (2001). Early auditory experience induces frequency-specific, adaptive plasticity in the forebrain gaze fields of the barn owl. *Journal of Neurophysiology*, 85, 2184–2194.

Miller, G. L., & Knudsen, E. I. (2003). Adaptive plasticity in the auditory thalamus of juvenile barn owls. *The Journal of Neuroscience*, 23, 1059–1065.

Miller, L. M., & Recanzone, G. H. (2010). Populations of auditory cortical neurons can accurately encode acoustic space across stimulus intensity. *Proceedings of the National Academy of Sciences of the USA*, 106, 5931–5935.

Mrsic-Flogel, T. D., Schnupp, J. W. H., & King, A. J. (2003). Acoustic factors govern developmental sharpening of spatial tuning in the auditory cortex. *Nature Neuroscience*, 6, 981–988.

Nodal, F. R., Kacelnik, O., Bajo, V. M., Bizley, J. K., Moore, D. R., & King, A. J. (2010). Lesions of the auditory cortex impair azimuthal sound localization and its recalibration in ferrets. *Journal of Neurophysiology*, 103, 1209–1225.

Nummela, S. U., & Krauzlis, R. J. (2010). Inactivation of primate superior colliculus biases target choice for smooth pursuit, saccades, and button press responses. *Journal of Neurophysiology*, 104, 1538–1548.

Phillips, S. L., Gordon-Salant, S., Fitzgibbons, P. J., & Yeni-Komshian, G. (2000). Frequency and temporal resolution in elderly listeners with good and poor word recognition. *The Journal of Speech, Language and Hearing Research*, 43, 217–228.

Rajan, R., Aitkin, L. M., Irvine, D. R., & McKay, J. (1990a). Azimuthal sensitivity of neurons in primary auditory cortex of cats. I. Types of sensitivity and the effects of variations in stimulus parameters. *Journal of Neurophysiology*, 64, 872–887.

Rajan, R., Aitkin, L. M., & Irvine, D. R. (1990b). Azimuthal sensitivity of neurons in primary auditory cortex of cats. II. Organization along frequency-band strips. *Journal of Neurophysiology*, 64, 888–902.

Rauschecker, J. P. (1998). Parallel processing in the auditory cortex of primates. *Audiology and Neuro-otology*, 3, 86–103.

Rauschecker, J. P., & Scott, S. K. (2009). Maps and streams in the auditory cortex: Nonhuman primates illuminate human speech processing. *Nature Neuroscience*, 12, 718–724.

Rauschecker, J. P., & Tian, B. (2000). Mechanisms and streams for processing of "what" and "where" in auditory cortex. *Proceedings of the National Academy of Sciences of the USA*, 97, 1180–11806.

Rauschecker, J. P., Tian, B., & Hauser, M. (1995). Processing of complex sounds in the macaque nonprimary auditory cortex. *Science*, 268, 111–114.

Rauschecker, J. P., Tian, B., Pons, T., & Mishkin, M. (1997). Serial and parallel processing in rhesus monkey auditory cortex. *The Journal of Comparative Neurology*, 382, 89–103.

1–Reale, R. A., Jenison, R. L., & Brugge, J. F. (2003). Directional sensitivity of neurons in the primary auditory (AI) cortex: Effects of sound-source intensity level. *Journal of Neurophysiology*, 89, 1024–1038.

Recanzone, G. H., & Beckerman, N. S. (2004). Effects of intensity and location on sound location discrimination in macaque monkeys. *Hearing Research*, 198, 116–124.

Recanzone, G. H., & Sutter, M. L. (2008). Biological basis of audition. *Annual Review of Psychology*, 59, 119–142.

Recanzone, G. H., Merzenich, M. M., Jenkins, W. M., Grajski, K. A., & Dinse, H. R. (1992). Topographic reorganization of the hand representation in cortical area 3b of owl monkeys trained in a frequency-discrimination task. *Journal of Neurophysiology*, 67, 1031–1056.

Recanzone, G. H., Schreiner, C. E., & Merzenich, M. M. (1993). Plasticity in the frequency representation of primary auditory cortex following discrimination training in adult owl monkeys. *The Journal of Neuroscience*, 13, 87–103.

Recanzone, G. H., Makhambra, S. D. D. R., & Guard, D. C. (1998). Comparison of relative and absolute sound localization ability in humans. *The Journal of the Acoustical Society of America*, 103, 1085–1097.

Recanzone, G. H., Schreiner, C. E., Sutter, M. L., Beitel, R. E., & Merzenich, M. M. (1999). Functional organization of spectral receptive fields in the primary auditory cortex of the owl monkey. *The Journal of Comparative Neurology*, 415, 460–481.

Recanzone, G. H., Guard, D. C., & Phan, M. L. (2000a). Frequency and intensity response properties of single neurons in the auditory cortex of the behaving macaque monkey. *Journal of Neurophysiology*, 83, 2315–2331.

Recanzone, G. H., Guard, D. C., Phan, M. L., & Su, T. K. (2000b). Correlation between the activity of single auditory cortical neurons and sound localization behavior in the macaque monkey. *Journal of Neurophysiology*, 83, 2723–2739.

Romanski, L. M., & Averbeck, B. B. (2009). The primate cortical auditory system and neural representation of conspecific vocalizations. *Annual Review of Neuroscience*, 32, 315–346.

Romanski, L. M., Tian, B., Fritz, J., Mishkin, M., Goldman-Rakic, P. S., & Rauschecker, J. P. (1999). Dual streams of auditory afferents target multiple domains in the primate prefrontal cortex. *Nature Neuroscience*, 12, 1131–1136.

5–Sabin, A. T., Macpherson, E. A., & Middlebrooks, J. C. (2005). Human sound localization at near-threshold levels. *Hearing Research*, 199, 124–134.

Shapley, R., & Lennie, P. (1985). Spatial frequency analysis in the visual system. *Annual Review of Neuroscience*, 8, 547–583.

Smith, A. L., Parsons, C. H., Lanyon, R. G., Bizley, J. K., Akerman, C. J., Baker, G. E., et al. (2004). An investigation of the role of auditory cortex in sound localization using muscimol-releasing Elvax. *European Journal of Neuroscience*, 19, 3059–3072.

Snell, K. B., & Frisina, D. R. (2000). Relationships among age-related differences in gap detection and word recognition. *The Journal of the Acoustical Society of America*, 107, 1615–1626.

Snell, K. B., Mapes, F. M., Hickman, E. D., & Frisina, D. R. (2002). Word recognition in competing babble and the effects of age, temporal processing, and absolute sensitivity. *The Journal of the Acoustical Society of America*, 112, 720–727.

Stevens, S. S., & Newman, E. B. (1936). The localization of actual sources of sound. *The American Journal of Psychology*, 48, 297–306.

Su, T. I., & Recanzone, G. H. (2001). Differential effect of near-threshold stimulus intensities on sound localization performance in azimuth and elevation in normal human subjects. *The Journal of the Association for Research in Otolaryngology*, 2, 246–256.

Thompson, G. C., & Cortez, A. M. (1983). The inability of squirrel monkeys to localize sound after unilateral ablation of auditory cortex. *Behavioral Brain Research*, 8, 211–216.

Tian, B., Reser, D., Durham, A., Kustov, A., & Rauschecker, J. P. (2001). Functional specialization in rhesus monkey auditory cortex. *Science*, 292, 290–293.

Tsao, D. Y., & Livingstone, M. S. (2008). Mechanisms of face perception. *Annual Review of Neuroscience*, 31, 411–437.

Ungerleider, L. G., & Haxby, J. V. (1994). 'What' and 'where' in the human brain. *Current Opinions in Neurobiology*, 4, 157–165.

Ungerleider, L. G., & Mishkin, M. (1982). Two visual cortical systems. In D. J. Ingle, M. A. Goodale, & R. J. W. Mansfield (Eds.), *Analysis of visual behavior* (pp. 549–586). Cambridge, MA: MIT Press.

Van Wanrooij, M. M., & Van Opstal, A. J. (2005). Relearning sound localization with a new ear. *The Journal of Neuroscience*, 25, 5413–5424.

Van Wanrooij, M. M., & Van Opstal, A. J. (2007). Sound localization under perturbed binaural hearing. *Journal of Neurophysiology*, 97, 715–726.

Westheimer, G. (1984) Spatial vision. *Annual Review of Psychology*, 35, 201–226.

Wightman, F. L., & Kistler, D. J. (1989a). Headphone simulation of free-field listening I: Stimulus synthesis. *The Journal of the Acoustical Society of America*, 85, 858–867.

Wighman, F. L., & Kistler, D. J. (1989b). Headphone simulation of free-field listening II: Psychophysical validation. *The Journal of the Acoustical Society of America*, 85, 868–878.

Woods, T. M., Lopez, S. E., Long, J. H., Rahman, J. E., & Recanzone, G. H. (2006). Effects of stimulus azimuth and intensity on the single neuron activity in the auditory cortex of the alert macaque monkey. *Journal of Neurophysiology*, 96, 3323–3337.

Zhang, H., Cai, R., Zhang, J., Pan, Y., & Sun, X. (2009). Environmental enrichment enhances directional selectivity of primary auditory cortical neurons in rats. *Neuroscience Letters*, 463, 162–165.

Chapter 5
Neural Correlates of Auditory Object Perception

Jan W.H. Schnupp, Christian Honey, and Ben D.B. Willmore

Keywords Acoustics • Categorical perception • Cortex • Electrophysiology • Invariance • Phonemes • Scene analysis • Segregation • Sound • Speech • Timbre • Vocalization

5.1 What Are Auditory Objects?

As you sit in front of this book, reading these words, pause for a moment, listen, and ask yourself: What do I hear? Perhaps you hear some conversations going on in the background, some devices making noises of various kinds (an almost omnipresent feature of the modern world). Or perhaps you are in an unusually quiet place, and there is effectively nothing to hear. Almost certain is that you would not describe your auditory experience as one of oscillating air pressure in your ear canals that gently wiggle your ear drums. Yet strictly speaking, on the surface of it, that is all there ever is to "hearing."

What makes hearing so rich is that brains display a remarkable ability to, in a sense, "reconstruct" the very sources that have produced the vibration patterns that arrive at the ears, and by the time an auditory event impinges on a listener's consciousness they are essentially unaware of sound waves, but instead hear—well—"something." What that thing is may be quite clear and easily put into words: a bicycle bell perhaps, or a voice. Note how readily people, when asked to identify a particular sound, might answer: "a telephone" or "a dripping tap" even though neither phones nor taps are, of course, sounds. Sounds are not things, yet people appear to try very hard to perceive them as such. Even when one cannot unambiguously identify a

J.W.H. Schnupp (✉) • C. Honey • B.D.B. Willmore
Department of Physiology, Anatomy and Genetics, University of Oxford, Parks Road, Oxford OX1 3PT, UK
e-mail: jan.schnupp@dpag.ox.ac.uk; christian.honey@dpag.ox.ac.uk; benjamin.willmore@dpag.ox.ac.uk

Y.E. Cohen et al. (eds.), *Neural Correlates of Auditory Cognition*, Springer Handbook of Auditory Research 45, DOI 10.1007/978-1-4614-2350-8_5,

© Springer Science+Business Media New York 2013

sound, perhaps when faced with unfamiliar "whooshing" or "whistling" noises, one is nevertheless usually quite clear about whether one heard a single "thing" or several, that is, the perception is tentatively structured into separate "entities."

Because, introspectively, humans perceive "things" and "events," auditory object recognition seems to be what hearing is all about. Yet so far there is no generally accepted consensus on what exactly constitutes an "auditory object." But a lack of clear definitions is often more of a problem for philosophers than for experimental scientists. Many of the most interesting concepts in experimental psychology lack clear definitions, but progress can still be made if one can devise experiments that measure phenomena that clearly relate to "the thing of interest." This chapter does not try to arrive at a precise definition of an auditory object. Instead it examines a number of tangible and intertwined questions, which seem to be of obvious relevance to auditory object perception, even if the definition of auditory objects remains vague. These questions include: How do sounds convey identifiable physical properties of sound sources? How does the brain decide whether a number of time-frequency components of acoustic events belong to one auditory object or several? What sort of processes allow the brain to try to identify auditory objects as being of a particular class or type? Where in the brain do these processes play out?

5.2 Physical Acoustics of Auditory Objects

5.2.1 Acoustic Fingerprints of Sound Sources

It seems likely that our ability to detect and identify auditory objects evolved for the purpose of informing our behavior, and to do so these perceptual auditory objects must reflect the physical properties of the sound sources in our immediate environment. So a good way to start our exploration of auditory objects is with a very quick review of how sound sources make sounds in the first place, and how the sounds they make betray some of the properties of the source.

Fundamentally, sounds stem from vibrations, most commonly of some solid object (the source), which are then transmitted through a medium, such as the surrounding air. Many introductory texts on hearing like to show images of vibrating tuning forks to illustrate this point. Tuning forks are a good example of mechanical resonators, being made out of a heavy material (steel) that exhibits a spring-like stiffness. For reasons that we do not have the space to discuss here (the interested reader is referred to Chapter 1 of Schnupp et al., 2010), mass-spring systems naturally enter into sinusoidal vibrations known as "simple harmonic motion," with a frequency that is proportional to the square root of the stiffness and inversely proportional to the square root of the mass. Consequently, heavy objects tend to make lower frequency sounds than lighter ones, and objects that exhibit high "spring forces," either because they are intrinsically stiff or because they are under a great deal of tension, make higher sounds than objects that are relatively "floppy." The sound frequencies emitted by an object thus betray some of its fundamental physical properties. Resonant cavities, such as an empty bottle, also behave as spring-mass

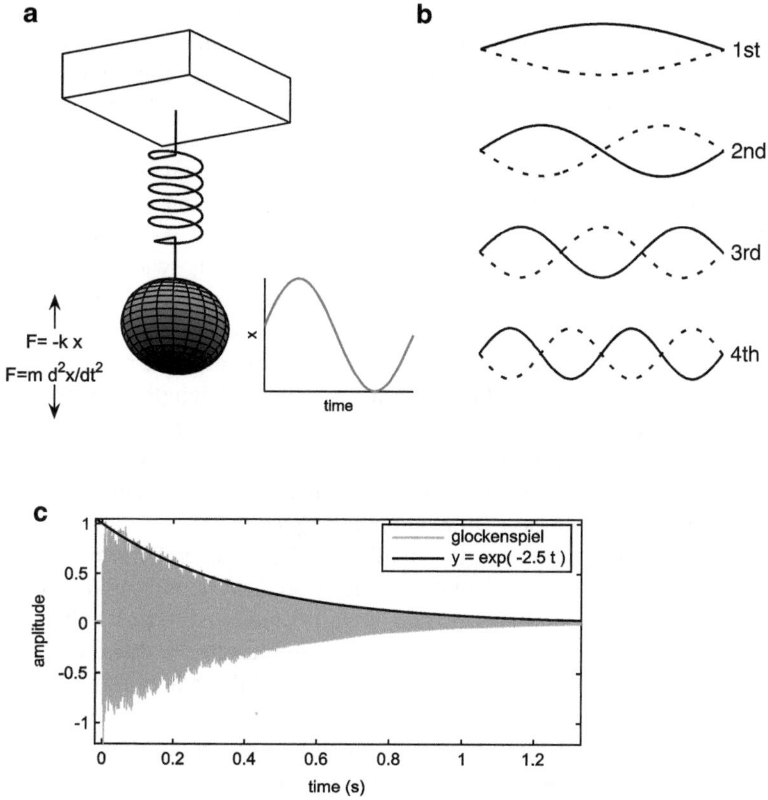

Fig. 5.1 (**a**) An object of a given mass hanging on the end of a spring—the "mass-spring system"—is a classic example of an oscillator. If you pull the object downward and release it, it returns to its original position, overshoots, and begins to oscillate up and down. Under ideal conditions, the oscillations are sinusoidal, and they have a characteristic frequency that depends only on the physical properties of the system: the mass of the object and the stiffness of the spring. Thus, the frequency of the oscillations tells you something about the physical nature of the system. (**b**) A taut string is a more complex oscillator. Because the ends are held in position, the string can only move in certain "modes" or patterns of vibration. These have wavelengths λ, $\lambda/2$, $\lambda/3$, ... and corresponding frequencies f, 2f, 3f, ... and are known as harmonics. When plucked, a string will oscillate in many of these modes at the same time. The relative strengths of the harmonics give the string a characteristic timbre. (**c**) The envelope of a sound—the variation of its loudness over time—also tells you something about the object that produced the sound. The more energy is converted to heat and sound by the oscillation, the more "damped" the sound source, and the faster its oscillations decay. In this case, the metal bar of the glockenspiel takes over a second to dissipate all its energy

oscillators, with the compressible air in the bottle acting as "spring" and the air in the bottle's neck acting as "mass." The effective spring constant of the air inside decreases with the volume of the bottle, so that large resonant cavities produce deeper hollow sounds than small ones.

The simplest mass-spring systems, when allowed to oscillate freely, exhibit purely sinusoidal vibrations at just their resonant frequency (see Fig. 5.1a). They should therefore emit pure tones. However, most real-world sound sources are physically more complex, and emit more complex sounds. A typical example is a taut string on a musical instrument, which may be thought of as a chain of string segments, each elastically coupled to the next. Such a structure can enter in transverse vibrations in a number of different "modes" (see Fig. 5.1b). The whole string can vibrate up and down as one, or each half (or third, fourth, … 1/*N*th) string can vibrate separately, in counterphase with the neighboring segment. Of course, each half string weighs only half as much as the whole string, but it is also effectively twice as stiff, as there are only half as many "elastic links." As a consequence, the second mode of vibration has a resonant frequency that is twice that of the first, fundamental mode. By the same argument, the third mode has a frequency three times that of the fundamental, and the *N*th mode one that is *N* times the fundamental. Air-filled tubes, such as organ pipes, can similarly exhibit multiple modes of vibration corresponding to longitudinal standing waves, with wavelengths corresponding to integer fractions of the length of the whole pipe. This is one of the reasons why many real-world objects do not emit pure tones, but complex tones composed of a series of frequencies that are often "harmonically related," meaning that they are integer multiples of a common "fundamental frequency."

Of course, sound sources do not normally oscillate forever. On every cycle of the vibration, a fraction of the mechanical energy stored in the vibration is lost. Some of it is radiated off as sound energy, and some is converted to heat. The amplitude of the vibrations of most sound sources therefore tend to die down exponentially (see Fig. 5.1c), but different sound sources differ greatly in how rapidly this decay occurs. "Highly damped" sound sources, such a wooden board, convert oscillations to heat much faster, and therefore have much shorter decay time constants and produce much shorter sounds than poorly damped sources, such as a steel bar, which may continue to "ring" for many seconds after being struck. Thus, the temporal envelope of a sound wave can tell us something about the materials that a sound source is made of.

And for any sounds to be emitted, something must mechanically excite the source by delivering the mechanical energy to set up the vibration in the first place. The manner of excitation of a sound source, perhaps unsurprisingly, greatly impacts the sounds that will be emitted. A very common and quite natural form of excitation occurs when the sound source is struck, that is, an isolated impulse is delivered. If excited in this manner, sound sources will invariably emit sounds at their own resonant frequencies (including higher harmonics if the corresponding modes of vibration were excited) with a very sharp onset and an exponential decay, as shown in Fig. 5.1c. However, a sound source might also be fed mechanical energy either repeatedly or continuously. Most sound sources can be well approximated as "linear systems," and therefore exhibit "sine wave in, sine wave out" behavior, meaning that if they are fed energy by coupling to a source of sustained sinusoidal vibration, then they will vibrate exclusively at the frequency that they are being supplied, not at their own resonant frequency. However, the amplitude of their vibration will be much larger if the supplied frequency is closer to the resonant frequency. Such sinusoidal excitation is very rare in nature, but it helps us understand what happens if a sound source is excited by a noise waveform, for example, a turbulent airflow,

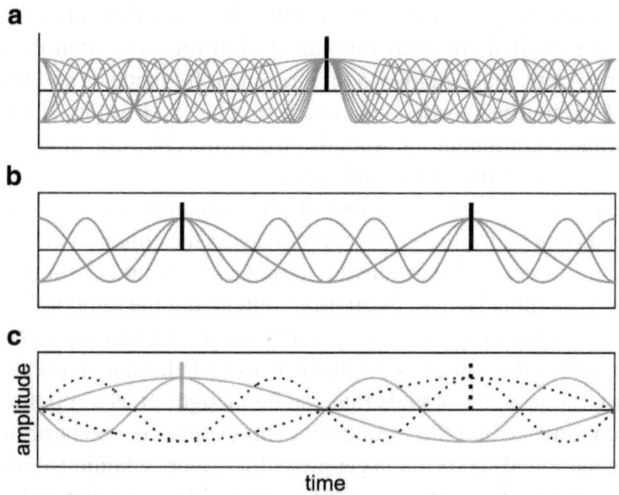

Fig. 5.2 Animal vocalizations are often composed of click trains—series of evenly spaced impulses. The frequency spectrum of a click train is a harmonic series. (**a**) According to Fourier theory, an impulse is equivalent to the superposition of an infinite series of sinusoids, whose phases are aligned so that they all have a peak at the time of the impulse. At every other time, the sinusoids mutually cancel to 0. (**b**) When clicks are evenly spaced with an interval *T*, successive clicks share sinusoidal components that are integer multiples of the repetition rate, 1/*T*. These components reinforce one another through constructive interference. (**c**) All other sinusoids mutually cancel one another through destructive interference. The result is a harmonic series, much like that of a bowed string

or some form of irregular friction. We can think of the supplied noise waveform as a sum of many sine waves, and the sound source will respond strongly, and thereby effectively amplify, only those frequencies that can excite its modes of vibration, that is, those that lie close to its resonant frequencies. A bowed violin string therefore vibrates at similar frequencies as a plucked violin string, but the emitted sound's temporal envelope is very different, as only the plucked string produces a transient sound characterized by a very sharp onset and rapid exponential decay.

Finally let us consider what happens when a sound source is excited by a steady stream of regularly repeating impulses. According to Fourier analysis, any one impulse can itself be thought of as a superposition of infinitely many sine waves, all canceling each other out everywhere except at the one point where all "sine wave components" become aligned in phase and the impulse becomes manifest (see Fig. 5.2a). If we now consider several impulses spaced at some regular time interval *T* (see Fig. 5.2b) then this series of impulses will share all the sine wave components that have a frequency of 1/*T*, 2/*T*, 3/*T*, and so forth, and only these shared sine components will become prominent through constructive interference, while the other sine components of each impulse in the train will be suppressed by destructive interference (Fig. 5.2c). A regular train of impulses or clicks is therefore equivalent to a harmonic tone complex with infinitely many harmonically related sine components whose frequencies are integer multiples of the inverse of the click interval. Sound sources that are excited by such regular click trains will then emphasize those harmonic components that lie close to their resonant frequencies and attenuate others. This mode of excitation through sustained, regular click trains is very commonly found in animal communication sounds, from the chirping of grasshoppers and cicadas to the voices of mammals and birds.

In summary, we have seen a number of physical attributes of sound sources that shape the sounds that they emit. High stiffness favors the emission of high frequencies, while large size or weight favors lower frequencies. Sounds from heavily damped sources decay very quickly. The frequency content and the temporal envelope of the sound emitted by a sound source depend partly on its resonance properties, partly on the manner in which the source is mechanically excited. Many natural sound sources emit "complex" sounds, i.e. sounds that can be thought of as a mixture of numerous frequency components, and the frequency components may be harmonically related either because the sound source has a geometry that favors vibration at several regularly spaced modes of vibration, or because it is excited by a regular pulse train. In this manner, each sound generating event or process has its own particular "acoustic fingerprint." The challenge of auditory object perception is to try to capture and analyze these fingerprints to recognize and classify the underlying processes.

5.2.2 Vocalizations as a Special Case of "Objects"

Perhaps unsurprisingly, animal communication sounds in general, and human speech in particular, are a class of auditory objects that hold a particular fascination for the scientific community. It seems reasonable to assume that discriminating different kinds of communication calls, for instance those emitted by conspecifics, is of particular importance to an animal, and conspecific calls, be they from song birds (Olveczky & Gardner, 2011), rodents (Geissler & Ehret, 2004), cats (Gourévitch & Eggermont, 2007), or primates (Romanski & Averbeck, 2009) have consequently been popular stimuli in electrophysiological experiments looking to elucidate functions of higher-order auditory centers. Implicit in this approach is the often unstated and unquestioned assumption that vocalizations by members of the listener's own species represent a particularly "natural" class of auditory object, and that much of the neural infrastructure of the higher-order auditory pathways is likely to exhibit specializations designed to deal with this class of stimuli (Ghazanfar & Santos, 2004).

But from the point of view of physical acoustics, vocalizations are, at first glance, not that special. Most vocalizations are generated from resonant cavities in the animal's vocal tract, which are excited by a series of impulses, the "glottal pulse train." These glottal pulses are generated when air is forced through the vocal folds, causing them to be pushed open and then flap shut again, repeatedly, at regular, short intervals. For reasons discussed in the preceding text and illustrated in Fig. 5.2, the glottal pulse train will impose countless harmonics onto the spectrum of the vocalization, but the resonances of the vocal tract will ensure that not all harmonics carry

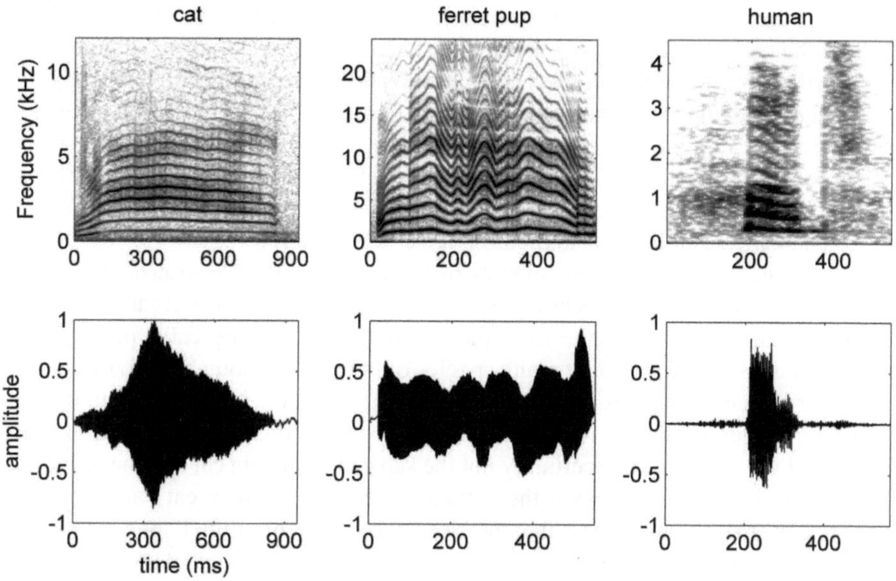

Fig. 5.3 Examples of vocalizations of three types of mammal: a domestic cat (left) a ferret pup (middle), and a human (right). The top panels show spectrograms of these vocalizations, and the bottom panels show the temporal waveform. In each case, the sound is produced by the repeated opening and shutting of the vocal folds, producing click trains containing many harmonics. The resonance of the animal's vocal tract then enhances some harmonics and suppresses others. The resulting frequency structure is visible as roughly horizontal bands in the spectrograms. Because the vocal apparatus varies in size and structure from animal to animal, the frequency content of the spectrograms carries information about the size of the animal (compare the cat and human vocalizations, for instance). Further, each animal can alter the repetition rate of the click trains and articulate the sound, resulting in dynamic variation in frequency content (most clearly visible in the ferret pup spectrogram)

equal energy. Rather, those harmonics that lie close to the resonant (so called "formant") frequencies of the vocal tract are strongly amplified. Pronounced harmonic structure that is subject to amplitude modulation by formant filtering is therefore a feature seen not just in human speech sounds, but a common feature of many animal vocalizations (for examples, see Fig. 5.3).

Some physical properties of a vocalizing animal will be reflected in the spectrum of its calls. Larger vocal tracts will produce lower resonances, and hence lower formants. They will also have larger, heavier vocal folds, which move at a slower rate, producing glottal pulse trains with a lower fundamental frequency. Consequently the bark of a 70-kg Newfoundland dog sounds much lower than that of a 1-kg Yorkshire terrier, and one instinctively recognizes the much lower voice as a hallmark of a more impressive physique. But unlike inanimate objects, animals can actively manipulate their own acoustic properties, and thus substantially increase the complexity and the diversity of sounds that they can produce. For example, animals can raise the pitch of their voice by increasing the tension on their vocal folds. Or they can change the configuration of their vocal tract resonators by moving a number of "articulators," including the tongue, jaw, lips, and soft palate. This will change the number and volume of the resonant cavities that are involved, and thus alter the formant structure of the vocalization. All of these aspects of a vocalization are controlled dynamically, and may change rapidly, producing amplitude and frequency modulations that generate a "syllabic structure" in the call, typically consisting of vowels framed by onset and offset consonants that result from a transient partial or complete interruption of the air flow at different points in the vocal tract. For further details on these aspects of vocalization and articulation, see also Steinschneider, Chapter 7.

As a consequence, any one species of animal may be able to produce a wide range of different types of vocalizations, and each may carry a different "meaning." Typical examples might range from the warning calls, mating calls, pup calls, or territorial display calls used by many species of mammal or song bird, to the spoken words of human languages. At this level of abstraction, the job of the auditory system is no longer just that of recognizing a sound source. It is said that a rose is a rose is a rose, but a hissing cat is certainly not the same as a purring cat or a meowing cat. Ignore the differences between these three types of "auditory cat" at your peril. Rather than merely answering the question "what type of animal" one also has to consider "what type of communication call." From a point of view of auditory object perception, this complicates matters greatly, and illustrates the need to adopt a more differentiated, "layered," hierarchical and context dependent view of the nature of auditory object perception. For a full understanding of auditory object processing one must grapple not only with the complexities of physical acoustics and auditory physiology, but also with those of semantics.

5.3 Perception of Auditory Objects

There seem to be at least two fairly uncontroversial ways in which the perception of auditory objects can be tackled experimentally: one presents subjects with sound waves, and then asks them either a "what question," or a "how many question." "What" experiments might require a subject to recognize sounds, perhaps by identifying the source, or by classifying or discriminating the sound in some other manner. "How many" experiments, in contrast, ask the subject whether she perceives a particular sound sample as containing sound from just a single auditory object or several. Thus, experimental psychologists treat auditory objects as "perceptual entities" that arise in the mind of the listener as a result of processes that include both a grouping or binding of constituent acoustic features, as well as sensory discrimination and categorization.

5.3.1 "What Type of Object?" Invariance in Categorical Perception

What goes on in the mind of a subject who is trying to identify or categorize a particular sound? One way to think about "what" experiments is that they map out the "topography" of a person's or animal's "perceptual space." The idea here is that the brain maps the continuum of all possible physical sound features onto some form of high-dimensional perceptual space, in which the "perceptual distances" between different sound stimuli are not necessarily directly proportionate to their dissimilarity in terms of physical acoustics. Thus, several different sound stimuli come to belong to the same auditory category (are the same type of auditory object) if the physical differences between sounds within one category are perceived as relatively small or unimportant. In contrast, changes in physical stimulus parameters that cross category boundaries in perceptual space are salient, and may lead to the sound being perceived as a "different thing."

In the 1950s, Liberman et al. (1957) asked human subjects "what" they heard when they listened to a set of artificial speech sounds that simulated a continuum between the syllables /ba/, /da/, and /ga/. The subjects never had much difficulty identifying the sounds as either /ba/, /da/, or /ga/. The stimulus set contained a number of intermediate sounds, but instead of a comparable number of intermediate percepts there appeared to be sharp boundaries between only three perceived stimulus categories. To one side of one such boundary humans perceive a clear /da/, on the other side they perceive a clear /ga/, and there is no perceptual intermediate or hybrid. This type of categorical perception of sounds is clearly different from that of, say, of a note played on a violin, the pitch of which can change smoothly and gradually, without encountering sharp perceptual boundaries. The stimulus dimension of pitch seems to follow a rather even, continuous line in perceptual space, while formant transitions in spoken syllables, in contrast, seem to inhabit dimensions of perceptual space, which are heavily "warped" so that continua of physical sounds fall into clusters of perceived sounds that correspond to distinct phoneme categories.

Another example of such perceptual categories is the distinction between syllables /ba/ and /pa/. The key distinguishing characteristic between these two phonemes is voice onset time, that is, the time at which the vocal folds begin to vibrate to voice the /a/ after the lips have separated to articulate the /p/ or /b/. In the case of /ba/, voicing starts quickly, after only about 20 ms, whereas in the case of /pa/, voice onset takes somewhat longer, typically around 60–70 ms. Voice onset time is the distinguishing "stimulus dimension" for the discrimination of /ba/ and /pa/ sounds, in the sense that other dimensions such as loudness, perceived sound source direction, fundamental frequency of the vowel, and so forth are all largely irrelevant for this discrimination: Our perception is invariant with respect to those other stimulus dimensions. The /ba/–/pa/ voice onset time distinction has an interesting electrophysiological correlate: recordings by Steinschneider et al. (2003) in the primary auditory cortex of rhesus monkeys (*Macaca mulatta*) and by Eggermont (1995) in cortex of cats, has shown that, in response to /p/, the low- to mid-frequency neurons typically produce two bursts of action potentials, whereas in response to /b/ they produce just one. In this particular case, the position of the category boundary along the relevant stimulus feature dimension may reflect fairly simple physiological constraints. The perceived phoneme category appears to depend on whether forward masking produced by the consonant suppresses the vowel onset response in a subset of the neurons. If voice onset time is long enough for cortical neurons to have recovered from forward masking, then a /p/ is heard. Otherwise, the perception is /b/. (For a more detailed discussion of the neural processing of such phonetic category boundaries, see Steinschneider, Chapter 7).

In some cases at least, category boundaries may therefore be "hard wired" into the mammalian auditory pathway. Alternatively, and perhaps more commonly, category boundaries may reflect a parcellation or warping of stimulus space that reflect the statistics of an individual's perceptual experience. An interesting example of the latter case comes from Kuhl's (1991) work on human vowel perception. Human vowels are characterized by a number of formants, in which each formant is due to a different resonance in the vocal tract. By changing the position of articulators such as the lips, tongue, and soft palate, speakers change these resonant frequencies, which allows them to produce different types of vowels. Although human vowels may contain as many as five or six formant frequencies, the bottom two formants, usually denoted F1 and F2, appear to be the chief determinants of which vowel is heard. Kuhl used artificial vowels in her experiments, and first asked her subjects to rate which F1/F2 combinations were "good" or "typical" exemplars of the vowel category /i/ as in "peep." She then picked one artificial vowel that had been judged consistently as the best or most typical /i/ and designated it as the "prototype," and she also chose another that was generally rated as a poor example of /i/ as a "nonprototype" vowel. Kuhl then proceeded to measure her subjects' abilities to detect small deviations in the formant frequencies from either the prototype or the nonprototype. Interestingly, subjects found it harder to detect small changes away from the prototype than from the nonprototype. It seemed as if to the subjects, "all more or less good /i/s sounded very much alike." In contrast, if sounds needed to be compared that did not fall right into the center of a well established perceptual category, such as a vowel class in the subject's mother tongue, then their differences were easier to perceive.

Kuhl named this tendency of failing to distinguish members of the same, highly trained, sound category the "perceptual magnet effect." So it seems that with experience, for example, when learning a language, perceptual differences between sounds may expand or shrink so as to facilitate the appropriate category distinctions.

It is now established that many nonhuman animals can learn to categorize all sorts of sensory stimuli, including vocalizations and human speech sounds, and, once trained, they will also show clear categorical boundaries. This has been seen in chinchillas (Kuhl & Miller, 1975), rats (Mercado et al., 2005; Eriksson & Villa, 2006), starlings (*Lamprotornis hildebrandti*) (Kluender et al., 1987), and macaques (Sinnott & Brown, 1997; Sinnott et al., 1998), and it has been argued that categorical boundary and perceptual magnet effects may be essential characteristics of auditory object perception (Schouten & van Hessen, 1992; Kluender & Lotto, 1994; Holt et al., 2001). Whether nonhuman animals experience the perceptual magnet effect is less clear. Kuhl (1991) observed the magnet effect in human infants and adults but not in adult rhesus monkeys that had learned the same vowel categorization as their human competitors. She concluded that the magnet effect may be limited to human speech acquisition. Han et al. (2007) reported that rats reared under strong exposure to particular tone frequencies develop impaired tone discrimination near the frequencies to which they were exposed, but better tone discrimination in distant frequency regions, compared to control animals. However, Han et al. argue that passive exposure to particular sounds outside a behavioral context is not an example of category learning and that decreased discriminability may contribute to the perceptual magnet effect in humans but does not represent such a magnet effect in and of itself. In addition, although Kluender et al. (1998) observed that starlings discriminate less within a prototypical sound category than in a nonprototypical category, the authors exercise caution in calling these results a perceptual magnet effect, arguing that the observed lack of sound discrimination does not necessarily entail a perceptual inability of discrimination. In summary, whether the perceptual magnet effect is limited to human speech recognition or represents a general aspect of sound category learning and cortical map formation (argued, for example, by Guenther et al., 1996) remains an open question.

Both the formation of categorical boundaries and the perceptual magnet effect can be thought of as manifestations of invariant perception: Perceived category membership remains constant both over very wide ranges of stimulus parameters along "nondistinguishing" dimensions, as well as over fairly wide ranges along distinguishing dimensions, as long as the parameter ranges do not straddle category boundaries. Invariance in categorical perception may provide a handle to investigate the neurophysiology of auditory object recognition. Presumably, the neural responses that underpin a particular, categorically perceived auditory object should exhibit invariances that parallel those observed psychoacoustically. Further, one would expect such invariant neurophysiological response properties to be shaped by experience, as an organism learns to distinguish, for example, an /a/ from an /i/.

5.3.2 "How Many Objects?" Auditory Stream Segregation

Experiments that ask "Do you hear one thing or several?" instead of "What do you hear?" are often referred to as "auditory scene analysis" experiments, and Bregman (1994) did much to pioneer and popularize this approach. Interestingly, Bregman himself avoided the use of the term "auditory object" in his writing, and preferred instead to talk about "auditory streams." He felt that the word "object" was more appropriate in the discussion of visual perception, which analyzes light reflected from material objects, than for audition, which analyzes sound emitted as a consequence of some mechanical process or event. Nevertheless, Bregman (1994) stated that, in his opinion, "the auditory stream plays the same role in auditory mental experience as the object does in the visual" (p. 11), so the concepts of "auditory stream" and "auditory object" are, if not synonymous, at least closely related.

In the real world, many sounds of interest occur against a background of noise or other sounds, and if the brain is to succeed in recognizing several simultaneously present auditory objects by their intermingled auditory fingerprints, it must solve two very difficult problems. One problem is to decide how many auditory objects contribute to the auditory scene, and to "unmix" the incoming signal so that each part of the auditory waveform can be assigned to one perceptual object or another. The second problem is to identify or categorize each of these parts of the scene. These problems seem to be conceptually separate, but are in fact intertwined, and it is unclear whether they can be solved separately.

At first glance, it might seem appropriate or necessary to separate out objects first, and only then to attempt to identify them. However, the decomposition of the received sound wave into the sum of waveforms emitted by each of multiple sources is an ill-posed problem. Mathematically, it is akin to a set of simultaneous equations with only two known quantities (the vibration of each eardrum) that needs to be solved to determine an a priori unknown, and possibly quite large, number of unknowns (the number of sound sources present and the vibration pattern of each). Such problems cannot be solved without additional assumptions, and it is therefore very helpful to know what sort of auditory objects to expect in a particular signal. The solving of the "how many" and the "what" questions therefore most likely proceed hand in hand, in a process of inference that involves heuristics and top-down–generated assumptions about the likely sound sources encountered in a particular context.

Gestalt psychology is an important source of ideas about the sort of heuristics the brain employs to solve the "how many" problem. One can formulate a set of expectations about how the different frequency components of a single auditory object might behave. For example, for reasons described in the first part of this chapter, harmonically related frequencies are likely to come from the same object, as are frequency components that share common onset and offset times or that are "comodulated," that is, their amplitude envelopes change in synchrony. Frequency components with a common interaural time difference (ITD) are also more likely to come from a single direction, and therefore from a single sound source, than frequency components whose ITDs differ. These rules are only heuristics—some particular object could defy any or all of these expectations—but they provide generally useful rules of thumb to create a first intelligent guess about which components may come from the same source. The grouping cues of common onset and harmonic structure are discussed further later.

5.3.3 Taking a Pragmatic Approach

The discussions in this section will have served to illustrate that a number of rather different things might be described as "auditory objects." Auditory objects may or may not be synonymous with either perceptual streams or with identifiable sound sources or acoustic events. If one takes the view that an object ought to be "recognizable" as belonging to a particular "class" or "category" of object, then one may quickly find oneself in a situation where it is not obvious which categorization is most "natural" or appropriate. Categories can be complex, hierarchical, or highly context dependent, as one can easily appreciate if one considers the process of recognizing a vocalization. Imagine that some listeners who overhear a conversation have recognized one of the acoustic signals as the spoken English word "please." Even in such a perfectly ordinary situation, the simple question "What did they hear?" has no single correct answer. One could say that he heard a plosive consonant followed by an approximant followed by a vowel followed by a fricative. That would be an accurate, but in most cases nevertheless an inappropriate answer. Saying that he heard "a pleading child" or "a sarcastic young lady" might be alternative answers that, depending on the context and on the pitch, timbre, and prosody of the acoustic stimulus, could be much closer to the mark.

This rich complexity and context dependence of auditory objects makes it almost impossibly challenging to study them in all their depth, but one can nevertheless scratch the surface by studying the physiological correlates of phenomena that must be linked to auditory object perception. Much of the rest of this chapter is dedicated to trying to illustrate this point, by discussing a number of recent neurophysiological studies that have been selected because of their relevance to the key concepts described here earlier, including streaming or foreground–background discrimination, or abstraction away from irrelevant stimulus features, or generalization within a category and sharp discrimination across category boundaries. Some of the literature that proposes that these functions may be carried out in a hierarchically organized, cortical object processing ("what") stream are also reviewed.

5.4 Neural Processing of Auditory Objects

5.4.1 Neural Correlates of Gestalt Grouping by Common Onset

If several frequency components have a common onset, then they are commonly perceived as belonging to the same auditory object. Conversely, if their onsets are asynchronous, then they are usually perceived as separate sounds. Darwin and Sutherland (1984) demonstrated this using the fine distinction between the vowels /I/ and /e/ in English. These vowels differ in the frequency of the first formant. A first formant frequency (F1) of less than approximately 450 Hz is normally heard as an /I/, while higher F1s are heard as /e/ (Fig. 5.4a, b).

Darwin and Sutherland generated synthetic vowels from harmonics of 125 Hz, and shifted the first formant by adjusting the relative levels of the third and fourth harmonics (375 and 500 Hz respectively). If the formant was right between these harmonics (i.e., the harmonics were of approximately equal amplitude), the listeners perceived the sound as an /I/ or as an /e/ with similar probabilities, but if the amplitude of the fourth harmonic was increased, the listeners' judgments reliably shifted toward /e/. However, increasing the amplitude of the fourth harmonic shifted the percept toward /e/ only if the onset and offset of the fourth harmonic coincided with that of the other harmonics. When the fourth harmonic started 240 ms earlier than the others (Fig. 5.4c), then it was heard as a separate 500-Hz tone, and the perceived vowel quality was no longer shifted toward /e/. (Interested readers can find a demonstration of these sounds at https://mustelid.physiol.ox.ac.uk/drupal/?q=topics/onsets-and-vowel-identity.)

These results very clearly demonstrate the importance of common onsets as cues to solving the "how many" problem, and also illustrate that solving the "what" and the "how many" problems must go hand in hand. In this scenario, whether you hear an /I/ or an /e/ depends on how much of the sound energy at 500 Hz you attribute to a formant of the vowel, and how much to a separate 500 Hz tone. The neural mechanisms that underpin this perceptual phenomenon are less clear. At first sight, the problem seems

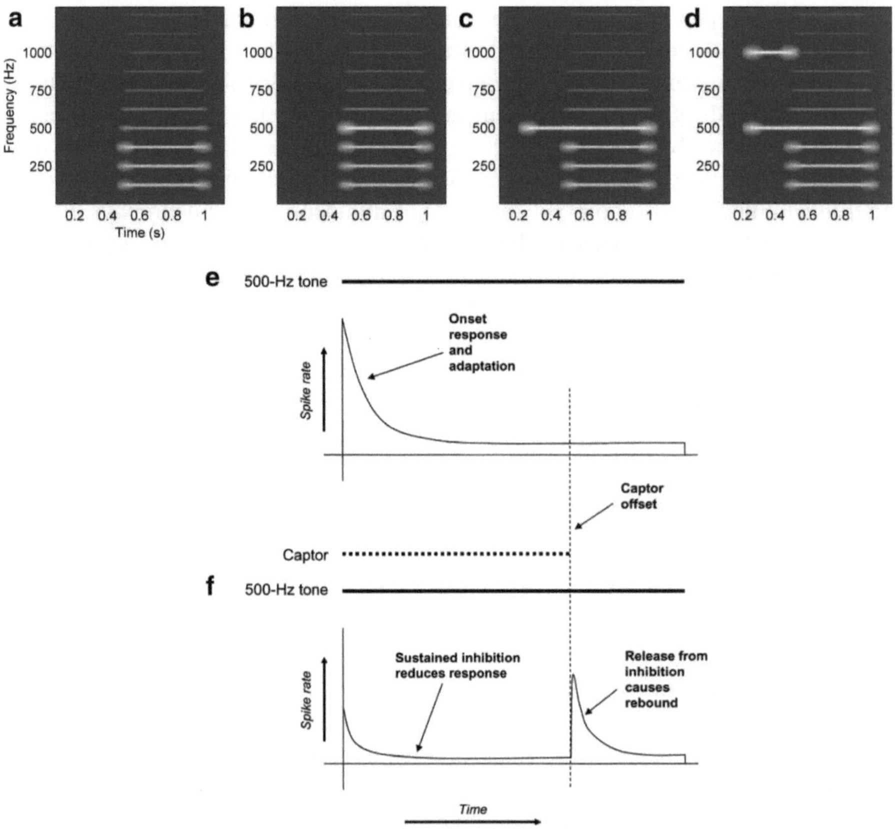

Fig. 5.4 Spectrograms of stimuli such as those used by Darwin and Sutherland (1984). (**a**) Artificial vowel with formant peak near 375 Hz. Normal native speakers of English hear this as an /I/-like vowel. (**b**) Same artificial vowel as in (**a**), but with a 500-Hz tone superimposed. The onset of the 500-Hz tone is synchronous with the vowel, and is perceived as belonging to the vowel. The vowel therefore appears to have a formant at 500 Hz, and is perceived as /e/-like. (**c**) Same as (**b**), but the 500-Hz tone now starts 250 ms before the vowel. Listeners perceive this as a tone pip preceding a separate, /I/-like vowel. (**d**) same as (**c**), but there is now also a 1000-Hz "captor" tone that starts 250 ms before the vowel and ends at vowel offset. Listeners perceive a brief harmonic complex followed immediately by an /e/-like vowel. (**e, f**) Schematic of neural responses to the 500-Hz component in the stimuli shown in (**c**) and (**d**) respectively, based on Fig. 2 of Holmes and Roberts (2006). (**e**) Owing to spike rate adaptation, responses to the 500-Hz tone are reduced at vowel onset, and no longer substantially enhance the activity in the frequency band associated with a formant characteristic for an /e/-like vowel. The vowel is therefore heard as /I/. (**f**) In the presence of a captor tone, wideband inhibition will suppress the sustained response to the 500 Hz even further, but this suppression ceases when the captor tone ends, leading to a "rebound" of activity that coincides with vowel onset and enhances activity in the 500-Hz band, biasing perception of the vowel toward /e/

to require some kind of sophisticated inference that might be performed by high-level, cortical mechanisms. However, an increasing body of research has suggested that bottom-up, relatively low-level mechanisms, such as adaptation in auditory nerve fibers, may play key roles in this process. If the fourth harmonic precedes the rest of the vowel, then the neural response to it will have decreased by the time the remaining harmonics of the vowel begin (Fig. 5.4e). Thus, the neural population will represent two consecutive onsets—the fourth harmonic followed by the rest of the sound. By the time the second sound begins, the response to the 500-Hz component will have declined, so that perception of the sound is less influenced by it.

However, not all of Darwin and Sutherland's results can be explained in these simple terms. In a further experiment they reasoned that, if the effect of onset asynchrony is due to perceptual grouping and segregation, then they should be able to reduce the effect of asynchronous onset by "capturing" the early, "lonely" 500-Hz harmonic into another harmonic complex, and they could then try to signal to the auditory system that this complex ends just before the start of the vowel. To do this, they added a "captor tone" at 1000 Hz, which started together with the early 500-Hz harmonic, and ended just at vowel onset (Fig. 5.4d). Indeed, the captor tone reversed, at least to some degree, the effect of onset asynchrony, in that it allowed the 500-Hz energy after the captor offset to be perceived as part of the vowel, causing it to shift the percept of the vowel back toward /e/. This indicates that more must be going on than merely spike rate adaptation at the level of the 500-Hz fibers of the auditory nerve, as the presence or absence of the 1000-Hz captor tones has no influence on that adaptation.

An alternative mechanism, which may be able to account for the "captor tone" result, relies on so-called wideband inhibitor neurons in the dorsal cochlear nucleus. These neurons respond poorly to pure tones, but vigorously to multiple tone complexes, or sounds with a wide bandwidth. Wideband inhibition could supply an alternative low-level account of the effects of the captor tone (Fig. 5.4f). When the captor tone is played, it could (together with the 500-Hz tone) activate wideband inhibition, which in turn would reduce the responses of cochlear nucleus neurons responding to 500 Hz. These would consequently fire less but therefore also experience less spike rate adaptation. When the 1000-Hz captor stops, the wideband inhibition ceases, and the 500-Hz neurons would generate a "rebound" burst of activity. Because the 1000-Hz captor ends when the vowel starts, the rebound burst in the 500-Hz neurons would coincide with the onset bursts of the various neurons that respond to the harmonics of the vowel, and the fact that these harmonics, including the 500-Hz tone, all fired a burst together will make it look as if they had a common onset. Recent experiments (Holmes & Roberts, 2006; Roberts & Holmes, 2006, 2007) have shown that the characteristics of the captor tone psychophysics are well matched to the response properties of wideband inhibitor neurons, and similarly Bleeck et al. (2008) have documented wideband inhibition, as well as the resulting rebound firing at captor offset, using harmonic complexes that were similar to those used in the human experiments by Darwin and Sutherland described in the preceding text.

5.4.2 *Neural Correlates of Stream Segregation*

The constituent harmonics of a vowel used by Darwin and Sutherland are a good example of grouping cues coming to play in a scenario with what might be called "ecological validity." Another, less ecological, but simpler, stimulus paradigm that has proven very popular and useful for studying the binding or segregation of acoustic features is the so-called "a b a" or "galloping rhythm" paradigm first investigated by van Noorden (1975). The stimulus is illustrated in Fig. 5.5. It consists of sequences of short pure tone pips of one of two possible frequencies, a and b, played in a repeating "a b a – a b a – …" pattern. The interest of this stimulus is that it can be perceived in one of two possible ways. When the frequencies a and b are close together (as in Fig. 5.5a) then listeners are likely to perceive the stimulus as rapidly galloping "da di da – da di da" triplets, but when the frequencies are more widely spaced (Fig. 5.5b), the percept breaks apart into two separate streams, one fairly rapid "da – da – da" stream against the background of a slower "- di – – – di – ." Whether the listener perceives one "auditory object" or two therefore depends on a gestalt grouping cue of proximity in frequency and time. Interestingly, for a rather wide range of intermediate frequency separations, the stimulus is ambiguous, and can flip back and forth between being perceived as just a one stream or two. Galloping rhythm "a b a" sound stimuli are therefore bistable, not unlike the well known "Necker cube" image (Fig. 5.5c), which can be perceived in one of two possible configurations: the top right corner appearing to be either part of the front or the back surface (Pressnitzer & Hupé, 2006; Schnupp, 2008). The initial percept is usually of just a single stream, but the wider the separation between frequencies a and b, the sooner the percept will break into two, and the more time it will spend being perceived as two streams rather than one.

The neurophysiological basis of perceiving one or several streams is not completely understood, but one very popular working hypothesis is that the number of perceived streams may correspond to the number of independently activated "ensembles" or populations of neurons. If the frequency separation between frequencies a and b is small relative to the typical tuning width of frequency tuned auditory neurons in the brain, then the population of neurons responding to the a tone will overlap heavily with that responding to the b tone. In other words, both tones activate, more or less, the same pool of neurons, and are therefore perceived as belonging to the same sort of object. However, when a and b are widely separated in frequency, there will be much less overlap in the neural populations they activate, which will lead to them

Fig. 5.5 "Galloping rhythm" or "a b a" stimulus used to investigate the binding and segregation of acoustic features. Tones of two different frequencies, a and b, are presented in an alternating pattern "a b a — a b a — …." (**a**) When the two frequencies are similar, the stimulus is perceived as a single stream of repeating triplets with a galloping rhythm: "da di da — da di da." (**b**) When the two frequencies are widely separated, the percept may break up into two separate streams, one rapid ("da — da — da") and one slower ("di — — — di — — — di"). Galloping rhythm stimuli are often perceptually bistable, starting off as a single stream, but then spontaneously breaking apart or merging again, and the proportion of time spent in the single or two stream perceptual mode depends on the frequency separation. (**c**) The Necker cube is an ambiguous bistable perspective drawing. The top-right corner can be seen as belonging to either the front or the back face of the cube, and the percept may spontaneously flip from one configuration to the other. (**d**) Responses of a macaque auditory cortex neuron to the first and the last triplet of galloping rhythm stimuli with frequency separations of either 1 semitone (black line) or 9 semitones (gray line).

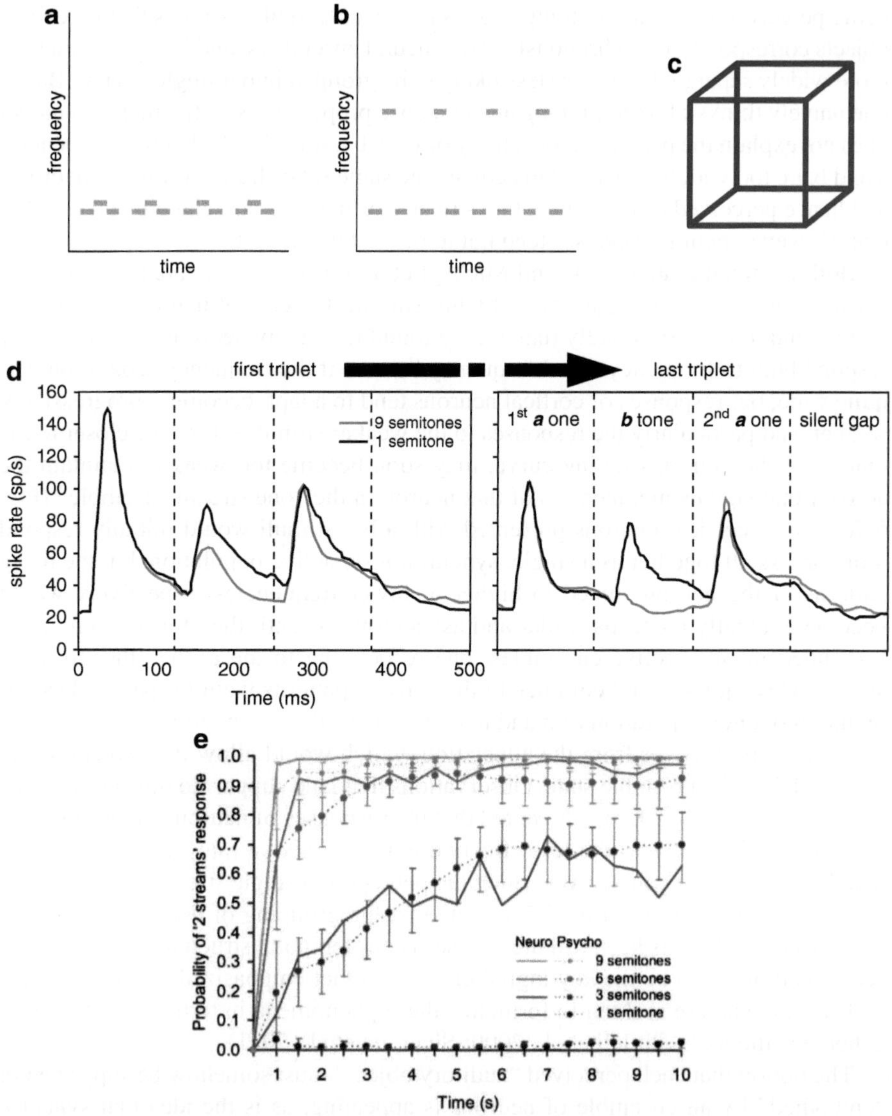

Fig. 5.5 (continued) The neuron responds more strongly to the a tone than to the b tone, but when the first triplet is presented (left), the neuron nevertheless responds strongly to each tone in the triplet regardless of initial frequency separation. However, owing to adaptation, the response to b is essentially nonexistent during the last triplet (right) in the 9 semitone separation condition. (**e**) Continuous lines: output of a "neurometric" model that attempts to predict the likelihood that a continuous a b a stimulus will be perceived as one or two streams at a time after stimulus onset given by the *y*-axis, based on whether adapted responses to one of the tones have fallen below a certain threshold, resulting in largely desynchronized cell assemblies. Different shades of gray give the results for different frequency separations between the a and b tones. Stippled lines, observed likelihoods of a "one-stream" or "two–stream" perception in normal human listeners subjected to behavioral ("psychometric") testing. (**d** and **e** based on Figs. 2 and 4 of Micheyl et al., 2005)

being perceived as "different things." This model essentially assumes that perceived objects correspond to synchronously active neural ensembles, and it can explain why more widely separated tones are less likely to be grouped into a single stream. But, if one naively thinks of neural tuning and response properties as static, then this model does not explain the perceptual bistability of the a b a stimulus. Either the frequencies a and b are too widely separated to activate the same set of frequency tuned neurons, and hence perceived as two perceptual streams, or they are not, and are perceived as one. However, neural responses tend not to be as static as that.

Both Fishman et al. (2004) and Micheyl et al. (2005) investigated responses of cortical neurons to a b a galloping rhythm stimuli. Given that neurons in auditory cortex tend to be very broadly tuned, they found that many neurons would initially respond both to frequency a and frequency b, even if the frequency separation was quite wide, but responses of cortical neurons tend to adapt, becoming progressively weaker, and particularly the responses to the weaker stimulus, the tone closer to the margin of the neuron's tuning curve, may soon become too weak to maintain the activity that signals membership of that neuron in the "one stream ensemble" (Fig. 5.5d). Thus, cortical neurons presented with a b a stimuli would initially respond more or less all together as a single, synchronously active population, but the firing patterns of the neurons tuned to higher or lower frequencies respectively would become gradually more dissimilar and asynchronous, until they form two more or less independently active ensembles, one responding to the a, the other to the b sounds. How quickly and completely the activity patterns would separate depends on the frequency separation of a and b tones. Once adapted, neurons fire somewhat less and partly recover from the adaptation, which would allow the percept to flip back and forth in a bistable state. Observations that lend support to this model come from Micheyl et al. (2005), who noted that the responses of neurons in the cortex of anesthetized monkeys to the less effective tone declined at rates that seemed well matched to the time course over which the percept of a single stream breaks up into two in human observers (Fig. 5.5e). Note that the grouping of neural responses in time, which highlights key attributes of the spectrotemporal structure of the auditory scene, is thought to be the key ingredient here, a fact emphasized by Shamma and colleagues, who are seeking to formalize these phenomena in terms of a "temporal coherence theory" (Elhilali et al., 2009; Shamma et al., 2011).

The notion that each perceived "auditory object" must somehow be supported or "embodied" by an ensemble of neurons is appealing, as is the idea that synchronized activity might provide the "glue" that holds these ensembles together, and it is encouraging that these notions appear supported by available evidence. However, there are still many open and unanswered questions. For example, the apparent correspondence between time courses of physiological adaptation in anesthetized monkey cortex and perceived, psychophysically assessed breakup of streaming in awake, behaving humans suggests that these two phenomena may be linked, but it does not prove that they are. Obtaining more compelling experimental evidence would be very challenging though, because ideally the activity of large numbers of individual neurons would need to be recorded in an awake, behaving animal trained to give a reliable indication of the number of streams it perceived. This would be technically extremely difficult. But there are also conceptual questions that need to be addressed. For example, is it better to assume that any and all neurons anywhere in the brain are equally eligible to form these ensembles that shape our conscious perception, or would it be more sensible to expect some anatomical levels, perhaps those in higher-order cortical regions, to be in a "privileged" position? Intuitively one might think that "higher" levels of processing may matter more than early ones, but this intuition was drawn into question when Pressnitzer et al. (2008) announced that they were able to replicate the results of Micheyl et al. (2005) using neurophysiological recordings obtained from the cochlear nucleus of anesthetized guinea pigs. It appears that even bushy cells of the anteroventral cochlear nucleus, which receive only minimally convergent inputs directly from auditory nerve fibers, and thus represent the very earliest input stages of the central auditory system, will exhibit patterns of adaptation when presented with a b a tone stimuli that should separate them out into largely asynchronously active ensembles over time courses that match the perceptual breakup of the galloping rhythm into two separate streams. As in the case of the Darwin and Sutherland experiments described in the previous section, here too it seems that very early and low-level physiological phenomena may make crucial contributions to the formation of auditory objects.

5.4.3 Auditory Object Representations in Primary Cortical Areas

The previous sections have illustrated that some at first sight apparently complex aspects of auditory scene analysis and object processing may be solved at least in part by relatively simple physiological mechanisms operating at low levels of the auditory pathway. Nevertheless, many researchers in the field suspect that higher, and particularly cortical, levels may have key roles to play in the formation of auditory objects. For example, Israel Nelken and his colleagues (Griffiths et al., 2004; Bar-Yosef & Nelken, 2007) have strongly argued that primary auditory cortex may make crucial contributions. The core of their argument is that most neurons in A1 fail to display selectivity for any one simple acoustic feature or set of features, such as falling or rising tone edges, patterns of frequency over time (STRFs), pitch, timbre, direction in space, and so forth. Rather cells in A1 appear to be tuned to combinations of features (Bizley et al., 2009) and to integrate information at different time scales (Nelken et al., 2003).

Experimental support for the special role of A1 comes from recordings by Israel Nelken and colleagues (Bar-Yosef & Nelken, 2007; Nelken, 2008) of cat A1 neurons responding to bird song stimuli recorded in the field. Because the recordings were made "in the real world," the songs were accompanied by additional acoustic components such as echoes and background noises. Using digital signal processing techniques, Nelken and colleagues separated out the foreground bird song from the background noise, and, in addition to the original sounds, also presented the cleaned "foreground only" song, or the much quieter remaining background sounds, to anesthetized cats. Figure 5.6 displays the responses of four cat auditory cortex neurons to these stimuli. The responses are displayed as rasters; there are 20 repeats of each

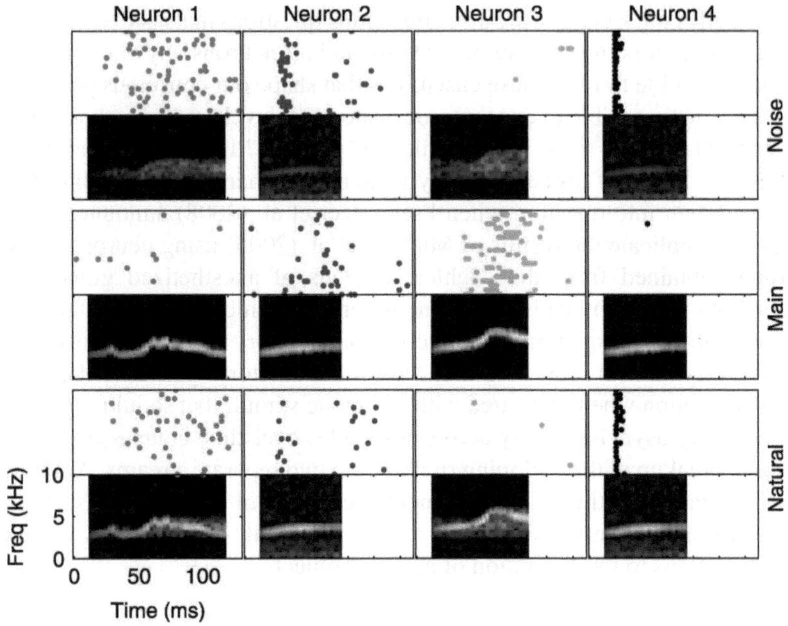

Fig. 5.6 Responses of four cat auditory cortex neurons to natural and manipulated bird chirp recordings. The responses are shown as dot rasters (each dot represents one action potential; each row of dots shows the action potential train evoked by a single presentation of the stimulus), and each dot raster is shown above a spectrogram of the corresponding stimulus. The bottom row shows the responses of the neurons to the original ("natural") bird chirp recording, the middle row shows responses to the isolated "foreground" (or "main") chirp, and the top row shows the response to the background noise and reverberation or echoes of the main signal. Note that, whereas the stimuli in the bottom row equal the sums of the stimuli in the top and middle row, the same cannot be said of the neural responses. Thus, the neurons seem to perform highly nonlinear operations that are suggestive of foreground–background separation. (Adapted from Fig. 13 of Bar-Yosef & Nelken, 2007, with kind permission of the authors)

sound and, in each repeat, a dot represents the time of occurrence of a spike. The stimuli are shown as spectrograms.

Three stimuli are shown for each neuron. The bottom stimulus is a segment from a natural sound recording, including the bird chirp plus all echoes (the "halo" around the chirps) and background noises that form part of a natural scene. The middle stimulus is the "foreground only" bird chirp, and the upper stimulus is the remainder, that is, just the echoes and background. Considering that most of the sound energy of the original, natural sound is contained in the bird chirp, the responses to the original recording and the cleaned chirp (bottom and middle rows) can be surprisingly different. In fact, in the examples shown here, the responses to the background, played alone, were often much more similar to the responses to the full natural stimulus than were the responses to the foreground-only stimulus.

The responses of these neurons can be interpreted as correlates of the end point of a process of scene analysis—they behave as if they responded either to foreground or to background objects present in an auditory scene, rather than the complete scene. However, these neurons responded more commonly to the background. Perhaps these neurons are really doing the hard part of auditory scene analysis—the foreground bird chirp is easy to hear. Analyzing the background is harder, but could be very important, for example, if a predator sought to take advantage of the foreground sounds to mask the noise of its own approach. Listening out for subtle background sounds could be a key function of auditory cortex. The same group presented data suggesting that responses in IC to these sounds are usually more closely related to the physical structure of the sounds, and therefore more strongly influenced by the high-intensity foreground. Thalamic responses, on the other hand, would appear to be more similar to the cortical responses (Chechik et al., 2006). This does suggest that the thalamocortical system may have a key role to play in scene analysis and auditory object processing.

Another, rather unusual but nevertheless very interesting, example of cortical neurons carrying out operations that appear related to auditory object processing comes from studies of echolocating pale spear-nosed bats (*Phyllostomus discolor*) by Firzlaff et al. (2007). Echolocation is of course in many ways very different from the cases discussed so far. The introductory sections of this chapter, which describe the acoustic fingerprints of objects and events, do not apply in the case of echolocation, because during echolocation, the bat does not listen out for sounds generated by external objects themselves. Instead the bat listens to acoustic reflections of a sound that it emitted itself. The delay between the emitted and the received sound provides valuable distance cues, and the intensity of the echo gives cues to the size and acoustic reflectance of the object. Some animals capable of echolocation even appear to be able to infer quite a great deal about the geometric structure of objects within their acoustic beam from the time signature of the reflected echo. Consider an object that presents several sound reflecting surfaces toward the bat, at different distances or depths. Each of these surfaces will send back a separate echo component, and these components will be delayed relative to each other in proportion to their distances. Consequently, each solid object, when hit by a brief echo pulse, will exhibit a characteristic impulse response.

An interesting feature of these impulse responses is that they scale with the size of the object. If the object increases in size, the reflective surfaces will increase and produce a stronger echo. The relative distances of the reflective surfaces would also increase, stretching the impulse response out in time. A bat that needs to recognize the impulse response of, say, a flower of a particular shape, regardless of whether the particular flower presented to it happens to be large or small, will need to be able to factor out the size scaling of the impulse response. Pale spear-nosed bats appear to be able to do just that to develop size-invariant auditory object representations. Firzlaff et al. (2007) were able to show not only that these bats were able to categorize echoes in a size-invariant manner, but also that a proportion (approximately 13%) of auditory cortical neurons investigated in these bats exhibited size invariant responses to echo stimuli. The anatomical parcellation of cortical areas within this particular species of bat is still not fully elucidated, but the size invariant, and therefore arguably "object specific," neurons observed by Firzlaff et al. (2007) did not appear to be anatomically clustered, and appeared to be found mostly in A1 and the anterior auditory field (AAF).

5.4.4 *Higher-Order Auditory Cortex: An Auditory "What" Stream?*

The last section presented two examples of phenomena that appear related to auditory object processing, and that both have been observed in primary auditory areas: foreground–background discrimination of stimuli recorded in the wild, and scale-invariant representations of classes of objects recognized by echo processing in bats. Nevertheless, primary auditory cortex is only the first station of an extensive cortical processing hierarchy that extends over large parts of temporal and parietal cortex and beyond (Poremba, 2003) and involves numerous higher order so-called "belt" and "parabelt" areas (Kaas & Hackett, 2000). All this higher-order cortex presumably has to serve a function, and, arguing by analogy with the presumed organization of the visual system, many researchers suspect that at least some of these higher-order auditory fields may make central contributions to the perception of auditory objects.

It has long been hypothesized that the visual system is organized along cortical streams that are specialized for the representation of either spatial attributes or object identity of stimuli respectively (Ungerleider & Haxby, 1994; Nassi & Callaway, 2009). Thus, the higher visual cortex comprises two anatomically distinct pathways (the ventral and dorsal stream), though they are richly interconnected. Moreover, neural responses in parts of the dorsal pathway, such as area MT, are highly sensitive to the position and motion of visual stimuli (Britten et al., 2009) while the ventral pathway is clearly involved in object recognition, as can be inferred from translation invariant responses to specific types of stimuli such as pictures of hands or faces (Tovee et al., 1994) or lesion studies that show that damage to inferiotemporal areas can lead to visual agnosias, that is, an inability to recognize certain objects by sight. Thus, even though there is continued controversy over whether these pathways truly represent separate "what" and "where" streams of visual analysis, the idea of parallel pathways is appealing, and can be supported by experimental evidence.

Romanski et al. (1999) have long argued, based on anatomical tracer studies, that similar specializations for spatial and object recognition specific processing streams may exist for the auditory system (Fig. 5.7a), and in 2001 Tian and colleagues (2001) published a study that set out to find physiological support for this idea. They presented seven different macaque communication calls from seven different spatial locations to anesthetized macaques and recorded responses from the caudolateral and anterolateral belt regions of the macaque auditory cortex respectively. They then quantified the spatial or call selectivity of the neurons simply by asking how many of the seven possible vocalizations or sound locations evoked responses with spike counts at least equal to half the maximum observed count for each neuron. Overall, neurons in the anterolateral belt appeared slightly more selective for vocalization type and less selective for, or invariant with respect to, spatial position, while the converse was observed in the caudolateral belt (Fig. 5.7b). These observations were taken as evidence that these respective belt areas were indeed parts of ventral "what" or dorsal "where" streams respectively.

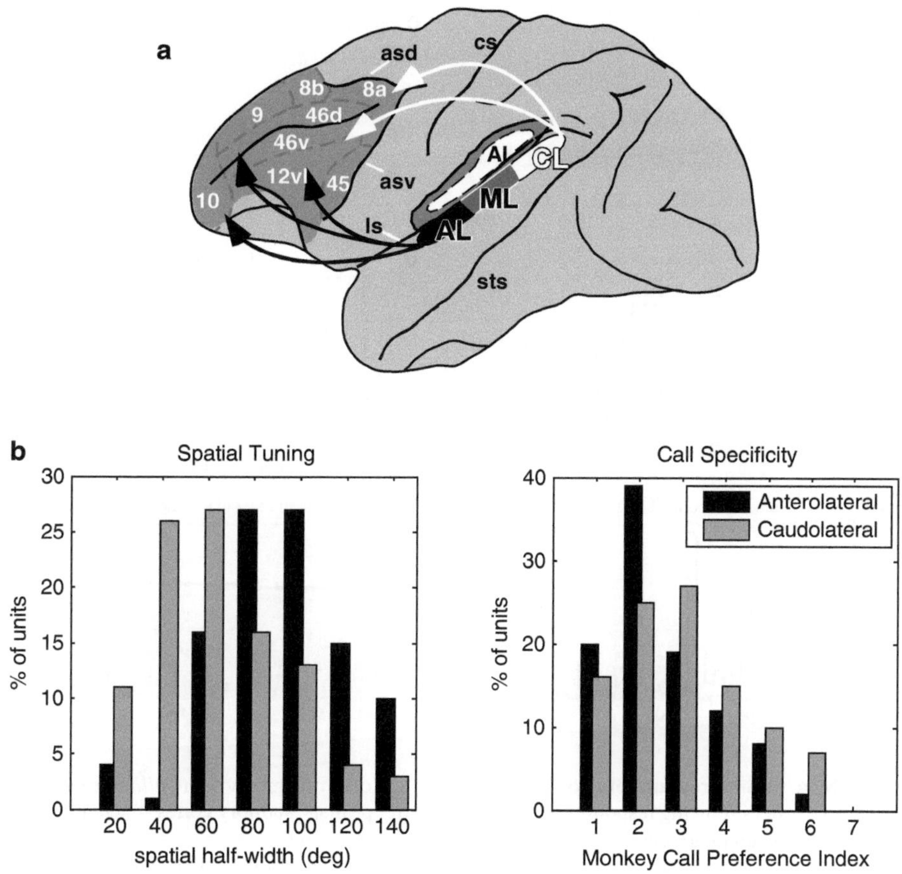

Fig. 5.7 (**a**) Putative auditory "what" and "where" processing streams in macaque cortex, based on anatomical connectivity. (Adapted from Fig. 3D of Romanski et al., 1999.) (**b**) Distributions of >50% maximal firing rate spatial tuning bandwidth (left) and "monkey call preference index" (number of calls out of seven which produced of >50% maximal firing rate, right) for neurons recorded in anterolateral (black bars) or caudolateral (white bars) belt of macaque auditory cortex, as observed by Tian et al. (2001). Although the distributions heavily overlap, there appears to be a trend for CL neurons to be more "narrowly tuned" for spatial position, and for AL neurons to be more narrowly tuned for call type

A shortcoming of that study lies in the very simplistic quantification of neural responses, which ignored the fact that temporal discharge patterns of cortical neurons can convey substantial information beyond that contained in spike counts accumulated over long time windows. For example, several studies showed that temporal discharge patterns or response latency distributions in many auditory cortical areas of the cat carry information about spatial location, vocalization call identity, or both (Brugge & Merzenich, 1973; Middlebrooks et al., 1994; Nelken et al., 2005; Bizley et al., 2009). To an extent this temporal pattern information carried by cortical discharges appears to reflect the fact that cortical neurons remain sensitive to the

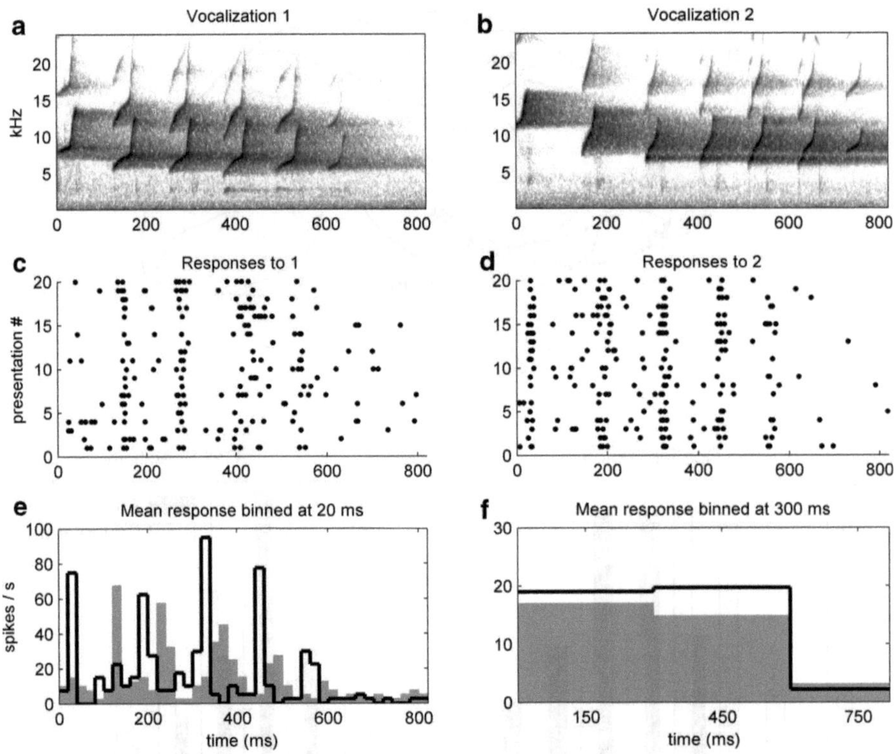

Fig. 5.8 (**a**, **b**) Spectrograms of two "twitter calls" of the common marmoset (*Callithrix jacchus*). (**c**, **d**) Dot rasters showing responses of a neuron in ferret primary auditory cortex. Each dot represents one nerve impulse, and each row of dots an impulse train fired in response to a single presentation of the corresponding stimulus. The neuron fires similar mean spike counts but with different temporal discharge patterns in response to each stimulus. (**e**, **f**) Responses shown in **c** and **d** as poststimulus time histograms. Responses to stimulus 1 shown in gray, those to stimulus 2 in black. At fine temporal resolutions (small histogram bin width, e.g., 20 ms shown in **e**) the differences in the response patterns are very clear and informative about stimulus identity, but at coarser temporal resolutions (e.g., 300 ms bin width, shown in **f**), the responses look very similar, and information about stimulus identity is lost. (Based on data from Schnupp et al., 2006)

temporal, "syllabic," or envelope structure of vocalization calls (Fig. 5.8), which appears to be reflected in the temporal discharge patterns at a resolution of about 10–20 ms (Schnupp et al., 2006; Walker et al., 2008). These temporal discharge patterns provide ample amounts of information for distinguishing different vocalization stimuli, even if overall mean spike counts do not. "Tuning width" estimates derived from mean spike counts over very long time windows such as those shown in Fig. 5.7b are therefore inadequate measures of the stimulus selectivity auditory cortical neurons.

As one continues along the cortical "what" stream through anterior belt areas along the anterior parts of the superior temporal gyrus (STG) and on to ventral prefrontal cortex (vPFC), one might expect responses to become less interested in physical features of sounds and more interested in their meaning. Indeed, recent experiments indicate that PFC activity is predictive of behavioral choices made by primates trained to categorize phonemes along a /b/-/d/ continuum (Lee et al., 2009). Consequently one might expect neurons in these higher-order areas to make less use of such temporal patterning of discharge at timescales of 10–20 ms to encode relatively fine grained detail of the stimuli. Meaning, after all, unfolds somewhat more slowly, over timescales that span several syllables. However, a recent set of experiments by Russ and colleagues (2008) suggest that this may not necessarily be the case. These investigators recorded activity from individual neurons in the STG and the vPFC of awake rhesus macaques, which were listening to ten very different types of rhesus communication calls, each of which is thought to carry a different meaning for the animals. Perhaps surprisingly, Russ et al. (2008) found that most STG and vPFC neurons responded vigorously (with >50% of their maximal firing rate) to more than half of the vocalizations tested.

Nor was there any trend for responses to become more specific as one ascends from STG to vPFC. But both STG and vPFC neurons carry a great deal of information about the vocalizations in their temporal discharge patterns. Using spike pattern classification techniques very similar to those used by Schnupp et al. (2006) to decode neural responses to vocalizations recorded in A1 of ferrets (*Mustella putorious*), Russ et al. (2008) were able to show that responses of neurons in macaque STG and vPFC also need to be decoded at a resolution of a few milliseconds if the individual vocalizations are to be correctly identified. Further, the reliance on precise temporal patterning of the discharges in monkey vPFC is also not obviously diminished compared to that seen in monkey STG or ferret A1. These observations seem somewhat at odds with the idea that vPFC neurons might be highly selective for the meaning of auditory stimuli, and much less so for their acoustic features. But it would be inaccurate to think that vPFC neurons are not interested in meaning at all. Indeed, another study from the same lab (Gifford et al., 2005) exploited that fact that the macaque vocal repertoire comprises several different vocalizations that appear to carry very similar meanings. For example, "warbles" and "harmonic arch" screams both appear to be used to signal the discovery of highly appetitive food items. It also appears that vPFC neural population responses do distinguish more clearly between two vocalization sounds that carry different meanings than between two acoustically different sounds that carry the same meaning. However, these "categorical" distinctions, although statistically significant at the population level, seem to be carried by a minority of vPFC neurons. Thus even at the very highest levels of the putative "what" stream many neurons remain clearly interested in acoustic features, and do not just signal abstract representations of auditory object categories, such as call type.

Another interesting question is whether responses in putative "what" processing areas become systematically more invariant with respect to changes in spatial location. Bizley et al. (2009) investigated this question by recording responses in five different cortical fields of the ferret to four different artificial vowel sounds ("a," "e," "i," and "u," respectively) which could occur at one of four different spatial positions (45° left, 15° left, 15° right, or 45° right) and with one of four possible fundamental frequencies or pitches (200, 336, 565, or 951 Hz). They then used a statistical variance decomposition

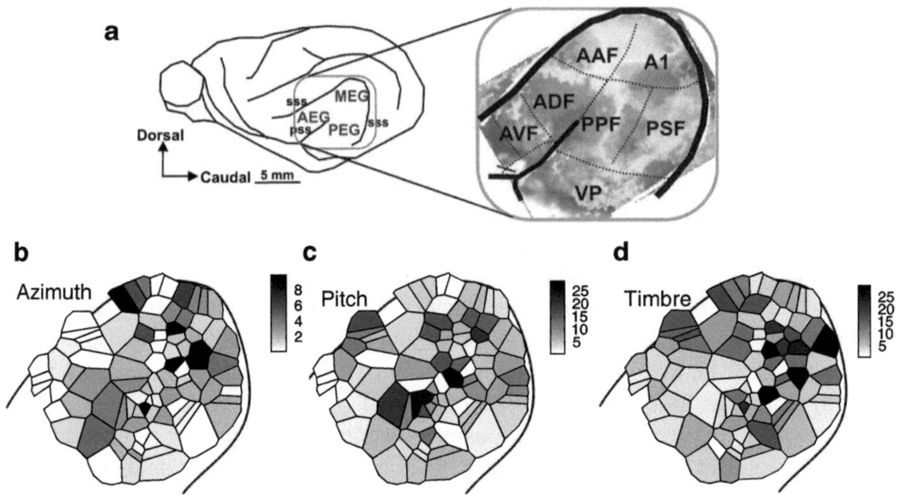

Fig. 5.9 (**a**) Schematic showing the organization of ferret auditory cortex. The subdivision of auditory cortex into physiological subfields is shown on the right, superimposed on a tonotopic mapping derived from previous optical imaging studies (Nelken et al., 2004). (**b**, **c**, **d**) Voronoi tessellations showing the distribution of sensitivity to sound source direction, pitch, and timbre of a set of artificial vowel sounds respectively. The grayscale of each tile shows the proportion of the variance in neural discharge patterns explained by variation in pitch, timbre, or azimuth respectively. (Adapted from Bizley et al., 2009)

to quantify how much of the observed changes in the neural discharge patterns in each cortical area was attributable to changes in stimulus formants (which are presumably a major cue to auditory object identity), how much to pitch, and how much to spatial location. They were also able to ask how many neurons exhibited substantial joint sensitivity to both formants and pitch or spatial location. Their results suggest a patchwork-like organization (Fig. 5.9), in which several areas appear to contain "hot spots" of high sensitivity to one feature or another, but none of the areas tested seems to sacrifice sensitivity to one stimulus attribute to specialize in the representation of another. Further, jointly formant- and pitch- or formant- and location-sensitive neurons formed the clear majority in all cortical areas tested. A clear division of labor, where some cortical fields clearly devote themselves to "what" (which vowel?) processing, and others concentrate on coding spatial location, is not readily apparent.

At first hand this may seem surprising when considered in the light of a recent study by Lomber and Malhotra (2008), who found that in cats, temporarily disabling a posterior belt area (PAF) by cooling with a cryoloop implant severely disrupted the animal's ability to perform a learned approach-to-target sound localization task, while cooling anterior cortical fields did not. Bearing in mind that many other studies have found high degrees of sensitivity to spatial cues in all parts of auditory cortex (Harrington et al., 2008; Bizley et al., 2009), this clearly localized effect of deactivation is surprising, but it may be possible to explain it if one assumes that posterior belt areas are privileged perhaps not so much in how they encode auditory information or what aspects of the auditory scene they represent, but in which other parts of the nervous systems they interact with. If one subscribes to the widely held belief that multisensory posterior parietal areas may be essential for directing attention to particular spatial locations (Andersen & Buneo, 2002) and one assumes, in line with current evidence (Romanski et al., 1999), that only posterior auditory belt areas project heavily to these areas, then Lomber and Malhotra's (2008) result is easy to explain even if anterior and posterior belt areas represent different aspects of auditory objects in very similar ways.

5.4.5 Are Auditory Objects Represented by Network States?

The cortical single unit data reviewed in the preceding text has provided a very mixed picture. On the one had there is some evidence for apparent foreground–background discrimination (Bar-Yosef & Nelken, 2007) and size-invariant responses in bat auditory cortex (Firzlaff et al., 2007), but on the other hand, most neurons right up to ventral prefrontal cortex appear preoccupied with acoustic detail rather than stimulus significance (Russ et al., 2008), and there is no evidence for neurons concerned with identifying stimuli becoming less interested in, or invariant to, spatial location (Bizley et al., 2009). But perhaps one needs to look at the network, rather than the individual neuron, level. Ohl et al. (2001) first presented evidence that primary auditory cortex overall may indeed respond in a category-specific manner, encoding auditory categories as invariant response patterns over the whole population of A1 neurons.

Ohl et al. (2001) trained gerbils to categorize tone sweeps according to whether the sweep moved upwards in frequency (the "up" category) or downwards (the "down" category). In the training cage one half of the cage floor was safe after an up-category stimulus was played and the other half was safe after a down-category stimulus was played. The gerbil's task was to move to the safe half of the cage to avoid small but unpleasant electric foot shocks. Animals had a surface electroencephalographic (EEG) grid implanted over their primary auditory cortex and cortical responses to each stimulus were recorded during training. Each response pattern recorded by the 18-channel EEG grid was analyzed as a 6–s-long trajectory in time, from 2 s prestimulus to 4 s poststimulus, through an 18-dimensional space. Each trajectory is the representation of the cortical response to one tone sweep from either category (up or down). Ohl et al. (2001) measured the Euclidean distance of each trajectory to the average trajectory for each of the two stimulus categories. This told them whether the global cortical response pattern, represented as a trajectory in time, could be assigned, based on similarity, to one of the two stimulus categories at any point in the 4 s after stimulus presentation.

During the first few training sessions, animals performed poorly in the task. During this learning phase cortical response patterns were very diverse, and did not distinguish sounds from each category. However, after only a few more training sessions each animal seemed to experience a "Eureka moment," after which performance dramatically improved, reaching more than 90% correct categorization, with a sharp category boundary. At the same time, the cortical activity patterns for stimuli within one stimulus category (upward or downward sweeps) became very similar after around 1 s poststimulus, while response patterns to one and the other stimulus category stopped to overlap. In summary, network states in primary auditory cortex were subject to category-learning: the acquired neural patterns now generalized over one physical stimulus dimension, that is, the particular rate or speed with which tone frequency swept up or down.

How individual A1 neurons are involved in forming such putative category-specific network states is uncertain, and sampling large portions of the population of auditory cortical neurons remains impossible, but simultaneous recordings from samples of only a few dozen neurons may be sufficient to generate new insights. For example, Luczak et al. (2009) simultaneously recorded responses from about 50 neurons distributed over 4 perpendicular array-electrode penetrations in the rat primary auditory cortex. Animals were either awake and passively listening or anesthetized. The novelty in this study was that the authors compared spontaneous events in A1 with events evoked by a number of tones and a number of cricket chirps, both in the anesthetized and the awake regime. First, they observed that spontaneous events (in both regimes) followed a stereotypical sequence of cell-by-cell activation, as had been shown earlier (Luczak et al., 2007). Further, the sequence in which neurons became active did not change considerably when A1 events were evoked by the external stimuli. So the temporal pattern of A1 activation seemed only partly determined by stimulus features. Second, they observed that stimulus-evoked response patterns depended on the stimulus category (tone, chirp, no stimulus) and these evoked patterns were a subset of the patterns observed during spontaneous activity. Tone-evoked patterns formed clusters that were separate from clusters for chirps, and within those clusters the response patterns for individual stimuli (e.g., tones of a specific frequency) formed subclusters.

In this study animals were naive and had not been trained to discriminate or categorize any of the tones and cricket chirps. However, against the extensive background of literature on learning-induced plasticity in primary auditory cortex (Weinberger, 2004; Kalatsky et al., 2005) in rats, cats, monkeys, and humans it appears highly likely that the response patterns observed by Luczak and colleagues in rats are subject to learning-induced changes. Perhaps such category-specific network states form the neural substrate of perceptual auditory object categories.

5.4.6 Objects in the Context of Action and Learning

The results discussed over the last few pages illustrate that, despite interesting and insightful studies over the last few decades, many aspects of the neural basis of auditory object perception remain poorly understood, and intuitions have proven to be an unreliable guide. Phenomena that one might think of as quite sophisticated and "high level," such as neural correlates of streaming, can be observed to some extent already at the very first relay station of the auditory pathway. Appealing notions of ever increasing abstraction of the neural representation as one ascends along a hierarchical cortical "what" stream remain only weakly supported by electrophysiological evidence, and many, perhaps most, neurons in all auditory belt regions and beyond remain stubbornly sensitive to relatively low-level features of sound stimuli, rather than exhibiting highly invariant, "object specific" responses. Also, the best examples of apparent figure–ground separation (Bar-Yosef & Nelken, 2007) or size-invariant shape sensitivity (Firzlaff et al., 2007) come from subsets of neurons recorded in primary areas, rather than further along the putative what stream. Similarly, studies that have had some success in identifying potential correlates of auditory categorization in neural population responses (Ohl et al., 2001; Luczak et al., 2009) have done so based on data recorded from primary fields. Much of this may seem counterintuitive, and difficult to tie together into a coherent conceptual framework.

As mentioned at the outset, the study of auditory object perception suffers to some extent from conceptual difficulties and a lack of uniformly accepted definitions of what constitutes an auditory object. That has not stopped experimental scientists from carrying out interesting and revealing experiments, but if one wishes to see a "big picture" emerging from these experiments then it is perhaps necessary to have another attempt at refining our notions of "auditory objects." One possible suggestion in this context is that perhaps it is time to move the emphasis away from "auditory," and ask simply, what is an "object"? And why do objects matter? If one asks a scholar of English grammar this question, then one might be told that an object is that which comes at the end of a sentence, after the verb. The reader may wonder whether bringing grammar into a discussion of neurobiology is really entirely a propos, but please bear with us for a moment. The potential insight here is that verbs describe actions, while objects tend to be manipulated by those actions.

If one thinks of objects as mental constructs that either inform or become instruments of actions, then it is perhaps unsurprising that the putative "what processing streams" appear to project to prefrontal cortical areas that have long been thought of as crucially involved action planning (Fig. 5.7a). This appears to be equally true for auditory and visual sensory processing streams. Particularly interesting in this context is the work of Earl Miller and colleagues carried out at MIT. They trained monkeys to discriminate computer-generated images of cats from images of dogs, and then recorded responses of neurons in inferotemporal as well as in prefrontal cortex (Miller et al., 2002; Freedman et al., 2003). The computer-generated images could be morphed, making images of imaginary animals that lay along the cat/dog continuum. After training, monkeys readily made categorical distinctions, generalized stimuli on either side of the category boundary, but were highly sensitive to changes near the boundary. Neurons in both inferotemporal and prefrontal cortical areas showed similar sharp distinctions between cats and dogs, while being much less sensitive to differences in stimuli within the dog or cat stimulus class. While neurons in the inferotemporal cortex appeared to be directly stimulus driven, and were active only while a stimulus was currently presented, those in the prefrontal cortex also exhibited what appeared to be working memory related behavior, for example, by firing in a sustained manner if the monkey had to hold in memory that the last stimulus presented was a cat or a dog to carry out a particular task.

It is worth bearing in mind that these experiments were carried out on laboratory monkeys that had no prior knowledge of either cats or dogs, and had no need to discriminate computer-generated images of these two types of carnivore before they found themselves in a situation in which making that distinction suddenly facilitated access to food or water rewards. The objects "cat" and "dog" emerged in the monkey's brain in the course of a trial and error learning procedure with positive reinforcement, during which the monkey's brain worked out that a particular parcellation of the sensory stimulus space, one that distinguished cat picture objects from dog picture objects, is highly advantageous for guiding efficient behaviors.

Sensory objects and category distinctions, be they visual or auditory, clearly serve to inform action selection processes that occur as neural activity reverberates through in the cognitive and motor loops that span prefrontal cortex and the underlying basal ganglia. But rather than merely being fed into these structures, they may even emerge from them. Evidence for this comes from another set of experiments carried out in Earl Miller's laboratory (Pasupathy & Miller, 2005; Loh et al., 2008; Histed et al., 2009) in which monkeys were trained with positive reinforcement to learn arbitrary mappings of particular visual stimulus categories onto particular actions. If you see a picture of a flower, look left; if you see a colored wallpaper, look right. Neurons in both the prefrontal cortex and in the caudate nucleus of the basal ganglia soon distinguished the visual stimuli, and fired in a manner that accurately predicted the animal's behavioral choice. But once the behavior was established, Miller and colleagues changed the rules. The correct behavior for seeing a flower was no longer "look left," but "look right." Interestingly, neurons in the caudate nucleus appeared to change their mapping of stimuli onto behavior very quickly, within fewer than 10 trials. Indeed, neurons in the caudate nucleus reflected the new set of appropriate rules for mapping stimuli to behavior much faster than either prefrontal cortex neurons or the monkey's behavior.

The idea that perceptual objects are largely or entirely shaped by our sensory experience is of course not new, but these experiments suggest where and how this shaping of perceptual objects may occur in the mammalian brain. Perhaps object categories are parcellations of stimulus space that emerge as our basal ganglia search for the most effective mappings of stimulus sets onto rewarding behavior patterns. Neurons in the striatum sit at the confluence of sensory, motor, and reward signals, and are therefore ideally placed to be among the first to discover the appropriate rules. They would then pass the learned rules back down the chain, first to prefrontal cortex and from there to higher-order visual and auditory areas in the temporal lobes, and so on. The rules that guide the mapping of stimuli onto behavior are of course highly context dependent. Whether listeners perceive a particular spoken utterance as a valuable piece of news or as an attractive voice depends on the context in which they find themselves, and whether it requires that attention be paid to the message or the messenger. In addition, the listeners' brains constantly have to learn new rules, as they struggle to make sense of new or as yet unfamiliar noises, new mobile phone ring tones, new words to add to their vocabulary, unfamiliar pronunciations encountered while traveling far from home, and so on.

If the striatum indeed plays a pioneering role in discovering these new rules and forming new "sound to meaning mappings," then that may help explain why many of the prefrontal cortex neurons described by Russ and colleagues (2008) show responses that appear clearly time-locked to relatively low-level acoustic features of incoming sounds. The discovery of new rules, or the updating of existing ones, does require access to a wealth of relatively raw, low-level data, which must be examined and reexamined to decide which details are important and which are not. This possibility may shift our expectations of how neurons along the putative "what" stream ought to behave. Different neurons may serve different functions. Perhaps only a subset of prefrontal and inferotemporal cortex neurons can be expected to exhibit "abstract" response properties that distinguish classes of objects, while others, perhaps many others, may be required to pass along large quantities of relatively raw, low-level acoustic information toward the basal ganglia, which need this raw information to test whether the rules and distinctions that govern perception and guide behavior still apply. Indeed, in the inferotemporal cortical areas of the ventral visual "what" processing stream, object specific responses are said to be confined to only a minority of neurons, while most neurons exhibit responses that are not obviously limited to any readily apparent object category (Desimone et al., 1984).

5.5 Concluding Remarks

At present, these are merely hypotheses, but they do illustrate that, to achieve a deep understanding of the neural basis of auditory object perception, one will probably need to consider areas and aspects of brain function that lie well beyond the parts that are classically considered part of the auditory pathway, such as prefrontal cortical areas, basal ganglia, and perhaps also parts of the hippocampal formation. There is also a clear need for more sophisticated experiments, which incorporate simultaneous recording of large numbers of neurons from multiple areas, ideally in awake, behaving animals that are actively learning and performing object recognition and discrimination tasks. Further, these studies will need to be complemented and supported by much more sophisticated and extensive data analysis and modeling techniques than those that have been deployed to date. Such ambitious studies remain extremely difficult to carry out, but with the technological advances that have emerged over recent years, they are no longer impossible.

References

Andersen, R. A., & Buneo, C. A. (2002). Intentional maps in posterior parietal cortex. *Annual Review of Neuroscience*, 25, 189–220.

Bar-Yosef, O., & Nelken, I. (2007). The effects of background noise on the neural responses to natural sounds in cat primary auditory cortex. *Frontiers in Computational Neuroscience*, 1(3), doi: 10.3389/neuro.10/003.2007.

Bizley, J. K., Walker, K. M. M., Silverman, B. W., King, A. J., & Schnupp, J. W. H. (2009). Interdependent encoding of pitch, timbre, and spatial location in auditory cortex. *Journal of Neuroscience*, 29(7), 2064–2075.

Bleeck, S., Ingham, N. J., Verhey, J. L., & Winter, I M. (2008). Rebound depolarization in single units of the ventral cochlear nucleus: A contribution to grouping by common onset. *Neuroscience*, 154(1), 139–146.

Bregman, A. S. (1994). *Auditory scene analysis: The perceptual organization of sound*. Cambridge, MA: MIT Press.

Britten, K. H., Newsome, W. T., Shadlen, M. N., Celebrini, S., & Movshon, J. A. (2009). A relationship between behavioral choice and the visual responses of neurons in macaque MT. *Visual Neuroscience*, 13(1), 87–100.

Brugge, J. F., & Merzenich, M. M. (1973). Responses of neurons in auditory cortex of the macaque monkey to monaural and binaural stimulation. *Journal of Neurophysiology*, 36(6), 1138–1158.

Chechik, G., Anderson, M. J., Bar-Yosef, O., Young, E. D., Tishby, N., & Nelken, I. (2006). Reduction of information redundancy in the ascending auditory pathway. *Neuron*, 51(3), 359–368.

Darwin, C. J., & Sutherland, N.S. (1984). Grouping frequency components of vowels: When is a harmonic not a harmonic? *Quarterly Journal of Experimental Psychology Section A*, 36(2), 193–208.

Desimone, R., Albright, T. D., Gross, C. G., & Bruce, C. (1984). Stimulus-selective properties of inferior temporal neurons in the macaque. *Journal of Neuroscience*, 4(8), 2051–2062.

Eggermont, J. J. (1995). Representation of a voice onset time continuum in primary auditory cortex of the cat. *Journal of the Acoustical Society of America*, 98(2 Pt 1), 911–920.

Elhilali, M., Ma, L., Micheyl, C., Oxenham, A. J., & Shamma, S. A. (2009). Temporal coherence in the perceptual organization and cortical representation of auditory scenes. *Neuron*, 61, 317–329.

Eriksson, J. L., & Villa, A. E. P. (2006). Learning of auditory equivalence classes for vowels by rats. *Behavioural Processes*, 73(3), 348–359.

Firzlaff, U., Schuchmann, M., Grunwald, J. E., Schuller, G., & Wiegrebe, L. (2007). Object-oriented echo perception and cortical representation in echolocating bats. *PLoS Biology*, 5(5), e100.

Fishman, Y. I., Arezzo, J. C., & Steinschneider, M. (2004). Auditory stream segregation in monkey auditory cortex: Effects of frequency separation, presentation rate, and tone duration. *Journal of the Acoustical Society of America*, 116(3), 1656–1670.

Freedman, D. J., Riesenhuber, M., Poggio, T., & Miller, E. K. (2003). A comparison of primate prefrontal and inferior temporal cortices during visual categorization. *Journal of Neuroscience*, 23(12), 5235–5246.

Geissler, D. B., & Ehret, G. (2004). Auditory perception vs. recognition: Representation of complex communication sounds in the mouse auditory cortical fields. *European Journal of Neuroscience*, 19(4), 1027–1040.

Ghazanfar, A. A., & Santos, L. R. (2004). Primate brains in the wild: The sensory bases for social interactions. *Nature Reviews Neuroscience*, 5, 603–616.

Gifford, G. W., MacLean, K. A., Hauser, M. D., & Cohen, Y. E. (2005). The neurophysiology of functionally meaningful categories: Macaque ventrolateral prefrontal cortex plays a critical role in spontaneous categorization of species-specific vocalizations. *Journal of Cognitive Neuroscience*, 9, 1471–1482.

Gourévitch, B., & Eggermont, J. J. (2007). Spatial representation of neural responses to natural and altered conspecific vocalizations in cat auditory cortex. *Journal of Neurophysiology*, 97(1), 144–158.

Griffiths, T. D., Warren, J. D., Scott, S. K., Nelken, I., & King, A. J. (2004). Cortical processing of complex sound: A way forward. *Trends in Neurosciences*, 27(4), 181–185.

Guenther, F. H., & Gjaja, M. N. (1996). The perceptual magnet effect as an emergent property of neural map formation. *Journal of the Acoustical Society of America*, 100(2Pt 1), 1111–1121.

Han, Y. K., Köver, H., Insanally, M. N., Semerdjian, J. H., & Bao, S (2007). Early experience impairs perceptual discrimination. *Nature Neuroscience*, 10(9), 1191–1197.

Harrington, I. A., Stecker, G. C., Macpherson, E. A., & Middlebrooks, J. C. (2008). Spatial sensitivity of neurons in the anterior, posterior, and primary fields of cat auditory cortex. *Hearing Research*, 240(1–2), 22–41.

Histed, M. H., Pasupathy, A., & Miller, E. K. (2009). Learning substrates in the primate prefrontal cortex and striatum: Sustained activity related to successful actions. *Neuron*, 63(2), 244–253.

Holmes, S. D., & Roberts, B. (2006). Inhibitory influences on asynchrony as a cue for auditory segregation. *Journal of Experimental Psychology*, 32(5), 1231–1242.

Holt, L. L., Lotto, A. J., & Kluender, K. R. (2001). Influence of fundamental frequency on stop-consonant voicing perception: A case of learned covariation or auditory enhancement. *Journal of the Acoustical Society of America*, 109(2), 764–774.

Kaas, J. H., & Hackett, T. A. (2000). Subdivisions of auditory cortex and processing streams in primates. *Proceedings of the National Academy of Sciences of the USA*, 97(22), 11793–11799.

Kalatsky, V. A., Polley, D. B., Merzenich, M. M., Schreiner, C. E., & Stryker, M. P. (2005). Fine functional organization of auditory cortex revealed by Fourier optical imaging. *Proceedings of the National Academy of Sciences of the USA*, 102(37), 13325–13330.

Kluender, K., Diehl, R., & Killeen, P. (1987). Japanese quail can learn phonetic categories. *Science*, 237(4819), 1195–1197.

Kluender, K. R., & Lotto, A. J. (1994). Effects of first formant onset frequency on [-voice] judgments result from auditory processes not specific to humans. *Journal of the Acoustical Society of America*, 95(2), 1044–1052.

Kluender, K. R., Lotto, A. J., Holt, L. L., & Bloedel, S. L. (1998). Role of experience for language-specific functional mappings of vowel sounds. *Journal of the Acoustical Society of America*, 104(6), 3568–3582.

Kuhl, P. K. (1991). Human adults and human infants show a "perceptual magnet effect" for the prototypes of speech categories, monkeys do not. *Perception & Psychophysics*, 50(2), 93–107.

Kuhl, P. K., & Miller, J. D. (1975). Speech perception by the chinchilla: Voiced-voiceless distinction in alveolar plosive consonants. *Science*, 190(4209), 69–72.

Lee, J. H., Russ, B. E., Orr, L. E., & Cohen, Y. E. (2009). Prefrontal activity predicts monkeys' decisions during an auditory category task. *Frontiers in Integrative Neuroscience*, 3(16), 1–12.

Liberman, A. M., Harris, K. S., Hoffman, H. S., & Griffith, B. C. (1957). The discrimination of speech sounds within and across phoneme boundaries. *Journal of Experimental Psychology*, 54(5), 358–368.

Loh, M., Pasupathy, A., Miller, E. K., & Deco, G. (2008). Neurodynamics of the prefrontal cortex during conditional visuomotor associations. *Journal of Cognitive Neuroscience*, 20(3), 421–431.

Lomber, S. G., & Malhotra, S. (2008). Double dissociation of 'what' and "where" processing in auditory cortex. *Nature Neuroscience*, 11(5), 609–616.

Luczak, A., Barthó, P., Marguet, S. L., Buzsáki, G., & Harris, K. D. (2007). Sequential structure of neocortical spontaneous activity in vivo. *Proceedings of the National Academy of Sciences of the USA*, 104(1), 347–352.

Luczak, A., Barthó, P., & Harris, K. D. (2009). Spontaneous events outline the realm of possible sensory responses in neocortical populations. *Neuron*, 62(3), 413–425.

Mercado, E., Orduña, I., & Nowak, J. M. (2005). Auditory categorization of complex sounds by rats (*Rattus norvegicus*). *Journal of Comparative Psychology*, 119(1), 90–98.

Micheyl, C., Tian, Biao, Carlyon, R. P., & Rauschecker, Josef P. (2005). Perceptual organization of tone sequences in the auditory cortex of awake macaques. *Neuron*, 48(1), 139–148.

Middlebrooks, J., Clock, A., Xu, L., & Green, D. (1994). A panoramic code for sound location by cortical neurons. *Science*, 264(5160), 842–844.

Miller, E. K., Freedman, D. J., & Wallis, J. D. (2002). The prefrontal cortex: Categories, concepts and cognition. *Philosophical Transactions of the Royal Society B: Biological Sciences*, 357(1424), 1123–1136.

Nassi, J. J., & Callaway, E. M. (2009). Parallel processing strategies of the primate visual system. *Nature Reviews Neuroscience*, 10(5), 360.

Nelken, I. (2008). Processing of complex sounds in the auditory system. *Current Opinion in Neurobiology*, 18(4), 413–417.

Nelken, I., Fishbach, A., Las, L., Ulanovsky, N., & Farkas, D. (2003). Primary auditory cortex of cats: Feature detection or something else. *Biological Cybernetics*, 89(5), 397–406.

Nelken, I., Bizley, J. K., Nodal, F. R., Ahmed, B., Schnupp, J. W. H., & King, A. J. (2004). Large-scale organization of ferret auditory cortex revealed using continuous acquisition of intrinsic optical signals. *Journal of Neurophysiology*, 92(4), 2574–2588.

Nelken, I., Chechik, G., Mrsic-Flogel, T. D., King, A. J., & Schnupp, J. W. H. (2005). Encoding stimulus information by spike numbers and mean response time in primary auditory cortex. *Journal of Computational Neuroscience*, 19(2), 199–221.

Ohl, F. W., Scheich, H., & Freeman, W. J. (2001). Change in pattern of ongoing cortical activity with auditory category learning. *Nature*, 412(6848), 733–736.

Olveczky, B. P., & Gardner, T. J. (2011). A bird's eye view of neural circuit formation. *Current Opinion in Neurobiology*, 21(1), 124–131.

Pasupathy, A., & Miller, E. K. (2005). Different time courses of learning-related activity in the prefrontal cortex and striatum. *Nature*, 433(7028), 873–876.

Poremba, A. (2003). Functional mapping of the primate auditory system. *Science*, 299(5606), 568–572.

Pressnitzer, D., & Hupé, J.-M. (2006). Temporal dynamics of auditory and visual bistability reveal common principles of perceptual organization. *Current Biology*, 16(13), 1351–1357.

Pressnitzer, D., Sayles, M., Micheyl, C., & Winter, I. M. (2008). Perceptual organization of sound begins in the auditory periphery. *Current Biology*, 18(15), 1124–1128.

Roberts, B., & Holmes, S. D. (2006). Grouping and the pitch of a mistuned fundamental component: Effects of applying simultaneous multiple mistunings to the other harmonics. *Hearing Research*, 222(1–2), 79–88.

Roberts, B., & Holmes, S. D. (2007). Contralateral influences of wideband inhibition on the effect of onset asynchrony as a cue for auditory grouping. *Journal of the Acoustical Society of America*, 121(6), 3655–3665.

Romanski, L. M., & Averbeck, B. B. (2009). The primate cortical auditory system and neural representation of conspecific vocalizations. *Annual Review of Neuroscience*, 32, 315–346.

Romanski, L. M., Tian, B., Fritz, J., Mishkin, M., Goldman-Rakic, P. S., & Rauschecker, J. P. (1999). Dual streams of auditory afferents target multiple domains in the primate prefrontal cortex. *Nature Neuroscience*, 2(12), 1131–1136.

Russ, B. E., Ackelson, A. L., Baker, A. E., & Cohen, Y. E. (2008). Coding of auditory-stimulus identity in the auditory non-spatial processing stream. *Journal of Neurophysiology*, 99(1), 87–95.

Schnupp, J. W. H. (2008). Auditory neuroscience: Sound segregation in the brainstem. *Current Biology*, 18(16), 705–706.

Schnupp, J. W. H., Hall, T. M., Kokelaar, R. F., & Ahmed, B. (2006). Plasticity of temporal pattern codes for vocalization stimuli in primary auditory cortex. *Journal of Neuroscience*, 26(18), 4785–4795.

Schnupp, J., Nelken, I., & King, A. (2010). *Auditory neuroscience: Making sense of sound.* Cambridge, MA: MIT Press.

Schouten, M. E., & van Hessen, A. J. (1992). Modeling phoneme perception. I: Categorical perception. *Journal of the Acoustical Society of America*, 92(4 Pt 1), 1841–1855.

Shamma, S., Elhilali, M., & Micheyl, C. (2011). Temporal coherence and attention in auditory scene analysis. *Trends in Neurosciences*, 34, 114–123.

Sinnott, J. M., & Brown, C. H. (1997). Perception of the American English liquid /ra-la/ contrast by humans and monkeys. *Journal of the Acoustical Society of America*, 102(1), 588–602.

Sinnott, J., Brown, C., & Borneman, M.A. (1998). Effects of syllable duration on stop-glide identification in syllable-initial and syllable-final position by humans and monkeys. *Perception & Psychophysics*, 60(6), 1032–1043.

Steinschneider, M., Fishman, Y. I., & Arezzo, J. C. (2003). Representation of the voice onset time (VOT) speech parameter in population responses within primary auditory cortex of the awake monkey. *Journal of the Acoustical Society of America*, 114(1), 307–321.

Tian, B., Reser, D., Durham, A., Kustov, A., & Rauschecker, J. P. (2001). Functional specialization in Rhesus monkey auditory cortex. *Science*, 292(5515), 290–293.

Tovee, M. J., Rolls, E. T., & Azzopardi, P. (1994). Translation invariance in the responses to faces of single neurons in the temporal visual cortical areas of the alert macaque. *Journal of Neurophysiology*, 72(3), 1049–1060.
Ungerleider, L. G., & Haxby, J. V. (1994). "What" and "where" in the human brain. *Current Opinion in Neurobiology*, 4(2), 157–165.
van Noorden, L. (1975). *Temporal coherence in the perception of tone sequences*. Doctoral thesis, Technische Hogeschool Eindhoven.
Walker, K. M. M., Ahmed, B., & Schnupp, J. W. H. (2008). Linking cortical spike pattern codes to auditory perception. *Journal of Cognitive Neuroscience*, 20(1), 135–152.
Weinberger, N. M. (2004). Specific long-term memory traces in primary auditory cortex. *Nature Reviews Neuroscience*, 5(4), 279.

Chapter 6
Phonemic Representations and Categories

Mitchell Steinschneider

Keywords Animal models • Context dependence • Heschl's gyrus • Human intracranial recording • Language networks • Multidimensional encoding • Place of articulation • Primary auditory cortex • Speech processing • Superior temporal gyrus • Superior temporal sulcus • Voice onset time

6.1 Introduction

The ease and automaticity with which we typically extract phonemic content from the acoustic stream of speech belies a high degree of underlying perceptual and physiological complexity. Phonemes are the smallest units of speech that can change word meaning within a given language (e.g., Holt & Lotto, 2010). Despite their humble rank in language processing, the ability to categorize a given phoneme successfully is fraught with numerous problems. Listeners must be able to identify speech in the face of marked variability in the acoustic signal. Dynamically changing environmental noise intermixes with the speech signal as it impinges on the ears. This obvious source of variability is compounded by many other more subtle, but equally important, sources, such as those based on vocal track differences across speakers, the rate of speech production, and the effects produced by the surrounding phonetic environment (Miller & Eimas, 1995; Holt & Lotto, 2010). As an example of the latter, the phoneme /p/ remains the same even though the acoustics for /p/ in the words [pill] and [spill] vary considerably (both allophones of phoneme /p/). On the other hand, an acoustically stable signal that is perceived as one phoneme (e.g., /d/) can change to the perception of another phoneme (e.g., /g/) depending on

M. Steinschneider (✉)
Department of Neurology, Albert Einstein College of Medicine, Rose F. Kennedy Center, 1300 Morris Park Avenue, Bronx, NY 10461, USA
e-mail: mitchell.steinschneider@einstein.yu.edu

Y.E. Cohen et al. (eds.), *Neural Correlates of Auditory Cognition*, Springer Handbook of Auditory Research 45, DOI 10.1007/978-1-4614-2350-8_6,

© Springer Science+Business Media New York 2013

whether the preceding sound is an /l/ or /r/ (Holt, 2006a). In the final analysis, perception cannot be based on an invariant transformation between a specific set of acoustic cues and a phonemic counterpart (e.g., Pisoni & Luce, 1987).

This lack of invariance has led some investigators to question whether phonemes are the most appropriate building block in spoken speech (Pisoni & Luce, 1987). Some models of sentence recognition utilize allophonic variations of phonemes to parse the syllables occurring in running speech (e.g., Church, 1987). For instance, the presence of a /t/ with aspiration noise following the onset of the sound indicates that the sound is in syllable initial position. Other models explicitly dismiss attempts to divide up running speech into separate phonemes, and instead initiate decoding by examining the signal for simpler phonologic features that in turn have their basis in the spectrotemporal analysis of the sound waves (McClelland & Elman, 1986; Stevens, 2002). As this chapter describes, these fundamental analyses are well accounted for by the outputs of core auditory cortical centers such as primary auditory cortex (A1), which in turn serve as inputs for later processing centers that perform the task of transforming acoustically based neural response patterns to one that represents phonemic information. Areas surrounding A1, such as regions of the posterior–lateral superior temporal gyrus (PLST) are shown to serve as an early stage in this process, with later refinement occurring within the upper bank of the superior temporal sulcus (STS). Higher-order transformations that require top-down processing to (1) match the outputs of these analyses with explicit knowledge of the phonotactic rules governing allowable phonological sequences for a given language and (2) match outputs with a language's lexicon are further steps in phonologic analyses but are not the focus of this chapter.

Instead, focus of this chapter is sixfold. First, physiological data are emphasized that support modern theories positing that phonemic perception is a categorization task involving the parsing of the incoming signal into a multidimensional acoustic space that allows extraction of phonemically relevant features (Hickok, 2009; Holt & Lotto, 2010). The output of this processing mechanism has been described as a probabilistically based decision that promotes categorization of the phonemic strings embedded within the acoustic signal (Obleser & Eisner, 2008). Second, the importance of temporal processing in auditory cortex is stressed (e.g., Shannon et al., 1995). Speech unfolds over time, and auditory cortex is shown to possess a temporal processing fidelity that can assist in both the parsing of the signal into discrete units amenable for speech decoding, and provide directly relevant information for phonemic classification. Third, the importance of basic rules of auditory cortical physiology for determining many contextual effects in phonemic perception is stressed. Speech-specific mechanisms need not be invoked for explaining fundamental aspects of phonetic perception such as trading relations (i.e., modifying the parameters of one acoustic cue can offset the perceptual effects of another) (Repp, 1982) and the consequences of the preceding acoustic environment on perception of succeeding phonemes. Fourth, the utility of physiological activity in auditory cortex as an independent method to directly assess specific hypotheses in speech perception is considered. Special reference is made to hypotheses proposing how stop consonants (e.g., /b/, /g/, and /d/) are differentially encoded, and those positing that developmental language abnormalities can be the result of abnormally prolonged temporal windows of integration. Fifth, this chapter outlines the transformations that occur in neural selectivity as processing extends out from A1 and other core areas into adjacent cortical regions. Homologies between auditory cortical areas across species become uncertain beyond core regions (e.g., Hackett, 2007), and thus discussions about response transformations are primarily restricted to human data. Emphasis is on summarizing the results of studies using higher spatial resolution techniques, such as those offered by intracranial recordings and functional magnetic resonance imaging (fMRI). Discussions of relevant work in nonhuman primates supplement those acquired in humans. Finally, the chapter's conclusion argues for the emerging notion that a modular model of phonemic categorization may not adequately explain speech encoding, and that the process can be better clarified by invoking more dynamic and interactive network mechanisms.

6.1.1 Terminology

To assist the less experienced reader in speech-related terminology, a brief description of both speech production and the resultant acoustic output is provided here (see Kent, 1993). The reader is encouraged to examine the sound spectrographs and schematic shown later in the chapter. The airstream emanating from the lungs passes within the lower vocal tract and through the vocal cords, which can be in one of two active states: either vibrating or nonvibrating and partially abducted. When vibrating, the acoustic energy has maximum power at the glottal pulsation rate (fundamental frequency, F_0), with remaining power that decays exponentially at integer multiples (harmonics) of the F_0. When the vocal folds are not vibrating and partially abducted, the constricted airflow becomes turbulent and the resultant acoustical power is one of wideband noise. Airflow enters the upper vocal tract, which can be broken down into an oral and nasal cavity. The latter can be open or closed to airflow. Cavities serve as resonators. Based on the positioning of the speech articulators (e.g., lips, teeth, tongue, roof of mouth), the upper vocal tract will have changing resonating frequencies, with some frequencies being accentuated and other frequencies being dampened. These changing spectral functions linearly convolve with the energy emanating from the lower vocal tract to produce the speech output. The radiated spectrum will have peaks of energy called formants. These are labeled from the lowest frequency peak (F1), to ever higher frequency peaks (e.g., F2, F3).

Sounds produced when the vocal folds are vibrating are termed voiced. Vowels are always voiced, and are classically characterized as having formants with steady-state frequencies. When one sound leads into another there will be a change in formant frequencies. These periods of frequency change are termed formant transitions. Consonants may be voiced (e.g., /b/, /v/, /z/) or unvoiced (e.g., /p/, /s/, "ch"). When unvoiced, the period between the onset of the sound and the start of glottal pulsations is called the voice onset time (VOT). For instance, American English stop consonants can be either voiced (/b/, /d/, /g/) or unvoiced (/p/, /t/, /k/). They are called stops because just before consonant onset there is a brief cessation of airflow due to blockage by the closing articulators with a resultant proximal buildup of pressure.

Stops have been of special interest in the phonetic literature because it has been difficult to identify invariant acoustic cues that correlate with their place of articulation. Place of articulation refers to the specific locations in the oral cavity where the articulators produce constrictions in airflow. For the bilabial stops (/b/ and /p/), constriction occurs at the lips, whereas for the alveolar stops (/d/ and /t/) constriction occurs just behind the upper teeth as the tongue tip contacts the roof of the mouth, while constriction for velar stops (/g/ and /k/) occurs at the back of the mouth with occlusion made by the tongue body. One reason for the invariance problem is that output frequencies depend on the preceding and succeeding sounds, and thus each phoneme is not independent of its neighbor. For instance, this coarticulation of speech sounds leads to variability in the formant transitions of /d/ when succeeded by different vowels (Liberman et al., 1967). Acoustic invariance and the effects of coarticulation remain major unsolved problems in the speech sciences (e.g., Miller & Eimas, 1995; Diehl et al., 2004).

Finally, manner of articulation refers to how speech sounds are generated. Stop consonants begin with a cessation of air flow, followed by a release. Fricatives (e.g., /f/, /v/) do not occlude airflow, but instead a tight constriction made by the articulators produces turbulent flow that results in a strong noise component. Nasal consonants are generated by allowing air flow through the nasal cavity (e.g., /m/, /n/). Note that these latter examples are also voiced and have the same place of articulation as /b/ and /d/, respectively. Acoustically, they differ in part from their equivalent stop by beginning with a prominent low frequency murmur. Additional categories will be discussed as needed.

6.2 Phonemic Representation within the Auditory System: General Principles

6.2.1 Utility of Animal Models

It is reasonable to question whether physiological data obtained from animal models may serve as a platform upon which useful conclusions can be drawn about human phonemic perception. Animals can learn phonetic categories and display categorical perception of conspecific vocalizations (Kluender et al., 1987; May et al., 1989). Basic features of phonemic perception in the human are paralleled by similar psychoacoustic findings in animals. These parallels run the gamut from perception of vowels (Le Prell et al., 2001), stop consonant place of articulation and VOT (Kuhl, 1986; Sinnott & Adams, 1987), consonant manner of articulation (Dent et al., 1997), context effects in perception (Lotto et al., 1997), and learned covariation of acoustic features (Holt et al., 2001). In summary, similarities in phonemic perception across species indicate that nonhuman animal models are a valuable tool for examining some of the key questions concerning physiological mechanisms of speech perception.

The success that animal models have for addressing issues related to phonemic perception may be based on analogous speech processing occurring early in development.

In contrast to adults, very young infants are "universalists" who (1) discriminate almost all phonemic contrasts used by the world's languages and (2) demonstrate categorical perception across phonemic boundaries (Werker & Tees, 1999; Kuhl, 2004; Gervain & Mehler, 2010). These capacities are presumed to represent infants' use of basic auditory processing mechanisms, abilities that animals clearly utilize (Kuhl, 2004). The manner in which this general capacity to parse speech becomes specific for the child's native language is a topic of major research, and is believed to represent a form of statistical learning (see Werker & Tees, 1999; Kuhl, 2004; Gervain & Mehler, 2010; Holt & Lotto, 2010). Further, this developmental process bears great resemblance to that occurring in songbirds, a model that is producing major advances in our understanding of language learning (e.g., Woolley et al., 2005; Prather et al., 2009; Hahnloser & Kotowicz, 2010). Unfortunately, both these exciting topics are beyond the scope of this chapter, and it is suggested that readers examine the aforementioned references for entry into these broad and fruitful literatures.

6.2.2 *Subcortical Physiology*

Phonemic processing begins at the ears, and continues through multiple subcortical stations in the auditory pathway. A detailed discussion of this highly and clinically important topic is a chapter unto itself (for review see Young, 2008). Suffice it to state several relevant points. Peripheral constraints such as the shape and bandwidth of auditory filters modify the way in which auditory nerve fibers represent the formants of the speech signal (e.g., Moore, 2008a). For instance, higher formants may be blurred into a single response peak across the array of auditory nerve fibers due to filter bandwidth considerations. Physiological issues that will be stressed at the cortical level are already significant considerations in the periphery. Thus, neural activity of auditory nerve fibers displays characteristics that support multidimensional analysis of the incoming speech signal. Encoding of stimulus spectra using both temporal (phase-locking) and rate codes are evident for vowels (e.g., Shamma, 1985; Recio et al., 2002) and consonants (e.g., Delgutte & Kiang, 1984; Carney & Geisler, 1986). Other features important for phonemic perception such as stimulus rise-time ("ch" vs. "sh") and manner of articulation are also represented by the patterns of auditory nerve fiber activity (Delgutte, 1980; Delgutte & Kiang, 1984). Temporal encoding of the VOT phonetic parameter is observed for stop consonants in word-initial, -medial, and -final positions (Sinex & McDonald, 1988; Sinex, 1993; Sinex & Narayan, 1994). Finally, context effects based on the acoustic environment are prominent in the peripheral encoding of speech (e.g., Delgutte & Kiang, 1984; Young, 2008). Although there is a large literature clarifying the encoding of other stimuli within subcortical auditory nuclei, animal studies directed at phonemic encoding are limited (see Young, 2008). However, it is generally assumed that the inferior colliculus collects and refines incoming acoustic information for later processing at more rostral auditory centers (Nelken, 2008). Finally, evidence is mounting that the brain stem auditory evoked response is a clinically relevant, aggregate index of normal and aberrant phonemic processing in subcortical structures (Chandrasekaran et al., 2009; Anderson et al., 2010).

6.2.3 Cortical Neural Networks

The problem of phonemic categorization cannot be solved by a single station in the auditory cortex. Instead, it requires the involvement of a network of highly interconnected structures that is slowly being elucidated. The goal of much of this research is to identify relevant structures, determine their role in phonemic perception, and clarify the mechanisms responsible for the computational transformations that occur across the network (e.g., Hickok, 2009). In addressing each of these concerns, it will become clear that only at the level of A1 will some mechanisms be clarified, in part because of the extensive animal model research that has occurred within this area. However, even here, there are divergent views (see Nelken, 2008 for review). Some models would propose that A1 promotes phonemic encoding by its organization along multiple overlapping parameter maps whose unique activation patterns would signify the presence of a specific phoneme (e.g., Mesgarani et al., 2008). A second set of observations stress the point that A1 serves as a "filter for specific combinations of acoustic features" (Nelken, 2008). Thus, some A1 neurons respond with sustained activity to a highly selective subset of sounds containing specific combinations of acoustic features (Wang et al., 2005). A third view espouses the hypothesis that A1 responds to auditory objects and not combinations of acoustic features. All perspectives would allow response patterns to be modified by attention and experience (see Fritz et al., 2007; Atiani et al., 2009). Finally, A1 can be viewed as a more complex auditory center that is modulated by multimodal stimulus cues, behavior, and reward (e.g., Brosch et al., 2005, 2011). While these views are not mutually exclusive, the first can be most directly related to the perceptual models noted previously that envision parsing the spectrotemporal patterns of speech as a first step in phonemic identification. That A1 can be viewed from such diverse perspectives is a testament to the profound complexity of response properties and neuronal computations occurring at even early cortical levels, and is indicative that processing contingencies will only become more intricate as information flow proceeds into higher cortical regions.

The basic outline for information flow relevant for phonemic processing has been well summarized (e.g., Hickok & Poeppel, 2007; Rauschecker & Scott, 2009). Two distinct streams are hypothesized. The dorsal stream can be divided into one subserving functions related to spatial location and a second related to mapping sensory and phonological inputs onto articulatory representations. Portions of the boundary between posterior temporal and inferior parietal cortex, often referred to as Wernicke's area, are envisioned to be part of this network, especially in the dominant hemisphere where there is a direct interface with Broca's area in the inferior frontal lobe. The ventral stream, which is the focus of this report, begins with activity in core auditory cortex and projects through PLST into ever more anterior regions of the STS and the anterior temporal lobe before interfacing with regions in ventral prefrontal cortex (e.g., Scott, 2005; Leff et al., 2008). Ever more abstracted representations of the speech signal are thought to occur as processing progresses from phonological representations to ones engaged in lexical interpretations.

6.2.4 *Hemispheric Asymmetries*

In contrast to prevalent views in decades past, the use of functional neuroimaging has demonstrated that only at higher stages of phonemic processing will activity along the ventral stream show pronounced degrees of hemispheric asymmetry (e.g., Scott, 2005; Leff et al., 2008; Hickok, 2009). Equal activation, however, should not be equated with equal function, and it is hypothesized that the left hemisphere has a greater role in processing rapidly changing temporal components of speech, while the right hemisphere is more specialized for identifying slower spectral features (Zatorre et al., 2002; Zatorre & Gandour, 2008). This asymmetry, in turn, is associated with the idea that left hemisphere encoding extracts lexical content, whereas the activity in the right hemisphere is optimized to characterize prosodic and affective components of speech (Boemio et al., 2005).

6.2.5 *Temporal Processing*

Within each spectral filter of the auditory nerve, a signal such as speech can be decomposed into an envelope of neural activity representing the sound's amplitude fluctuations as seen by that filter, and a temporal fine structure characterized by phase-locking to the individual spectral components that occur within the filter's frequency band (e.g., Moore, 2008b). It is generally acknowledged that temporal envelope encoding is more important than fine structure encoding for the perception of speech (e.g., Drullman et al., 1994a, b). The importance of temporal envelope encoding is exemplified by the classic study performed by Shannon and colleagues (1995). The authors severely degraded speech by replacing discrete spectral components with band-limited noise. Speech with from one to four band-limited noise bands was constructed. Amplitude and temporal cues were preserved. Vowels could be accurately encoded using three noise bands. Voicing and consonant manner of articulation were accurately perceived using only two bands. Only consonant place of articulation was poorly identified, and this was due to the limited spectral encoding required for accurate discrimination. Clinically, temporal waveform encoding is the principal manner in which people using cochlear implants process speech, and it is a fundamental deficit in auditory neuropathy (Zeng & Liu, 2006). Finally, disruption of envelope following may be a comorbid feature of clinical states affecting speech perception in conditions as diverse as aging (Kumar, 2011), dyslexia (e.g., Giraud et al., 2005; Goswami et al., 2011), poor perception after cochlear implantation (Roman et al., 2004), autism (Kwakye et al., 2011), and schizophrenia (Brenner et al., 2009).

Integrated with the notion that each hemisphere specializes in either temporal or spectral aspects of speech is the hypothesis that these differences are based on distinct windows of integration that govern neuronal computations (Poeppel et al., 2008).

At least two time scales have been suggested. The first is between 20 and 80 ms and operates to sample the detailed spatiotemporal patterns of speech to extract phonological features (see also Wallace & Blumstein, 2009; Yrttiaho et al., 2010). The second is between 150 and 300 ms and samples speech at a syllabic level.

Experimental support for these shorter and longer temporal windows of integration is observed in an influential fMRI study that modulated the duration of repetitive bandpass noise segments that varied in whether the noise within each segment was either steady-state or frequency-modulated (FM) (Boemio et al., 2005). Activation was centered in PLST and the STS and became greater as the repetitive noise segments increased in duration from 12 ms to 300 ms (i.e., syllabic length, approximately 3–4 Hz). This preferential activation to the longer duration noise segments is consistent with the longer temporal window of integration of about 300 ms. PLST was also sensitive to segment type, indicating that this region is sensitive to the spectral composition of sounds. The maximal difference between responses evoked by steady-state and FM noise segments occurred at durations of approximately 45 ms, leading the authors to propose that this differential pattern was based on a window of integration of 25–50 ms. STS was not differentially activated by segment type, but did show an asymmetric increase to the longest duration segments in the right hemisphere. This was interpreted as evidence for the preferential role of more distal structures in the nondominant hemisphere to process slowly changing segments in the speech stream with integration windows in the 3- to 4-Hz range. In turn, these longer windows would be better suited for spectral analysis versus one requiring rapid decoding of temporally dynamic speech features. This parsing of the speech stream using multiple temporal integration windows would allow efficient processing that can simultaneously encode complementary phonemic and supraphonemic information occurring over different time scales (Panzeri et al., 2010).

Another important concern is the time frame over which phonemic information is extracted. Addressing this question is critical for assessing the temporal sequence by which acoustic, phonemic, and then higher-order lexical and semantic information is processed by the brain. It can best be addressed using physiological methods that possess high temporal resolution, such as auditory evoked potentials (AEPs) and magnetoencephalographic (MEG) signals. Using these methodologies, it is evident that phonemic representation begins at the earliest stages in cortical processing (Obleser & Eisner, 2008). MEG signals that reflect consonant place of articulation occur within 50 ms (Tavabi et al., 2007), and by 100 ms, activity becomes more specific for speech as compared to syllables made unintelligible by spectrally inverting the sounds (Obleser et al., 2006). Invasive studies of PLST in patients undergoing surgical evaluation for medically intractable epilepsy have shown that AEPs and high γ responses embedded within the electrocorticogram (ECoG) reflect consonant place of articulation and VOT in the slightly later time frame of approximately 100–150 ms (Chang et al., 2010; Steinschneider et al., 2011). The timing of activation within the STS has yet to be defined.

6.3 Phonemic Representation Within A1

6.3.1 Tonotopic Organization

It is now well established that speech-evoked activity is differentially distributed across A1 in a temporally dynamic manner that is determined by the field's underlying tonotopic organization (e.g., Merzenich & Brugge, 1973; Morel et al., 1993) and the spectral content of the sounds (e.g., Steinschneider et al., 2003; Engineer et al., 2008; Mesgarani et al., 2008). This interaction means that A1 neurons will generally fire if the spectral content of the phonemes overlaps with the spectral sensitivity of the cells. Further, tonotopic gradients in A1 are spatially organized. For instance, in A1 of the macaque monkey, low frequencies are best represented in anterior–lateral portions of A1 while progressively higher frequencies are best represented in more posterior–medial portions. This organization suggests that the representation of different phonemes, each possessing unique spectral characteristics, will generate distinct, spatially distributed maps across the tonotopic array of A1 space (see Schreiner, 1998).

An example of this tonotopic organization of speech-evoked activity is illustrated in Fig. 6.1 (adapted from Steinschneider et al., 2003). Awake monkeys were presented with the synthetic syllables /da/ and /ta/ that varied in their VOT from 0 to 60 ms in 20-ms intervals. Fig. 6.1a illustrates the spectrograms of /da/ with a 0 ms VOT, and /ta/ with a 60 ms VOT. For /da/, low-frequency energy centered between 800 and 1200 Hz and dominated by F1 and F2 persists throughout the stimulus. There is a relative paucity of higher frequency energy centered between 2000 and 4000 Hz, due to the characteristic low amplitude of F3. In contrast, /ta/ begins with a relative paucity of low-frequency energy due to the attenuation of F1 before voicing onset, and a relative predominance of higher frequency energy before voicing onset due to the presence of aspiration noise. Multiunit activity (MUA) across multiple electrode penetrations was averaged according to the best frequency (BF) of the sites (Fig. 6.1b). MUA evoked by /da/ begins with a transient response to stimulus onset and is followed by sustained activity in the two lowest BF regions. This pattern is based on excitation evoked by the sustained low frequency energy. In contrast, MUA within the 2- to 4-kHz BF band contains only a lower amplitude transient response evoked by stimulus onset, followed by sustained suppression of activity throughout the duration of the stimulus. This pattern is based on the relative paucity of high-frequency energy in the stimulus. A fundamentally different pattern occurs for /ta/. Now, a transient response evoked by both stimulus onset and the delayed voicing onset, based on the accompanying introduction of low-frequency energy, is observed in the two lowest BF regions. Further, sustained excitatory activity occurring throughout the duration of the high-frequency aspiration noise is observed in the 2– to 4-kHz BF band, followed by a rapid change to sustained suppression of MUA when the aspiration noise terminates and the low-frequency, high-amplitude F1 energy dominates. This dynamic and tonotopically precise representation of the syllables follows the sound's changing spectral content in real time, demonstrating

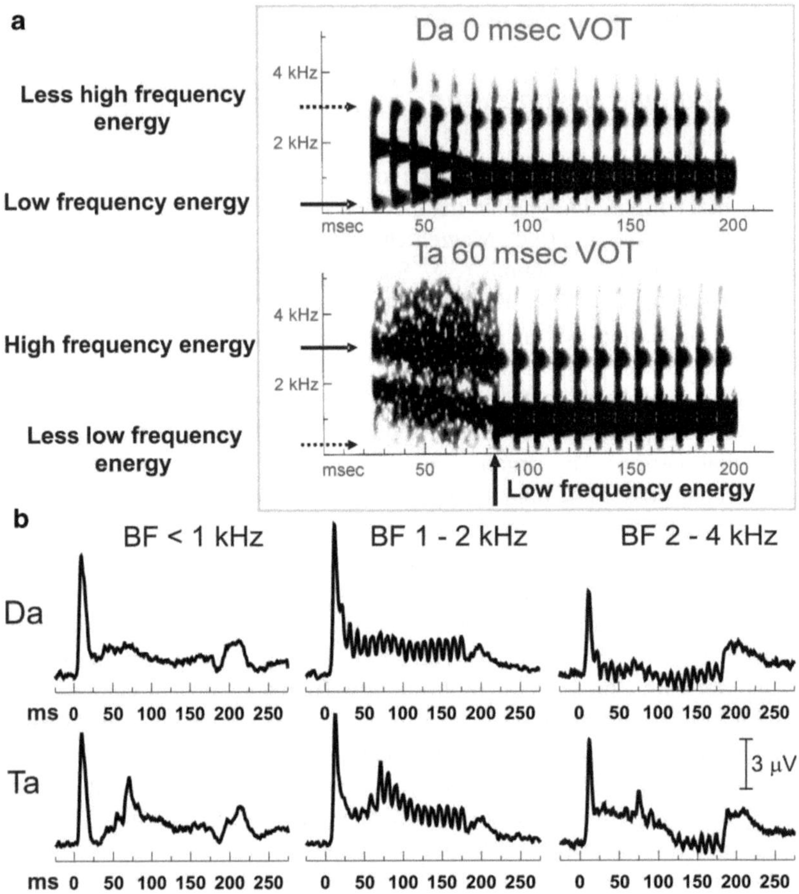

Fig. 6.1 Real-time encoding of the spectral and temporal structure of phonemes. (**a**) Sound spectrographs of the synthetic syllables /da/ and /ta/. (**b**) Multiunit activity (MUA) recorded from multiple electrode penetrations and averaged according to the best frequency (BF) of the recording sites, as determined by responses evoked by 60 dB SPL tones. See text for details. (Adapted from Steinschneider et al., 2003)

that speech sounds are encoded in A1 along spectral, temporal, and given an orderly representation of BF in A1, spatial dimensions.

The importance of the tonotopic organization in determining phonemic encoding in A1 is not restricted to stylized, synthetic syllables, but helps govern the representation of real speech with more complex spectral features. A prominent example is the work of Engineer and colleagues (2008), who recorded neuronal activity in rats evoked by 20 consonants that varied in their manner and place of articulation as spoken by an English-language speaker. Onset responses to the various consonants were predicted by the spectral sensitivity of the neurons. For instance, the fricative

/s/ with its higher spectral content activated high-BF neurons, while the fricative "sh" with its lower spectral content activated cells with a lower BF. Similar results were obtained for both consonants and vowels in A1 of ferrets, with vowel spectral patterns (e.g., Hillenbrand et al., 1995) mirrored by excitation patterns of neurons whose BFs overlap the major peaks of spectral energy (Mesgarani et al., 2008).

Relevance of A1 tonotopicity is not restricted to animal models, but is also applicable in human primary auditory cortex located on more medial aspects of Heschl's gyrus (e.g., Hackett, 2007). Single units recorded from this region are selectively activated by specific tone frequencies, and can have a specificity that approaches the limit of just noticeable differences in sound frequency (Howard et al., 1996; Bitterman et al., 2008). Spectral tuning derived from responses evoked by one segment of a movie soundtrack could reliably predict the responses occurring within a different segment of the soundtrack that included dialogue (Bitterman et al., 2008).

6.3.2 Multidimensional Representation

Neuronal sensitivity to the spectral components of phonemes is only one of several mechanisms that shape the representation of speech. Many phonemic distinctions cannot be made by discriminating the spectral composition of the speech sounds. Overall, the diverse sensitivities of A1 neurons are capable of capturing the essential features of all spoken phonemes. In this manner, dynamically changing ensembles of activated A1 neurons can uniquely represent the acoustic characteristics of each phoneme as they sequentially occur during ongoing streams of speech. These discrete ensembles would, in turn, transmit this ever changing pattern of activity to noncore areas of auditory cortex for further processing and characterization.

This principle is exemplified by the acoustic consequences resulting from differences in the manner of articulation for /ba/ and /wa/. The primary difference between the stop (/b/) and the glide (/w/) lies in the duration of the formant transitions, with those of /b/ being rapid (<50 ms), and those of /w/ being prolonged (e.g., Dent et al., 1997). A1 neurons can be characterized by their sensitivity to the rate of spectral change (Mesgarani et al., 2008). Those neurons with higher best rates were preferentially activated by /ba/, while /wa/ evoked more pronounced firing in neurons that responded best to slow spectral modulation rates. The authors were also able to demonstrate differentiation in speech-evoked responses based on whether neurons responded preferentially to broad (e.g., /d/) or narrow spectrum (e.g., "sh") sounds. Finally, some A1 neurons are selective for the direction of FM sounds as they sweep across a cell's excitatory bandwidth (Godey et al., 2005; Atencio et al., 2007). In principle, these FM-sensitive cells would be better activated by formant transitions moving in either an upward or downward direction, and enhance discrimination of phonemes with transitions that occur in characteristic patterns (e.g., /b/, /g/, /d/).

6.3.3 *Temporal Processing*

6.3.3.1 Windows of Integration

What is the duration of neural activity in A1 required to represent human speech accurately? This question is not trivial, as in the past, most studies would integrate the number of spikes occurring across the entire duration of a sound to estimate the specificity in the representation of one stimulus over another. Further, certain "preferred" stimuli evoke more sustained activity, whereas other "less preferred" stimuli evoke either a more transient response immediately after stimulus onset or no response at all (Wang et al., 2005). Thus, it is not immediately clear what physiological window of temporal integration best allows accurate phonemic discrimination. Further, it is unclear whether an optimal physiological temporal window of integration is parsimonious with the perceptual finding that windows of integration for speech can be as short as 20–50 ms.

Short latency responses after stimulus onset are able to discriminate phonemes and other complex vocalizations using temporal windows of integration compatible with those perceptually observed in humans. Optimal single-cell discrimination of marmoset vocalizations occurs over a temporal integration window of 10–50 ms (Wang et al., 1995; Schnupp et al., 2006). A computer algorithm classifier is able to discriminate 20 consonants presented in syllable-initial position using comparable integration windows when provided with onset responses recorded from rat A1 (Engineer et al., 2008).

Integrating tonotopic specificity with short windows of integration serves as a powerful example of how physiology can evaluate specific perceptual hypotheses attempting to define mechanisms underlying phonemic perception. An important illustration is based on the proposal that the short-term spectrum within the first 20 ms after consonant release is a major determinant for the differential perception of stop consonants (Stevens & Blumstein, 1978; Blumstein & Stevens, 1979, 1980): diffuse spectra maximal at lower and higher frequencies promote the perception of /b/ and /d/, respectively, while a more compact spectrum at intermediate frequencies support the perception of /g/. While other speech attributes such as F2 onset frequency and the direction of formant transitions are also key features in the perception of stops, onset spectra remains an important cue for their perception, especially during conversational speech occurring in noisy environments and in the hearing impaired (see Alexander & Kluender, 2008, 2009).

A1 physiology supports the fundamental hypothesis that onset spectra sampled within a short window of integration accurately discriminates stop consonants (Steinschneider et al., 1995; Steinschneider & Fishman, 2011). A set of the three syllables /ba/, /ga/, and /da/ were constructed that characterized the spectra at syllable onset described in the preceding paragraph (Fig. 6.2a). Figure 6.2b illustrates the spectral tuning of three representative sites in monkey A1 within a 20-ms window of integration beginning at response onset. BFs ranged from low (left), middle (center), and high (right) frequencies. Responses evoked by the syllables at the

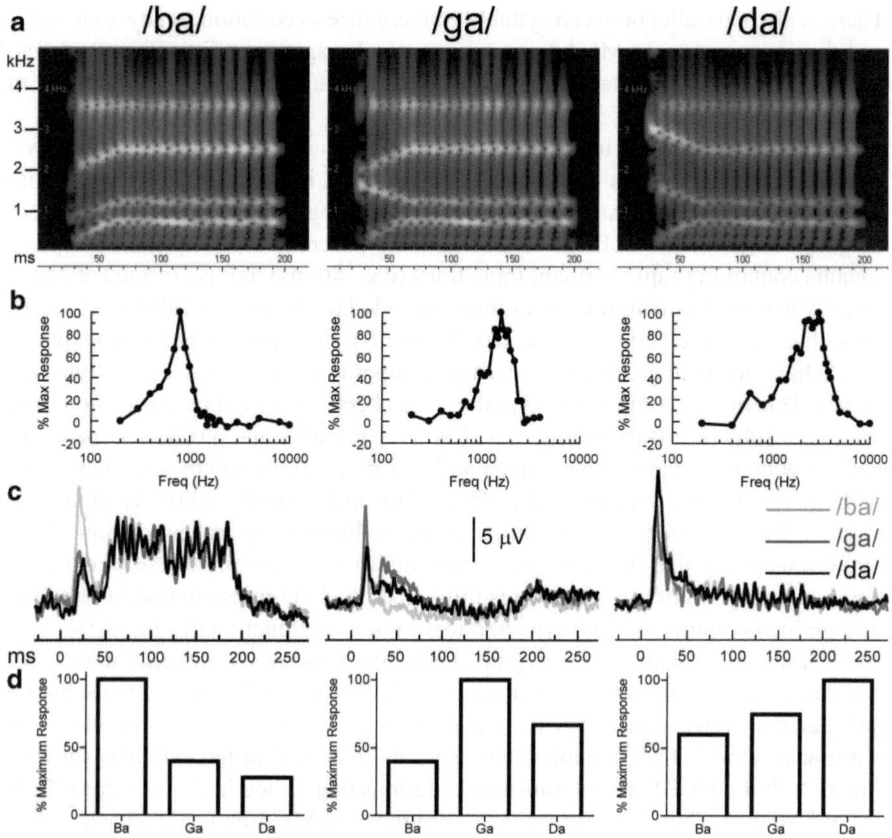

Fig. 6.2 Short windows of temporal integration are required to reliably differentiate rapidly changing phonemic structures. (**a**) Sound spectrographs and temporal waveforms of three highly stylized synthetic syllables varying in their consonant place of articulation. Sounds were constructed to have a high degree of overlap in their formant configuration. Spectral content at consonant onset is maximal at 800 Hz for /ba/, 1600 Hz for /ga/, and 3000 Hz for /da/. (**b**) Tuning curves at three representative recording sites with BFs of 800, 1600, and 3000 Hz, respectively. (**c**) MUA recording from the same three sites as in **b**. Activity is most dissimilar near stimulus onset, and then converges to a common pattern as the syllables become more similar to each other. (**d**) Amplitude of MUA evoked by each of the three syllables as recorded within the first 20 ms of the responses. Relative amplitudes are reliably predicted by the spectral tuning at each site. Thus, at the site with a BF of 800 Hz (left column), responses are largest to /ba/, while at the sites with BFs of 1600 Hz and 3000 Hz, responses are largest to /ga/ and /da/, respectively. Response differences related to consonant place of articulation would be obscured if a long temporal window of integration was used to calculate differences in activity across the three syllables

same sites shown in Fig. 6.2b are shown in Fig. 6.2c. Responses evoked by the three different syllables are depicted in grey-scale, with MUA elicited by /da/, /ga/, and /ba/ depicted as the darkest to lightest waveforms, respectively. Relative amplitudes of MUA within a 20-ms window of integration evoked by the three syllables were reliably predicted by the relative amplitudes of responses to pure tones (Fig. 6.2d).

There is also a parallel between syllable convergence to common steady-state values and the convergence of MUA to common neural responses (Fig. 6.2c). Prolonged temporal windows of integration would have obscured the differentiation of the initial stop consonants.

Expanded windows of integration may have profound ramifications for the development of normal speech processing. This proposal is exemplified by the seminal finding of Tallal and colleagues, who found that a subgroup of children with specific language impairment (SLI) had difficulty in the differential perception of stop consonants containing rapid formant transitions (e.g., 40 ms), but performed at control levels when the formant transitions were extended to 80 ms (see Tallal et al., 1993). This observation helped lead to the development of the temporal processing hypothesis, which posits that deficits in auditory temporal processing are causally related to SLI. This highly contentious hypothesis has been supported by replication of the original results (see Burlingame et al., 2005 for review), and has been expanded to include deficits in many nonlinguistic auditory perceptions and in some individuals with dyslexia (e.g., Wright et al., 1997; Hari and Renvall, 2001; Vandermosten et al., 2010). A recent interpretation of the auditory temporal processing deficit hypothesis posits that the underlying abnormality is based on the use of expanded temporal windows of integration, resulting in degraded representation of phonemic fine-grained patterns and subsequent phonological dysfunction (Tallal, 2004).

Neural activity within A1 is degraded at longer windows of integration in a manner that parallels the perceptual findings of Tallal and colleagues. Steinschneider and Fishman (2011) recorded from A1 of awake monkeys while presenting the consonant–vowel (CV) syllables /ba/, /ga/, /da/ identical to the syllables shown in Fig. 6.2a, but with F2 and F3 transition durations that varied in 20-ms steps (20, 40, 60, and 80 ms). They ranked response amplitudes of MUA evoked by each syllable based on the relative amplitudes of the responses evoked by 800-, 1600-, and 3000-Hz tones. These tones were centered at the maxima of the initial consonant spectrum preceding the formant transitions. Comparisons were made at multiple physiological windows of integration ranging from 10–30 ms after stimulus onset to 10–110 ms in 20-ms expansion increments. When short windows of integration were used (e.g., 20 ms), there was accurate concordance between the tone-evoked and the syllable-evoked activity for all four sets of formant transition duration stimuli. However, as windows of integration were progressively increased, there was a corresponding decrease in the concordance between the tone-evoked and speech-evoked activity for the syllables with rapid formant transitions. By an integration window of 90 ms, concordance was observed only for the syllables with the longest formant transitions.

At short windows of integration, syllables containing both rapid and prolonged formant transitions can be discriminated accurately. However, if an individual "chunks" time in abnormally long integration windows, then only syllables with extended formant transition durations will be discriminated accurately. Expanded windows of integration are also compatible with other proposed acoustically based abnormalities in dyslexia (see Goswami et al., 2011) or those suggesting that attention mechanisms are dysfunctional based on "sluggish" attentional switching

(Hari & Renvall, 2001). Further, it is possible that remediation therapies for SLI are effective in part by sharpening abnormally long integration windows (e.g., Tallal, 2004; Gillam et al., 2008).

Finally, the current discussion has focused on A1, where short windows of integration (e.g., 20 ms) are the norm. The other principal region of core auditory cortex, area R, has been shown to have significantly longer integration windows (e.g., 100 ms) (Scott et al., 2011). These parallel windows may promote the envisioned dual time scales thought important for phonemic perception (Poeppel et al., 2008).

6.3.3.2 Temporal Envelope Processing

At the level of A1, temporal encoding of slowly modulated rates typical of sentences is a general processing rule, and is observed for both acoustic stimuli (e.g., Wang et al., 1995, 2008; Kayser et al., 2009) and electrical stimuli generated by cochlear implants (e.g., Middlebrooks, 2008). In humans, temporal encoding of the slow oscillations embedded within complete sentences is observed in both AEPs and in theta band activity recorded using MEG (e.g., Ahissar et al., 2001; Luo & Poeppel, 2007; Nourski et al., 2009). The importance of this temporal encoding is emphasized by its disappearance as sentences became increasingly time compressed and the speech became incomprehensible (Ahissar et al., 2001; Nourski et al., 2009).

The fidelity of temporal encoding in human auditory cortex is illustrated in Fig. 6.3. Trajectories of two intracranial electrodes traversing Heschl's gyrus are shown in Fig. 6.3a. Fig. 6.3b depicts the synthetic syllables /da/ (VOT = 5 ms) and /ta/ (VOT = 40 ms) that were presented to the subject. AEPs (left column) and event-related band power (ERBP, right column) recorded from electrode L3 are shown in Fig. 6.3c. ERBP is a powerful method to examine neural activity occurring at higher frequencies often lost in the AEP. AEPs faithfully represent both stimulus VOT and F_0 in the temporal response patterns. After an initial response evoked by stimulus onset that is similar for both syllables, only /ta/ evokes a second response burst time-locked to the delayed VOT (red arrowhead). Superimposed oscillations are time locked to the F_0. This real-time following of the F_0 is more clearly seen in the ERBP, which contains a strong band of high γ activity whose center frequency parallels that of the stimulus F_0. The delayed VOT for /ta/ is seen as a relative diminution of high γ activity in the time period between consonant release and voicing onset (black arrowhead). The phase-locked nature of the high γ paralleling the stimulus F_0 is supported by estimating the non–phase-locked portion of the ERBP, determined by subtracting the AEP from individual trials and then calculating the residual time-frequency profiles (Crone et al., 2001; Steinschneider et al., 2008). This residual is depicted in the left-hand column of Fig. 6.3e. High γ activity no longer follows the F_0.

A markedly different temporal pattern is observed at a more anterolateral site (Fig. 6.3d). AEPs are of longer latency and no longer manifest an oscillatory response phase-locked to the F_0. A time-locked response to the expanded VOT for /ta/, however, is still generated (red arrowhead). ERBPs are also delayed relative

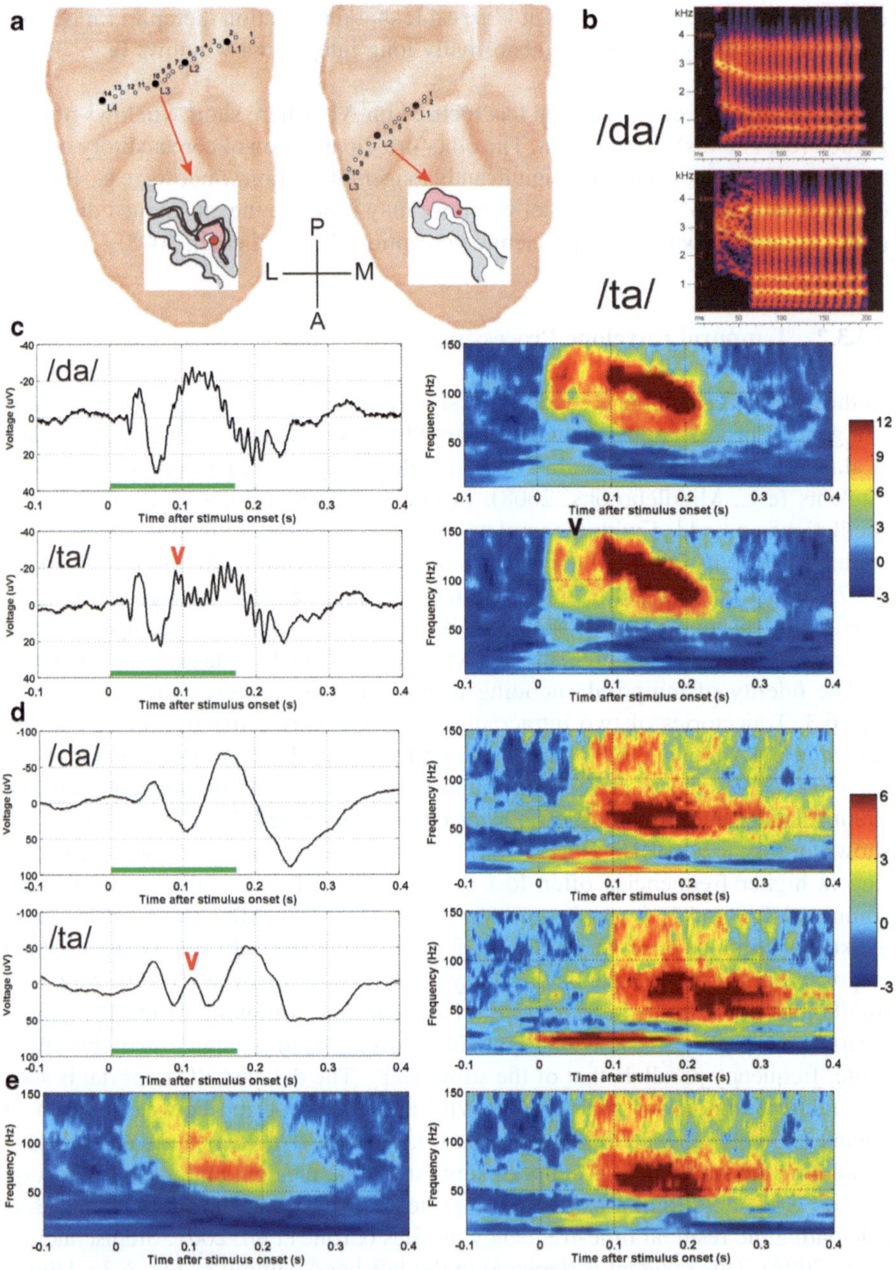

Fig. 6.3 Representation of speech in human auditory cortex shows high temporal fidelity. (**a**) Trajectories of two intracranial electrodes traversing Heschl's gyrus in a single subject. Larger filled circles represent locations of low-impedance electrode contacts; smaller unfilled circles represent locations of high-impedance contacts. Coronal reconstructions reveal the positions of two low-impedance contacts as within the gray matter (insets). Contact L3 is more posterior than contact L2.

to those seen in Fig. 6.3c, and there is also a delay in the onset of ERBP for /ta/ relative to /da/ that mirrors the VOT. In contrast to the previous example, non–phase-locked ERBP changes at this site elicited by /da/ (Fig. 6.3e, right column) appear nearly identical to the total power ERBP shown in Fig. 6.3d, indicating its non–phase-locked nature. This transformation of activity across the two sites illustrates that while different loci in auditory cortex can have markedly different response patterns, each site provided temporally precise information relevant for phonemic identification.

6.3.4 Context Dependence

Context dependence is an additional cornerstone of phonemic processing. This critical phenomenon will be emphasized by citing three classic examples that demonstrate how context dependence can be explained by basic principles of auditory cortex physiology that do not require the invoking of higher-order perceptual processes. Language-specific context dependence, such as observed when infants progressively become attuned to the sound patterns specific for a native language at the expense of non-native patterns, is another critical feature of phoneme perception, but is not discussed further (see Werker & Yeung, 2005; Kuhl, 2010).

Preceding phonemes, or for that matter, nonlinguistic sounds, alter the perception of subsequent phonemes. For instance, categorical boundaries shift such that /g/ is perceived more readily when preceded by the syllable /al/, while /d/ is more consistently perceived when preceded by the syllable /ar/ (Mann, 1980; Lotto & Kluender, 1998). A schematic of the formant structure that has been often used to demonstrate this context dependence is shown in Fig. 6.4. Steady-state formants of the vowel /a/ are succeeded by F2 and F3 transitions that typify /l/ (falling F2, rising F3) and /r/ (rising F2, falling F3). The following /ga/ and /da/ differ in the starting F3 position and in the direction of the subsequent transition converging on the steady-state frequency of F3. Synthetic /ga/ and /da/ syllables are generated with both typical and

---

Fig. 6.3 (continued) (**b**) Sound spectrographs of /da/ and /ta/ that were presented to the subject. (**c**) Auditory evoked potentials (AEPs) and event-related band power (ERBP) recorded from site L3. Responses elicited by /da/ are above those evoked by /ta/. Green bars under the AEPs depict the durations of the syllables. AEPs and ERBP represent the VOT and F_0 of the syllables with high fidelity. The red arrowhead denotes an AEP component evoked by the delayed VOT for /ta/, while the black arrowhead depicts decreased power in the interval between syllable onset and voicing onset for the same syllable when compared to the response elicited by /da/. (**d**) AEPs and ERBP elicited at site L2. Responses at this more anterior site have longer latencies than those recorded at L3. An AEP component is still evoked by the delayed VOT for /ta/ (red arrowhead), and the ERBP shows a delayed onset approximating the prolonged VOT when compared to the ERBP elicited by /da/. (**e**) Non–phase-locked ERBP elicited by /da/ at the two sites. The response recorded from L3 is shown in the left column. Much of the power at this site was phase-locked, as determined by the decreased response when compared to the response shown in **c**. This contrasts with the activity at L2. See text for details

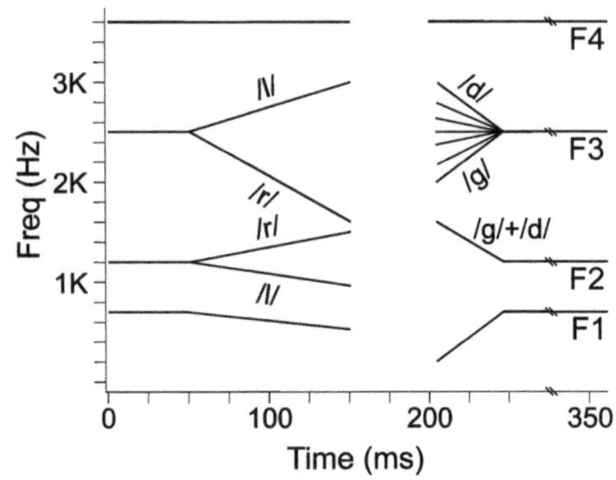

Fig. 6.4 Schematic of the formant structure for the syllables /al/ and /ar/ preceding those for /da/ and /ga/. These syllable sequences demonstrate how phonemic perception (e.g., /d/ vs. /g/) is context dependent and varies as a function of the preceding phonemic environment. See text for details

intermediate F3 starting frequencies. Intermediate F3s, that when presented without the preceding vowel-consonant (VC) lead to the perception of /d/, elicit shifts to the perception of /g/ when the preceding VC is /al/. Reverse effects occur when the preceding VC is /ar/. Animals demonstrate these context effects, implicating basic auditory system mechanisms, instead of previously suggested higher-order constructs invoking articulatory models of speech perception (Lotto et al., 1997). An auditory processing based model is further supported by the finding that both FM tone glides mirroring the trajectories of the changing F3 for /l/ and /r/, or steady-state tones mirroring the termination frequencies of the glides, produce similar perceptual results (Lotto & Kluender, 1998). These observations have been extended to paradigms where /ga/ and /da/ are preceded by a series of tones with variable frequency, but with means that centered either at higher frequencies, promoting the perception of /g/, or at lower frequencies, promoting the perception of /d/ (Holt, 2006b).

As discussed by Holt (2006b), this form of perceptual modulation can be explained by invoking stimulus specific adaptation (SSA). In the simplest paradigm, a tone centered within a neuron's excitatory spectral receptive field will inhibit responses to subsequent tones of similar frequency (Brosch & Schreiner, 1997). Conversely, when the first tone presented is outside the center region of the excitatory receptive field, the response to a second tone centered in the excitatory receptive field can be enhanced (Brosch & Schreiner, 2000). In more complex paradigms that mirror those used to elicit the human event-related mismatch negativity response, a novel or deviant tone interspersed among repeating standard tones will evoke an enhanced response, even when both tones lie within the excitatory receptive field of a cell (Ulanovsky et al., 2003). This observation was extended to show that the long term spectral history of the acoustic environment modulated the response to subsequent tones (Ulanovsky et al., 2004).

SSA can readily explain the perceptual result of why /d/ is more likely heard when preceded by /ar/, and /g/ more likely when preceded by /al/. When /al/ is presented, neurons usually more responsive to /d/ (i.e., higher frequencies) are adapted while those usually responsive to /g/ do not show this adaptation, and instead show enhanced responses (e.g., as a deviant stimulus would). This shift in response strengths would bias activity toward the mid-frequency range, which, as we have shown, would promote the perception of /g/ (e.g., Steinschneider & Fishman, 2011). A reverse situation would occur if /ar/ was the lead sound, and the high frequency range was accentuated at the expense of the mid-range. The findings by Ulanovsky and colleagues (2004) that the long-term spectral history affects responses to subsequent sounds explains why in the experiment of Holt (2006b), the statistical mean of the spectral environment biased responses toward either /da/ (lower frequency history) of /ga/ (higher-frequency history). Context dependence of phonemic categorization can thus be based on low-level rules of A1 physiology determined by the preceding spectral environment.

Context dependence is a likely mechanism for promoting categorical perception along the VOT dimension. Among the world's languages, differences along the VOT dimension are commonly used to distinguish one phoneme from another. In American English, short positive VOTs (consonant release precedes voicing) promote the perception of /b/, /d/, and /g/ (voiced stops), while long positive VOTs promote the perception of /p/, /t/, and /k/ (unvoiced stops) (see Figs. 6.1 and 6.3 for example). The change from, for instance, /d/ to /t/, is categorical, such that small changes in VOT generally do not produce a change in the phonemic percept. However, there is a temporal region of VOT, usually between +20 and +40 ms, where the same small change will evoke a pronounced shift in perception from one phoneme to another. This nonlinear phenomenon exemplifies categorical perception, one of the most well studied aspects of speech perception. Categorical perception of consonants varying along the VOT dimension is observed in animals with boundaries similar to those seen in humans (e.g., Kuhl, 1986; Sinnott & Adams, 1987), indicating that VOT categorical perception partly relies on basic auditory system mechanisms.

Categorical differentiation between voiced from unvoiced stop consonants appears to result from early synaptic events after stimulus onset (Steinschneider et al., 2003). This sequence is illustrated in Fig. 6.5, which depicts short-latency current source density (CSD) profiles recorded from two sites in A1 (center columns) that were evoked by the syllables /da/ and /ta/ with VOTs of 0, 20, 40, and 60 ms (rows). MUA recorded from cortical depths marked by the horizontal lines in the CSD profiles are shown adjacent to the contour plots. These depths correspond to lower lamina 3, as determined by their co-location with the initial current sink in A1 generated by inputs from thalamocortical fibers (Steinschneider et al., 1992; Metherate & Cruikshank, 1999). The initial current sink, which represents net depolarization within the region, initiates large-amplitude responses in the MUA. Following the initial sink is a large amplitude current source at the same recording depth. This source is associated with co-located decreases in the MUA, indicating that the source is a marker for net hyperpolarization. The initial current sink and following current source associated with increases and decreases in MUA are seen in the responses to all syllables. However, syllables with a 40- or 60-ms VOT (i.e., those that lead to perception of /ta/) evoke an additional current sink in lower lamina 3, indicating a new volley of depolarization, and with it, large-amplitude, time-locked MUA responses to voicing (arrows). In contrast, MUA evoked by the syllable with a 20-ms VOT fails to elicit this second time-locked response. Instead, the MUA

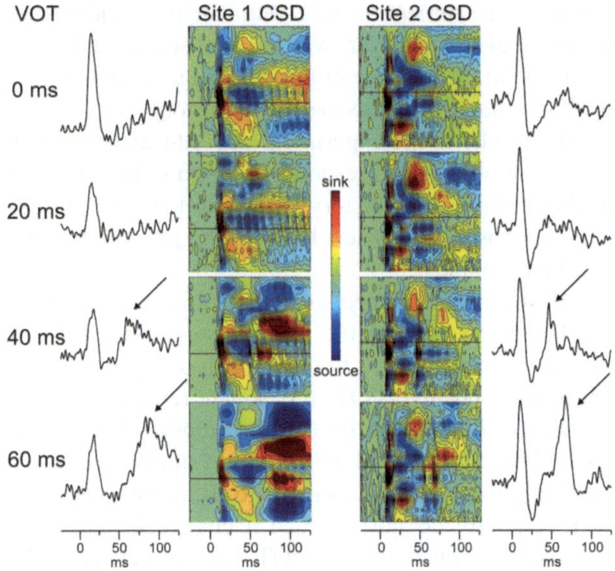

Fig. 6.5 Categorical responses for the VOT phonetic parameter occur in A1 of experimental animals. Middle columns depict contour plots of current source density (CSD) recorded simultaneously across the laminar depths using linear-array multicontact electrodes. Horizontal lines denote the recording depths where the MUA activity, shown in the lateral columns, was recorded. Rows denote the VOT of the syllables that varied between /da/ and /ta/. See text for details

pattern parallels that evoked by the syllable with a 0 ms VOT (0 and 20 ms VOT perceived as /da/). For the syllable with the 20-ms VOT, the response to voicing onset falls near the peak of the hyperpolarization, and is suppressed. Longer VOTs occur after this hyperpolarization, and are able to overcome this transient neuronal suppression and drive a second response. In total, the brain response pattern of a "single" onset (promote perception of /d/) versus a "double" onset (promote perception of /t/) response profile parallels the timing of categorical perception for the voiced versus unvoiced distinction of stop consonants (for additional details see Steinschneider et al., 2003; Sakai et al., 2009).

While other mechanisms could be invoked to explain the differential perception of voiced from unvoiced stop consonants (e.g., Mesgarani et al., 2008), the temporal parameter of VOT has been shown to be the principal determinant of this phonetic perception (Lisker and Abramson, 1964; Shannon et al., 1995; Faulkner and Rosen, 1999). Auditory cortex in humans exhibit similar temporal categorical boundaries in subjects classifying sounds as either /da/ or /ta/ (Steinschneider et al., 1999, 2005). Syllables generating a "single on" or "double on" response pattern could be rapidly and simply placed into a phonemic category based on "gross" temporal response patterns, whereas more subtle timing discriminations would be required for differentiating syllables residing on the same side of the perceptual boundary (e.g., Kewley-Port et al., 1988). This temporal processing scheme is consistent with the

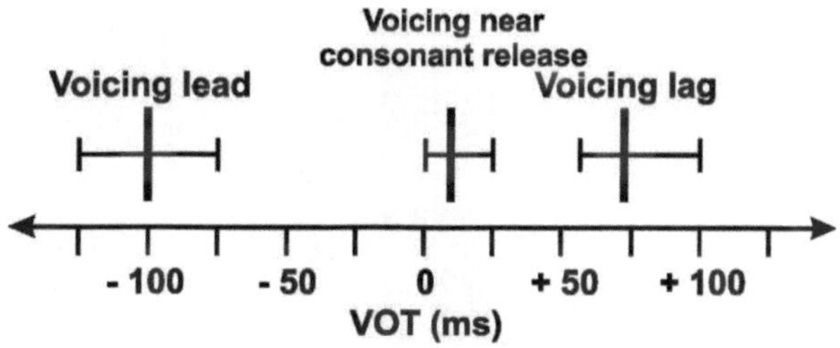

Fig. 6.6 Mean and range of VOT values for 11 languages. VOT for many languages of the world segregate into a trimodal distribution with temporal regions that are scrupulously avoided. (Adapted from Lisker & Abramson, 1964)

hypothesis that phonemic encoding is the product of a warped representation of acoustical properties, wherein some differences located along a specific portion of a physical continuum are readily distinguished from one another (e.g., VOT of 20 vs. 40 ms), while similar magnitude differences at other locations along the continuum (e.g., VOT of 40 vs. 60 or 0 vs. 20 ms) are difficult to discriminate (Stevens, 1981).

The idea of a "warped" representation of temporal space is consistent with the nonoverlapping trimodal distribution of VOTs used by many of the world's languages for stop consonants in syllable initial position (Fig. 6.6) (Lisker & Abramson, 1964; see also Cho & Ladefoged, 1999). VOTs are either strongly negative (voicing leads consonant release), near 0 ms, or strongly positive (voicing lags consonant release). The physiological boundary of between +20 and +40 ms in both monkeys and humans lies squarely in a temporal region of acoustic space avoided by world languages. Ambiguity in temporal brain patterns between a "single on" and a "double on" response that would result if VOT values fell within this region of acoustic space serves as a reasonable explanation as to why this temporal segment is avoided. While the presented physiological studies have focused on positive VOT values, similar results have been reported for negative VOT values used, for instance, in the French language (Liégeois-Chauvel et al., 1999; Trébuchon-Da Fonseca et al., 2005). In this case, voiced stops (e.g., /b/) begin with a low-amplitude, low-frequency murmur before consonant release while the unvoiced stops (e.g., /p/) have a short VOT without aspiration. As this difference between English and French exemplifies, temporal patterns alone cannot differentiate between voicing lead or lag. Both would generate "double on" response patterns. The addition of spectral cues, such as the presence of low-frequency energy before consonant release, or high-frequency aspiration noise after consonant release, could facilitate this discrimination. Further, given that in the two exemplifying languages, a voiced or unvoiced stop could display either temporal pattern indicates that the phonemic categorization must occur downstream of primary cortex.

Instead of being a speech-specific temporal boundary, the +20 to +40 ms VOT interval represents a general psychoacoustic boundary that is utilized by some languages (Keating et al., 1981). Infants can discriminate both speech and nonspeech sounds varying across this interval, even when it is not phonemically relevant for the infant's native language (e.g., Eilers et al., 1979; Jusczyk et al., 1989). It has been proposed that VOT perception is partly based on the ability to determine whether consonant release and voicing onset are sequential in time (Pisoni, 1977; see also

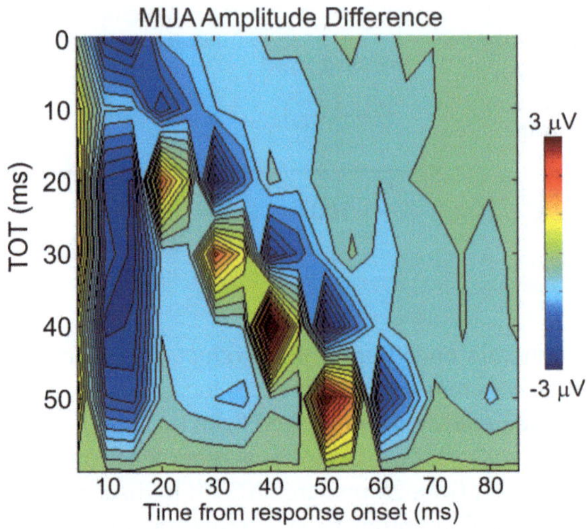

Fig. 6.7 Contour plot of MUA recorded from monkey A1 in response to two temporally overlapping tones in a paradigm that paralleled the psychoacoustic study of Pisoni (1977). Onsets of the two tones were separated by 0–50 ms in 10-ms increments. Differences in MUA are shown between responses occurring at a given time with that occurring 10 ms earlier. Only tones separated by 20 ms or more evoke a second response time locked to the delayed sound component. This pattern mirrors psychoacoustic results. (Adapted from Steinschneider et al., 2005)

Hirsh, 1959; Jusczyk et al., 1989). Pisoni (1977) demonstrated that there was a categorical tone onset time (TOT) boundary of about 20 ms when subjects were asked to judge whether a 1500-Hz tone preceding a 500-Hz tone were either simultaneous or sequential events. A physiological basis for this perceptual observation was provided by Steinschneider and colleagues (2005). MUA was recorded in monkey A1 while presenting two temporally overlapping tones separated in onset by 0–50 ms, thus mirroring the paradigm of Pisoni (1977). They found significant increases of activity to the second tone only when the TOT was 20 ms or greater (Fig. 6.7). TOTs of 0 and 10 ms evoked only an isolated response. Thus, a "single on" versus a "double on" temporal response pattern serves as a neurophysiological cue for whether two sounds occur simultaneously or sequentially and, in certain languages, promotes phonemic perception along the VOT continuum.

To this point, discussion has focused on how a preceding spectral or temporal history modulates phonemic perception. A third highly important context dependence is termed trading relations. Trading relations means that a phonemic perception generated by one acoustic attribute (e.g., VOT) can be changed by modulating another acoustic attribute (e.g., spectrum). One of the most studied examples of this principle is the shift in VOT boundaries with a shift in place of articulation. Shortest VOT boundaries are observed for differentiating /b/ from /p/ (approximately 20 ms), longest for differentiating /g/ from /k/ (approximately 40 ms), and intermediate for /d/ and /t/ (approximately 30 ms) (Lisker & Abramson, 1967). Similar boundary shifts have been observed in chinchillas, indicating that this trading relation has a nonlinguistic basis (Kuhl & Miller, 1978). The key acoustic attribute is the F1 frequency at voicing onset: it is highest for /b/ and /p/ with the shortest boundary, lowest for /g/ and /k/ with the longest boundary, and intermediate for /d/ and /t/ (Summerfield & Haggard, 1977; Soli, 1983). The acoustic basis for this phenomenon is highlighted by Japanese quail (*Coturnix coturnix japonica*), which are more likely to identify a stop consonant as unvoiced when there is a higher frequency F1 (Kluender, 1991; Kluender & Lotto, 1994). Thus, in this third example of context dependence, spectral attributes modulate temporal attributes.

A physiological correlate of this effect has been demonstrated (Steinschneider et al., 2005). Intracranial AEPs were acquired from Heschl's gyrus in response to stop CV syllables varying in both their F1 frequency and VOT. The subject placed the VOT perceptual boundary for the lowest F1 value (424 Hz) at between 40 and 60 ms, whereas higher F1 values (600 and 848 Hz) led to a boundary of 20–25 ms. When the F1 was at its lowest value, AEPs showed a "double on" pattern only for the syllable with a 60-ms VOT and not for the syllables with a shorter VOT. In contrast, a "double on" pattern was observed down to VOT values of 20–25 ms when the F1 was higher. Thus, there was a physiological/perceptual trading relationship in that the spectral attribute of F1 frequency modulated the temporal attribute of VOT encoding.

Further insight into this relationship is guided by the work of Parker (1988). She observed that as frequency differences between two temporally overlapping tones increased, the likelihood of judging that the tones occurred simultaneously increased as well. This result parallels that of the trading relations between F1 frequency and VOT perception. Physiological support for this work was provided by demonstrating that as two tones increase in spectral separation, a larger tone onset time is required to generate a "double on" response pattern (Steinschneider et al., 2005). In summary, trading relations for complex spectral and temporal interactions modulating phonemic perception can be clarified by basic features of A1 physiology.

6.3.5 Relationship Between Physiology and Perception

Although fundamental rules of A1 physiology are helpful for identifying mechanisms modulating phonemic perception, the resultant activity is inadequate to elucidate fully the means by which speech sounds are categorized. Several studies emphasize this point. One study compared the morphology of the N1 component of the scalp-recorded AEP between Hindi and English speakers in response to stop CV syllables varying in their VOT from –90 to 0 ms (Sharma & Dorman, 2000). Progressive increases in latency of the N1 component of the AEP paralleled the VOT as the syllables became more prevoiced. The morphology of N1 did not differ between the two groups, even though the Hindi speakers heard prolonged prevoiced syllables as /ba/ and the short negative VOT syllables as /pa/ whereas the English speakers heard all syllables as /ba/. Physiology followed the physical attributes, not the phonemic categorization.

Dissociation between physiology and perception is emphasized further by results obtained from intracranial recordings within Heschl's gyrus. Two studies examined high γ activity present in the ECoG in response to time-compressed speech (Nourski et al., 2009; Mukamel et al., 2010). The high γ activity phase-locked to the speech envelopes. However, phase-locking was present at all rates of compression, and occurred even when the speech was incomprehensible. Thus, temporal following in primary auditory cortex exceeded perceptual capacities. Further, firing rates of neurons were invariant across all time compressions, indicating that rate codes in core auditory cortex do not parallel perceptual encoding and that processing at the primary cortical level is not the rate-limiting step in speech perception (Mukamel et al., 2010).

6.4 Phonemic Representation and Categorization in Non-A1

6.4.1 General Principles

A1 represents the temporal and spectral details of speech with high fidelity. Phonemic categorization, however, entails a more complex set of processing demands. For instance, the same vowel can have different speaker-specific acoustic cues (e.g., Hillenbrand et al., 1995). Cues differ based upon the surrounding phonetic environment, exemplified by differences seen when a consonant is in syllable-initial, -medial, or -final position (e.g., Sussman et al., 1997), or simply when it is followed by a different vowel (e.g., Kewley-Port, 1982). There are developmental differences in the weights given to different acoustic cues when differentiating one phoneme from another (e.g., Mayo & Turk, 2004), as well as differences seen in normal hearing and hearing-impaired listeners (e.g., Hedrick & Younger, 2007).

These problems are beyond the capacity of A1 to resolve, and instead, require the activation of a large network of interconnected brain regions (e.g., Nelken, 2008). The distributed nature of this organization is exemplified by the work of Formisano and colleagues (2008). They examined fMRI patterns in response to three vowels spoken by three different speakers. Vowel discrimination utilized a widely distributed brain activation pattern bilaterally encompassing anterior-lateral portions of Heschl's gyrus and sulcus and much of the PLST and STS, with additional contributions from the left planum temporale (PT). The goal of this section is to highlight the putative role each region plays in phonemic categorization.

6.4.2 Posterior–Lateral Superior Temporal Gyrus

Posterior–lateral superior temporal gyrus (PLST) is the most thoroughly studied of all noncore areas with respect to speech processing. Nonhuman primate models define the organization of this region as partly based on convergence of inputs from lower level core regions (e.g., Rauschecker & Scott, 2009). This convergence allows more complex acoustic attributes to be encoded by a combination sensitivity derived from these inputs encoding more basic acoustic features. A straightforward example is the increased sensitivity of cells in monkey PLST to bandpass noise over pure tones (Rauschecker & Tian, 2004). This combination sensitivity is thought to explain why PLST neurons, especially those in the anterior lateral belt field termed AL, are highly responsive to conspecific vocalizations and show response specificity to vocalization type (e.g., Tian et al., 2001; Russ et al., 2008a).

Data from nonhuman primates also suggest PLST functions as an early stage in the phonemic categorization process (Tsunada et al., 2011). Single neurons in field AL were recorded while two macaque monkeys made a leftward eye saccade when two sequential speech sounds (/bad/ and /dad/) were perceived as same, and a rightward eye saccade when the two syllables were perceived as different. Prototypic syllables were systematically altered so that each took on the acoustic characteristics of the other. A behavioral categorical boundary at the 50% midpoint stimulus was established. Neuronal activity evoked by the second of the two syllables was consistently greater when the second stimulus straddled the categorical boundary, despite being compared to activity of equal acoustic change but which remained on the same side of the perceptual boundary.

Human electrophysiology also indicates that PLST is an early stage in phonemic categorization. Parametric modification of the starting frequency of F2 in 14 equal steps led to categorical perceptual changes that ranged from /ba/, to /da/, to /ga/ (Chang et al., 2010). These perceptual boundaries were compared with the amplitude and distribution of AEPs acquired from subdural grid electrodes placed over PLST while the subjects passively listened to the syllables. In the time range of 110–150 ms, responses to each of the 14 stimuli spatially clustered across the electrode grids in a "categorical-like" manner that paralleled the perceptual categorization. At a single-trial level, field potential distributions could accurately predict the perceptual category of the stop consonants, indicating that responses were not simply based on acoustic differences produced by the 14 stepwise changes used to generate the syllables. Although there was much overlap in the overall activation of PLST, categorization was a function of spatially discrete, local selectivity in the responses. Multidimensional scaling was used to identify the two major factors responsible for the spatially discrete response selectivity. Both factors were based on acoustic attributes. The first factor was the onset frequency of F2, while the second factor was the difference between the F2 onset frequency and the vowel F2 frequency. This multidimensional scaling used to classify the syllables is highly reminiscent of that proposed by Mesgarani and colleagues (2008) for differentiating phonemes at the level of A1 (see Section 6.3.2). Thus, one role of PLST may be to integrate spectrotemporal cues identified by neurons in core auditory cortex and integrate them into spatially and functionally differentiated maps.

In a related study, differences in the amplitude of high γ activity recorded in the 100–150 ms time frame and elicited by the syllables /ba/, /ga/, and /da/ could predict differences in high γ elicited by their unvoiced counterparts /pa/, /ka/, and /ta/ (Steinschneider et al., 2011). Despite variation in VOT and concomitant acoustic differences, responses in PLST categorized the consonants according to their place

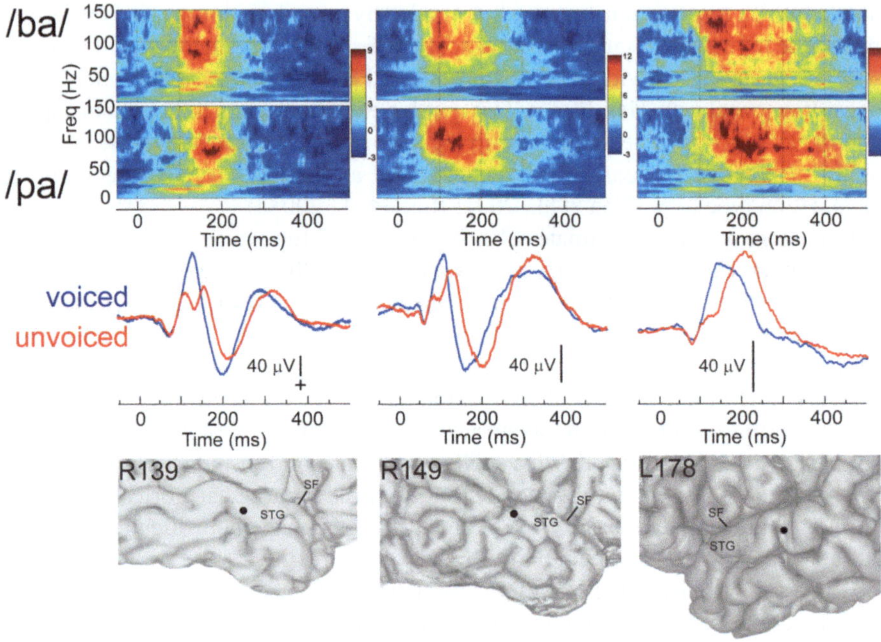

Fig. 6.8 VOT is represented by temporal response patterns in PLST. The top row depicts ERBP elicited by the syllables /ba/ and /pa/. There is either a delay in the high γ ERBP for /pa/ relative to /ba/, or enhanced activity at both syllable and voicing onset. The middle row depicts AEPs evoked by voiced syllables (/ba/, /ga/, and /da/) shown in blue, with the superimposed waveforms evoked by the AEPs evoked by unvoiced syllables (/pa/, /ka/, and /ta/) shown in red. There are time-locked differences in the AEPs reflecting the extended VOTs for the unvoiced syllables. The bottom row depicts the locations of the subdural recording sites on PLST for three subjects. STG: superior temporal gyrus. SF, sylvian fissure. (Adapted from Steinschneider et al., 2011)

of articulation. In parallel, consonants could also be categorized as being either voiced or unvoiced based on the amplitude of the high γ activity.

The relationship between increases in high γ activity and increases in cellular firing (e.g., Steinschneider et al., 2008) suggests that the categorization process involving consonants differing in their place of articulation or VOT are partly represented by a rate code of neural activity. VOT, however, is represented also by a temporal code similar to that seen in Heschl's gyrus; short VOT syllables have a "single on" response whereas syllables with a more prolonged VOT have a "double on" response (Steinschneider et al., 2011). This temporal patterning is exemplified in Fig. 6.8, which depicts ERBP elicited by the syllables /ba/ and /pa/, in the upper panels, AEPs evoked by the averaged responses to the voiced (blue) and unvoiced (red) CV syllables shown in the middle panels, and the anatomical locations of the recording sites in the lower panels. At all sites, AEP waveforms diverge based upon the temporally delayed VOT of the unvoiced syllables. Either a "double on" response is seen for the unvoiced syllable responses, (e.g., R139), or there is a trough in the rise and peak delay of the large amplitude negativity (e.g., L178). There is also either a delay in the high γ ERBP for /pa/ relative to /ba/, or enhanced activity at both syllable and voicing onset.

The aforementioned data indicate that PLST continues to track detailed spectral and temporal characteristics of speech sounds. fMRI data support the electrophysiological results. In one such representative study, naturally produced stop consonants /d/, /t/, /g/, and /k/ (i.e., vary along lines of place of articulation and VOT) placed in different vowel contrasts were presented both normally and as spectrally rotated sounds (Obleser et al., 2007). Rotated sounds are unintelligible, but have the same spectrotemporal complexity. Sites along PLST were activated by the spectral content occurring at consonant onset (e.g., responses to /d/ and /t/ greater than /g/ and /k/). At other sites, the voiced stops evoked greater activation than their unvoiced counterparts.

In one of the most conclusive fMRI investigations regarding the intermediate stage that PLST plays in phonemic categorization, subjects were presented with multiple exemplars of several classes of sounds, including human speech, and those produced by musical instruments, songbirds and other animals (Leaver & Rauschecker, 2010). Detailed analyses were performed on multiple acoustic attributes of all the stimuli, such as their spectral content, pitch strength, harmonics-to-noise-ratio, and temporal variability. When accounting for these and other variables, areas along PLST that appeared to be relatively selective for human speech were shown instead to be selective for acoustic attributes not specific to the presented syllables.

Despite this intermediate role, PLST is still crucial for speech perception. While undergoing electrocortical stimulation for evaluation of medically intractable epilepsy, subjects were required to respond in a same–different task when presented with CV syllable pairs (Boatman and Miglioretti, 2005). Syllable discrimination was disrupted with electrical stimulation of PLST. Concurrent deficits were also noted for discrimination of FM tones, but not for pure tones, nor for other tasks including picture naming and word reading. Further, in keeping with the known effects that visualizing articulatory mouth and lip movements have on auditory speech perception, it has been shown that visualizing speech production modulates PLST responses elicited by concurrently presented syllables (Reale et al., 2007; Besle et al., 2008).

6.4.3 *Planum Temporale*

A remarkable transformation has occurred with regard to the putative role of the planum temporale (PT) in speech and language processing. Anatomically, this region is generally larger in the language-dominant hemisphere (Geschwind & Levitsky, 1968; Galaburda & Sanides, 1980; Shapleske et al., 1999), concordant with the specialized role classic neurology places on this area for phonological processing (see Dronkers et al., 2000). However, modern anatomical studies fail to confirm significant deficits in speech perception with strokes restricted to this region (Dronkers et al., 2004). Anatomical studies are supported by multiple fMRI studies, which also fail to identify PT as being specifically involved with phonemic perception (e.g., Griffiths & Warren, 2002; Hickok, 2009). Clarification for the roles of the PT is complicated by it having multiple subdivisions which have yet to be classified (Shapleske et al., 1999). In one fMRI study, the percept of sound motion better activated medial portions of PT, whereas the percept of frequency modulation led to more activation within lateral segments (Hart et al., 2004). Limited electrophysiological studies have shown that nondominant PT responds with latencies only slightly later than those on medial Heschl's gyrus to speech and musical sounds in a nonspecific manner (Steinschneider et al., 1999; Fishman et al., 2001). Involvement of PT in multiple aspects of hearing have led to the hypothesis that PT serves as an early-stage sound processing "computational hub," directing inputs representing different acoustic attributes into diverse processing streams for further analysis (Griffiths & Warren, 2002; Kumar et al., 2007). In recent models, the PT on the dominant hemisphere is hypothesized to serve as an integral part of a bidirectional sensory–motor integration loop that involves the intraparietal sulcus, and premotor and motor regions to modulate articulatory representations and commands with acoustic inputs and feedback (e.g., Hickok, 2009; Rauschecker & Scott, 2009).

6.4.4 Superior Temporal Sulcus

Over a decade of research has implicated the superior temporal sulcus (STS) as the most likely candidate region specifically involved in phonemic categorization. Earlier work described (1) how the STS was preferentially activated by human vocalizations over other types of sounds and (2) defined acoustic characteristics that led to these enhanced responses. For instance, the STS was activated more strongly by words, pseudowords, and reversed speech than by FM tones or white noise (Binder et al., 2000). Passive listening to speech or nonspeech sounds such as laughs, cries, or coughs, produced greater bilateral activation than that elicited by diverse, energy-matched environmental stimuli such as animal vocalizations or mechanical sounds (Belin et al., 2000). Activation was also greater when compared to scrambled speech, generated by randomly intermixing the magnitude and phase of each spectral component while preserving the amplitude envelope of the original vocal signals (Belin et al., 2000). Findings were interpreted as indicating that frequency structure was more important than amplitude envelope in generating the enhanced responses. In support of the latter conclusion, Belin and colleagues (2000) also showed that activation was marked decreased when vocal stimuli were spectrally filtered so that either low- or high-frequency components of the sounds were removed. Overall, it appears that spectral complexity may be one factor in determining the degree of STS activation.

More recent work has highlighted a more direct role in phonemic categorization. Responses elicited by synthetic CV syllables whose formant values varied along a continuum from /ba/ to /da/ were compared with those elicited by nonphonemic sounds generated by spectrally inverting the onset frequencies of the formants and thus preserving the acoustic characteristics of the syllables while disrupting their phonemic quality (Liebenthal et al., 2005). Behavioral performance showed that only the syllables were categorically perceived. During the fMRI scanning, subjects performed a two-alternative forced-choice ABX discrimination task for both the phonemic and nonphonemic conditions. Comparisons across the two conditions revealed that the phonemic discrimination task preferentially activated the left middle and anterior segments of the STS. Similar results using similar stimuli have been obtained during passive listening experiments, indicating that speech activation in STS is partly an automatic process (Spitsyna et al., 2006; Obleser et al., 2007). More recently, greater activation in the upper banks of the left middle STS to speech was found when compared to multiple other types of complex sounds, even after controlling for effects produced by acoustic similarities and differences across sound categories (Leaver & Rauschecker, 2010). Finally, a meta-analysis examining 23 fMRI studies studying sublexical speech perception implicated the left middle and posterior portions of the STS in phonological processing (Turkeltaub & Coslett, 2010).

The literature, however, has been inconsistent with regard to laterality effects. Both bilateral and left-dominant results have been reported. In trying to sort out reasons for this discrepancy, Obleser and colleagues (2008) modified real words by either temporally or spectrally degrading the signals. Subjects were instructed to rate the intelligibility of each sound during fMRI scanning. Intelligibility based on spectral detail positively correlated with the degree of bilaterally enhanced STS activation. However, there was a greater effect for enhanced temporal detail to activate the left anterolateral STS and enhanced spectral detail to activate the homologous region in the right hemisphere. Thus, STS of both hemispheres appears important for phonemic perception, but with subtle variations in the degree to which it emphasizes either spectral or temporal components of speech. Another interpretation for left/right differences in the processing role of the STS posits that each hemisphere analyzes speech using different windows of integration, with the left side examining incoming signals at a faster rate (25–50 Hz) than the right hemisphere (4–8 Hz) (e.g., Hickok, 2009). The faster processing rate of the left hemisphere would be better equipped to evaluate rapid spectrotemporal changes (e.g., consonants), whereas the slower rate would better discriminate speech with slower acoustic changes (e.g., vowels). Finally, discrepancies across studies reflecting both left/right differences and those reflecting differences between the three subdivisions of the STS (anterior to Heschl's gyrus, subjacent to Heschl's gyrus, and posterior to Heschl's gyrus) may be partially based on multiple factors including behavioral task requirements, specifics of the stimuli used and functional imaging comparisons made (Hickok & Poeppel, 2007), and quite likely, on technical and neural response variability.

Given the variable results and the fact that fMRI provides only an indirect measure of neural activity, it is likely that major advancements will require integration with other methodologies. To date, human intracranial recordings in STS have been lacking. Scalp recordings lack the spatial specificity. However, animal models may shed light on the specific mechanisms by which the STS encodes speech signals. A potentially homologous region in macaque monkey has been identified that is located on the anterior portion of the superior temporal plane (Petkov et al., 2008;

see also Poremba & Mishkin, 2007). This "voice region" is anterior to auditory cortex, and responds preferentially to species-specific vocalizations when compared against heterospecific animal vocalizations and diverse natural sounds. This area, a higher-level nexus of the ventral or "what" pathway of auditory sound processing (Rauschecker & Scott, 2009), contains neurons that respond to a progressively more restricted number of sound stimuli when compared with core areas and appears to prefer species-specific vocalizations (Kikuchi et al., 2010). Further electrophysiological exploration of this region will undoubtedly provide invaluable insights into the mechanisms by which STS becomes selective for human vocalizations and subserves phonemic categorization.

6.4.5 Supramarginal Gyrus

Several studies have identified the left supramarginal gyrus (SMG) as being involved in phonemic categorical perception (Turkeltaub & Coslett, 2010). Dehaene-Lambert and colleagues (2005) reported this finding using sine wave speech, a stimulus whose frequencies correspond to the center frequencies of formants. People usually begin hearing these stimuli as whistles and later switch to a phonemic mode of perception. fMRI activation patterns were compared when frequencies were changed such that in a speech mode of perception, the sounds were perceived as either a repeating /b/ or /d/, or as a categorical change in place of articulation. The left SMG was selectively activated when there was a categorical change in perception, and only when subjects heard the sounds in a phonemic mode of hearing. This finding was replicated when comparing responses produced by acoustic changes that did not alter perception from /b/ to /d/, versus when the same degree of acoustic change crossed the perceptual boundary (Raizada & Poldrack, 2007). However, recent work questions the interpretation that the SMG is specifically involved with phonemic perception, and instead suggests that this region is more generally activated by salient stimulus changes (Zevin et al., 2010). This observation does not mean that the area is not involved with phonemic perception, but that its function must be viewed in a larger context. Thus, a role in both change detection and phonological processing implies that the supramarginal gyrus is involved with phonological short-term memory (Turkeltaub & Coslett, 2010).

6.4.6 Ventral Prefrontal Cortex

Within both humans and nonhuman primates, the ventral prefrontal cortex (vPFC) serves as the final station in the ventral ("what") processing stream of sound (Rauschecker & Scott, 2009; Romanski & Averbeck, 2009). Neurons in monkey vPFC respond to multiple complex stimuli, including species-specific vocalizations (e.g., Cohen et al., 2007; Romanski & Averbeck, 2009). Both citations stressed that responses were not a reflection of sensitivity to basic acoustic features, but instead, were modulated by both higher-order, categorical features of the sounds as well as the behavioral choices made by the animals (e.g., Nelken, 2008; Russ et al., 2008b). As discussed earlier, an identical paradigm to that of Russ and colleagues (2008b) found that neurons in field AL on PLST were already responsive to the categorical nature to the syllables (Tsunada et al., 2011). Thus, it would appear that, at least in monkeys, the vPFC may be more involved with associating a given stimulus with a given motor response (Romanski & Averbeck, 2009). If parallels exist, as the previous authors suggested, between monkey vPFC and the human inferior frontal gyrus (i.e., Broca's area), then the neuronal response patterns observed could reflect some interaction between articulatory movements and the speech signal. A direct link between phonemic perception and human vPFC remains to be established (see Hickok, 2009 for discussion).

6.5 Conclusions

6.5.1 Serial Processing versus a Bidirectional and Interactive Neural Network

Implicit in this chapter's description of phonemic encoding is a modular and serial organization wherein each succeeding station performs an ever more complex and abstract task. Core auditory cortex follows the detailed spectrotemporal details of the speech stream in real time. Fundamental properties of A1 physiology explain many contextual effects of phonemic perception. Convergence of inputs from core auditory cortex onto secondary fields such as PLST permits a complex, multidimensional scaling of its neuronal responses that begins the categorization process. This process becomes further refined in the STS, SMG, and possibly, in vPFC.

Although this scenario fits much of the data, serial processing does not account for the distributed pattern of neural responses occurring when phonemes are categorized. Thus, identification of different vowels requires bilaterally distributed patterns of activation within anterior Heschl's gyrus and sulcus, STG and STS, and the left PT (Formisano et al., 2008). Ambiguous consonants are correctly identified based on activation patterns predominantly within left Heschl's gyrus, Heschl's sulcus, and anterior PT (Kilian-Hütten et al., 2011). Similar findings have been observed in auditory belt regions of the right hemisphere (Raizada et al., 2010). The involvement of regions immediately surrounding primary areas suggest a degree of stimulus abstraction not typically ascribed to belt auditory cortex. Even within A1, abstraction beyond simple analysis of spectrotemporal patterns is indicated by modulation of neural responses based on stimulus valence and task requirements (e.g., Atiani et al., 2009; Frita et al., 2010; Brosch et al., 2011). Overall, this higher level processing suggests that these regions are also representing auditory objects, in addition to their being an analyzer of spectrotemporal details (Nelken, 2008).

This degree of sophistication is incompatible with a simple feedforward mechanism, and instead indicates a dynamic and bidirectional interaction between higher processing centers and lower centers via reentrant pathways (Obleser & Eisner, 2008; Fritz et al., 2010). Mechanisms of this interaction are just now being clarified. Functional and anatomical pathways between primary, secondary, and higher level regions of auditory responsive cortex are being identified using Granger causality mapping and diffusion tensor probabilistic mapping (e.g., Upadhyay et al., 2008). Temporal coherence in the underlying brain rhythms between lower and higher auditory centers may promote temporal binding of activity related to the abstract representation of speech (Bonte et al., 2009; see also Kayser & Logothetis, 2009). The later timing of the coherence (e.g., 300 ms as observed by Bonte et al., 2009) between areas suggests that the early activity of lower centers performs basic stimulus attribute computations whereas later activity driven by feedback relates to more abstract representations of the speech signal. Reliable relationships between the phase of ECoG rhythms and neuronal firing patterns add much information about the nature of the stimulus, helps disambiguate stimuli that might evoke similar firing rates but at different phases of the ongoing ECoG rhythms, and facilitates information transfer in high noise conditions that are typically encountered in a normal hearing environment (Kayser et al., 2009).

6.5.2 Future Directions

Further clarification of the mechanisms underlying the dynamic interactions between the various nodes and pathways engaged in phonemic encoding is clearly on the horizon. This clarification will undoubtedly include detailing how perception occurs when there are simultaneous and competing speech streams. Identifying the interplay between dorsal and ventral streams will likely promote some degree of reconciliation between the extreme views of phoneme perception that stress either basic acoustic processing mechanisms or the importance of reacquisition of articulatory gestures (see Diehl et al., 2004). The manner in which written speech processing enters into pathways originally shaped by audition is another topic that will receive great interest. Finally, emphasis must be given to the development of these nodes and pathways, and how infants transform from speech "universalists" to experts restricted to their native languages. Only then will we more fully understand the scope that learning and plasticity have on sculpting the unique properties that each speech-responsive region possesses in the processing scheme underlying phonemic perception.

Acknowledgments This work was supported by NIH grants DC-00657 and DC-04290.

References

Ahissar, E., Nagarajan, S., Ahissar, M., Protopapas, A., Mahncke, H., & Merzenich, M. M. (2001). Speech comprehension is correlated with temporal response patterns recorded from auditory cortex. *Proceedings of the National Academy of Sciences of the USA*, 98, 13367–13372.

Alexander, J. M., & Kluender, K. R. (2008). Spectral tilt change in stop consonant perception. *Journal of the Acoustical Society of America*, 123, 386–396.

Alexander, J. M., & Kluender, K. R. (2009). Spectral tilt change in stop consonant perception by listeners with hearing impairment. *Journal of Speech, Language, and Hearing Research*, 52, 653–670.

Anderson, S., Skoe, E., Chandrasekaran, B., & Kraus, N. (2010). Neural timing is linked to speech perception in noise. *Journal of Neuroscience*, 30, 4922–4926.

Atencio, C. A., Blake, D. T., Strata, F., Cheung, S. W., Merzenich, M. M., & Schreiner, C. E. (2007). Frequency-modulation encoding in the primary auditory cortex of the awake owl monkey. *Journal of Neurophysiology*, 98, 2182–2195.

Atiani, S., Elhilali, M., David, S. V., Fritz, J. B., & Shamma, S. A. (2009). Task difficulty and performance induce diverse adaptive patterns in gain and shape of primary auditory cortical receptive fields. *Neuron*, 61, 467–480.

Belin, P., Zatorre, R. J., Lafaille, P., Ahad, P., & Pike, B. (2000). Voice selective areas in human auditory cortex. *Nature*, 403, 309–312.

Besle J., Fischer, C., Bidet-Caulet, A., Lecaignard, F., Bertrand, O., & Giard, M.-H. (2008). Visual activation and audiovisual interactions in the auditory cortex during speech perception: Intracranial recordings in humans. *Journal of Neuroscience*, 28, 14301–14310.

Binder, J. R., Frost, J. A., Hammeke, T. A., Bellgowan, P. S. F., Springer, J. A., Kaufman, J. N., & Possing, E. T. (2000). Human temporal lobe activation by speech and nonspeech sounds. *Cerebral Cortex*, 10, 512–528.

Bitterman, Y., Mukamel, R., Malach, R., Fried, I., & Nelken, I. (2008). Ultra-fine frequency tuning revealed in single neurons of human auditory cortex. *Nature*, 451, 197–201.

Blumstein, S. E., & Stevens, K. N. (1979). Acoustic invariance in speech production: Evidence from measurements of the spectral characteristics of stop consonants. *Journal of the Acoustical Society of America*, 66, 1001–1017.

Blumstein, S. E., & Stevens, K. N. (1980). Perceptual invariance and onset spectra for stop consonant vowel environments. *Journal of the Acoustical Society of America*, 67, 648–662.

Boatman, D. F., & Miglioretti, D. L. (2005). Cortical sites critical for speech discrimination in normal and impaired listeners. *Journal of Neuroscience*, 25, 5475–5480.

Boemio, A., Fromm, S., Braun, A., & Poeppel, D. (2005). Hierarchical and asymmetric temporal sensitivity in human auditory cortices. *Nature Neuroscience*, 8, 389–395.

Bonte, M., Valente, G., & Formisano, E. (2009). Dynamic and task-dependent encoding of speech and voice by phase reorganization of cortical oscillations. *Journal of Neuroscience*, 29, 1699–1706.

Brenner, C. A., Krishnan, G. P., Vohs, J. L., Ahn, W.-Y., Hetrick, W. P., Morzorati, S. L., & O'Donnell, B. F. (2009). Steady state responses: Electrophysiological assessment of sensory function in schizophrenia. *Schizophrenia Bulletin*, 35, 1065–1077.

Brosch, M., & Schreiner, C. E. (1997). Time course of forward masking tuning curves in cat primary auditory cortex. *Journal of Neurophysiology*, 77, 923–943.

Brosch, M., & Schreiner, C. E. (2000). Sequence sensitivity of neurons in cat primary auditory cortex. *Cerebral Cortex*, 10, 1155–1167.

Brosch, M., Selezneva, E., & Scheich, H. (2005). Nonauditory events of a behavioral procedure activate auditory cortex of highly trained monkeys. *Journal of Neuroscience*, 25, 6797–6806.

Brosch, M., Selezneva, E., & Scheich, H. (2011). Representation of reward feedback in primate auditory cortex. *Frontiers in Systems Neuroscience*, 5, article 5.

Burlingame, E., Sussman, H. M., Gillam, R. B., & Hay, J. F. (2005). An investigation of speech perception in children with specific language impairment on a continuum of formant transition duration. *Journal of Speech, Language, and Hearing Research*, 48, 805–816.

Carney, L. H., & Geisler, C. D. (1986). A temporal analysis of auditory-nerve fiber responses to spoken stop-consonant vowels. *Journal of the Acoustical Society of America*, 79, 1896–1914.

Chandrasekaran, B., Hornickel, J., Skoe, E., Nicol, T., & Kraus, N. (2009). Context-dependent encoding in the human auditory brainstem relates to hearing speech in noise: Implications for developmental dyslexia. *Neuron*, 64, 311–319.

Chang, E. F., Rieger, J. W., Johnson, K., Berger, M. S., Barbaro, N. M., & Knight, R. T. (2010). Categorical speech representation in human superior temporal gyrus. *Nature Neuroscience*, 13, 1428–1432.

Cho, T., & Ladefoged, P. (1999). Variation and universals in VOT: Evidence from 18 languages. *Journal of Phonetics*, 27, 207–229.

Church, K. W. (1987). Phonological parsing and lexical retrieval. *Cognition*, 25, 53–69.

Cohen, Y. E., Theunissen, F., Russ, B. E., & Gill, P. (2007). Acoustic features of rhesus vocalizations and their representation in the ventrolateral prefrontal cortex. *Journal of Neurophysiology*, 97, 1470–1484.

Crone, N. E., Boatman, D., Gordon, B., & Hao, L. (2001). Induced electrocorticographic gamma activity during auditory perception. *Clinical Neurophysiology*, 112, 565–582.

Dehaene-Lambertz, G., Pallier, C., Serniclaes, W., Sprenger-Charolles, L., Jobert, A., & Dehaene, S. (2005). Neural correlates of switching from auditory to speech perception. *NeuroImage*, 24, 21–33.

Delgutte, B. (1980). Representation of speech-like sounds in the discharge patterns of auditory-nerve fibers. *Journal of the Acoustical Society of America*, 68, 843–857.

Delgutte, B., & Kiang, N. Y. S. (1984). Speech encoding in the auditory nerve: IV. Sounds with consonant-like dynamic characteristics. *Journal of the Acoustical Society of America*, 75, 897–907.

Dent, M. L., Brittan-Powell, E. F., Dooling, R. J., & Pierce, A. (1997). Perception of synthetic /ba/-/wa/ speech continuum by budgerigars (*Melopsittacus undulatus*). *Journal of the Acoustical Society of America*, 102, 1891–1897.

Diehl, R. L., Lotto, A. J., & Holt, L. L. (2004). Speech perception. *Annual Review of Psychology*, 55, 149–179.

Dronkers, N. F., Pinker, S., & Damasio, A. (2000). Language and the aphasias. In E. R. Kandel, J. H. Schwartz, & T. M. Jessell (Eds.), *Principles of neural science* (pp. 1169–1187). New York: McGraw-Hill.

Dronkers, N. F., Wilkins, D. P., Van Valin, R. D., Jr., Redfern, B. B., & Jaeger, J. J. (2004). Lesion analysis of the brain areas involved in language comprehension. *Cognition*, 92, 145–177.

Drullman, R., Festen, J. M., & Plomp, R. (1994a). Effect of temporal envelope smearing on speech reception. *Journal of the Acoustical Society of America*, 95, 1053–1064.

Drullman, R., Festen, J. M., & Plomp, R. (1994b). Effect of reducing slow temporal modulations on speech reception. *Journal of the Acoustical Society of America*, 95, 2670–2680.

Eilers, R. E., Gavin, W., & Wilson, W. R. (1979). Linguistic experience and phonetic perception in infancy: A crosslinguistic study. *Child Development*, 50, 14–18.

Engineer, C. T., Perez, C. A., Chen, Y. H., Carraway, R. S., Reed, A. C., Shetake, J. A., et al. (2008). Cortical activity patterns predict speech discrimination ability. *Nature Neuroscience*, 11, 603–608.

Faulkner, A., & Rosen, S. (1999). Contributions of temporal encodings of voicing, voicelessness, fundamental frequency, and amplitude variation to audio-visual and auditory speech perception. *Journal of the Acoustical Society of America*, 106, 2063–2073.

Fishman, Y. I., Volkov, I. O., Noh, M. D., Garell, P. C., Bakken, H., Arezzo, J. C., et al. (2001). Consonance and dissonance of musical chords: Neural correlates in auditory cortex of monkeys and humans. *Journal of Neurophysiology*, 86, 2761–2788.

Formisano, E., De Martino, F., Bonte, M., & Goebel, R. (2008). "Who" is saying "what"? Brain-based decoding of human voice and speech. *Science*, 322, 970–973.

Fritz, J. B., Elhilali, M., & Shamma, S. A. (2007). Adaptive changes in cortical receptive fields induced by attention to complex sounds. *Journal of Neurophysiology*, 98, 2337–2346.

Fritz, J. B., David, S. V., Radtke-Schuller, S., Yin, P., & Shamma, S. A. (2010). Adaptive, behaviorally gated, persistent encoding of task-relevant auditory information in ferret frontal cortex. *Nature Neuroscience*, 13, 1011–1019.

Galaburda, A. M., & Sanides, F. (1980). Cytoarchitectonic organization of the human auditory cortex. *Journal of Comparative Neurology*, 190, 597–610.

Gervain, J., & Mehler, J. (2010). Speech perception and language acquisition in the first year of life. *Annual Review of Psychology*, 61, 191–218.

Geschwind, N., & Levitsky, W. (1968). Human brain: Left-right asymmetries in temporal speech region. *Science*, 161, 186–187.

Gillam, R. B., Loeb, D. F., Hoffman, L. M., Bohman, T., Champlin, C. A., Thibodeau, L., et al. (2008). The efficacy of Fast ForWord language intervention in school-age children with language impairment: a randomized controlled trial. *Journal of Speech, Language & Hearing Research*, 51, 97–119.

Giraud, K., Démonet, J. F., Habib, M., Marquis, P., Chauvel, P., & Liégeois-Chauvel, C. (2005). Auditory evoked potential patterns to voiced and voiceless speech sounds in adult developmental dyslexics with persistent deficits. *Cerebral Cortex*, 15, 1524–1534.

Godey, B., Atencio, C. A., Bonham, B. H., Schreiner, C. E., & Cheung, S. W. (2005). Functional organization of squirrel monkey primary auditory cortex: Responses to frequency-modulation sweeps. *Journal of Neurophysiology*, 94, 1299–1311.

Goswami, U., Fosker, T., Huss, M., Mead, N., & Szűcs, D. (2011). Rise time and formant transition duration in the discrimination of speech sounds: The Ba–Wa distinction in developmental dyslexia. *Developmental Science*, 14, 34–43.

Griffiths, T. D., & Warren, J. D. (2002). The planum temporale as a computational hub. *Trends in Neurosciences*, 25, 348–353.

Hackett, T. A. (2007). Organization and correspondence of the auditory cortex of humans and nonhuman primates. In J. H. Kaas (Ed.), *Evolution of the nervous system* (pp. 109–119). Oxford: Elsevier.

Hahnloser, R. H. R., & Kotowicz, A. (2010). Auditory representation and memory in birdsong learning. *Current Opinion in Neurobiology*, 20, 332–339.

Hari, R., & Renvall, H. (2001). Impaired processing of rapid stimulus sequences in dyslexia. *Trends in Cognitive Sciences*, 5, 525–532.

Hart, H. C., Palmer, A. R., & Hall, D. A. (2004). Different areas of human non-primary auditory cortex are activated by sounds with spatial and nonspatial properties. *Human Brain Mapping*, 21, 178–190.

Hedrick, M. S., & Younger, M. S. (2007). Perceptual weighting of stop consonant cues by normal and impaired listeners in reverberation versus noise. *Journal of Speech, Language and Hearing Research*, 50, 254–269.

Hickok, G. (2009). The functional neuroanatomy of language. *Physics of Life Reviews*, 6, 121–143.

Hickok, G., & Poeppel, D. (2007). The cortical organization of speech processing. *Nature Reviews Neuroscience*, 8, 393–402.

Hillenbrand, J., Getty, L. A., Clark, M. J., & Wheeler, K. (1995). Acoustic characteristics of American English vowels. *Journal of the Acoustical Society of America*, 97, 3099–3111.

Hirsh. I. J. (1959). Auditory perception of temporal order. *Journal of the Acoustical Society of America*, 31, 759–767.

Holt, L. L. (2006a). Speech categorization in context: Joint effects of nonspeech and speech precursors. *Journal of the Acoustical Society of America*, 119, 4016–4026.

Holt, L. L. (2006b). The mean matters: Effects of statistically defined nonspeech spectral distributions on speech categorization. *Journal of the Acoustical Society of America*, 120, 2801–2817.

Holt, L. L., & Lotto, A. J. (2010). Speech perception as categorization. *Attention, Perception, & Psychophysics*, 72, 1218–1227.

Holt, L. L., Lotto, A. J., & Kluender, K. R. (2001). Influence of fundamental frequency on stop-consonant voicing perception: A case of learned covariation or auditory enhancement? *Journal of the Acoustical Society of America*, 109, 764–774.

Howard, M. A. III, Volkov, I. O., Abbas, P. J., Damasio, H., Ollendieck, M. C., & Granner, M. A. (1996). A chronic microelectrode investigation of the tonotopic organization of human auditory cortex. *Brain Research*, 724, 260–264.

Jusczyk, P. W., Rosner, B. S., Reed, M. A., & Kennedy, L. J. (1989). Could temporal order differences underlie 2-month-olds. discrimination of English voicing contrasts? *Journal of the Acoustical Society of America*, 85, 1741–1749.

Kayser, C., & Logothetis, N. K. (2009). Directed interactions between auditory and superior temporal cortices and their role in sensory integration. *Frontiers in Integrative Neuroscience*, 3, article 7.

Kayser, C., Montemurro, M. A., Logothetis, N. K., & Panzeri, S. (2009). Spike-phase coding boosts and stabilizes information carried by spatial and temporal spike patterns. *Neuron*, 61, 597–608.

Keating, P. A., Mikoś, M. J., & Ganong, W. F., III. (1981). A cross-language study of range of voice onset time in the perception of initial stop voicing. *Journal of the Acoustical Society of America*, 70, 1261–1271.

Kent, R. D. (1993). Vocal tract acoustics. *Journal of Voice*, 7, 97–117.

Kewley-Port, D. (1982). Measurement of formant transitions in naturally produced stop consonant-vowel syllables. *Journal of the Acoustical Society of America*, 72, 379–389.

Kewley-Port, D., Watson, C. S., & Foyle, D. C. (1988). Auditory temporal acuity in relation to category boundaries; speech and nonspeech stimuli. *Journal of the Acoustical Society of America*, 83, 1133–1145.

Kikuchi, Y., Horwitz, B., & Mishkin, M. (2010). Hierarchical auditory processing directed rostrally along the monkey's supratemporal plane. *Journal of Neuroscience*, 30, 13021–13030.

Kilian-Hütten, N., Valente, G., Vroomen, J., & Formisano, E. (2011). Auditory cortex encodes the perceptual interpretation of ambiguous sound. *Journal of Neuroscience*, 31, 1715–1720.

Kluender, K. R. (1991). Effects of first formant onset properties on voicing judgments result from processes not specific to humans. *Journal of the Acoustical Society of America*, 90, 83–96.

Kluender, K. R., & Lotto, A. J. (1994). Effects of first formant onset frequency on [-voice] judgments result from auditory processes not specific to humans. *Journal of the Acoustical Society of America*, 95, 1044–1052.

Kluender, K. R., Diehl, R. L., & Killeen, P. R. (1987). Japanese quail can learn phonetic categories. *Science*, 237, 1195–1197.

Kuhl, P. (1986). Theoretical contributions of tests on animals to the special-mechanisms debate in speech. *Experimental Biology*, 45, 233–265.

Kuhl, P. K. (2004). Early language acquisition: Cracking the speech code. *Nature Reviews Neuroscience*, 5, 831–843.

Kuhl, P. K. (2010). Brain mechanisms in early language acquisition. *Neuron*, 67, 713–727.

Kuhl, P. K., & Miller, J. D. (1978). Speech perception by the chinchilla: Identification functions for synthetic VOT stimuli. *Journal of the Acoustical Society of America*, 63, 905–917.

Kumar, A. U. (2011). Temporal processing abilities across different age groups. *Journal of the American Academy of Audiology*, 22, 5–12.

Kumar, S., Stephan, K. E., Warren, J. D., Friston, K. J., & Griffiths, T. D. (2007). Hierarchical processing of auditory objects in humans. *PLoS Computational Biology*, 3(6), e100.

Kwakye, L. D. et al., Kwakye, L. D., Foss-Feig, J. H., Cascio, C. J., Stone, W. L., & Wallace, M. T. (2011). Altered auditory and multisensory temporal processing in autism spectrum disorders. *Frontiers in Integrative Neuroscience*, 4, article 129.

Leaver, A. M., & Rauschecker, J. P. (2010). Cortical representation of natural complex sounds: Effects of acoustic features and auditory object category. *Journal of Neuroscience*, 30, 7604–7612.

Leff, A. P., Schofield, T. M., Stephan, K. E., Crinion, J. T., Friston, K. J., & Price, C. J. (2008). The cortical dynamics of speech. *Journal of Neuroscience*, 28, 13209–13215.

Le Prell, C. G., Niemiec, A. J., & Moody, D. B. (2001). Macaque thresholds for detecting increases in intensity: Effects of formant structure. *Hearing Research*, 162, 29–42.

Liberman, A. M., Cooper, F. S., Shankweiler, D. P., & Studdert-Kennedy, M. (1967). Perception of the speech code. *Psychological Review*, 74, 431–461.

Liebenthal, E., Binder, J. R., Spitzer, S. M., Possing, E. T., & Medler, D. A. (2005). Neural substrates of phonemic perception. *Cerebral Cortex*, 15, 1621–1631.

Liégeois-Chauvel, C., de Graaf, J. B., Laguitton, V., & Chauvel, P. (1999). Specialization of left auditory cortex for speech perception in man depends on temporal coding. *Cerebral Cortex*, 9, 484–496.

Lisker, L., & Abramson, A. S. (1964). A cross-language study of voicing in initial stops: Acoustical measurements. *Word*, 20, 384–422.

Lisker, L., & Abramson, A. S. (1967). Some effects of context on voice onset time in English stops. *Language and Speech*, 10, 1–28.

Lotto, A. J., & Kluender, K. R. (1998). General contrast effects in speech perception: Effect of preceding liquid on stop consonant identification. *Perception & Psychophysics*, 60, 602–619.

Lotto, A. J., Kluender, K. R., & Holt, L. L. (1997). Perceptual compensation for coarticulation by Japanese quail (*Coturnix coturnix japonica*). *Journal of the Acoustical Society of America*, 102, 1134–1140.

Luo, H., & Poeppel, D. (2007). Phase patterns of neuronal responses reliably discriminate speech in human auditory cortex. *Neuron*, 54, 1001–1010.

Mann, V. A. (1980). Influence of preceding liquid on stop-consonant perception. *Perception & Psychophysics*, 28, 407–412.

May, B., Moody, D. B., & Stebbins, W. C. (1989). Categorical perception of conspecific communication sounds by Japanese macaques, *Macaca fuscata*. *Journal of the Acoustical Society of America*, 85, 837–847.

Mayo, C., & Turk, A. (2004). Adult-child differences in acoustic cue weighting are influenced by segmental context: Children are not always perceptually biased toward transitions. *Journal of the Acoustical Society of America*, 115, 3184–3194.

McClelland, J. L., & Elman, J. L. (1986). The TRACE model of speech perception. *Cognitive Psychology*, 18, 1–86.

Merzenich, M. M., & Brugge, J. F. (1973). Representation of the cochlear partition on the superior temporal plane of the macaque monkey. *Brain Research*, 50, 275–296.

Mesgarani, N., David, S. V., Fritz, J. B., & Shamma, S. A. (2008). Phoneme representation and classification in primary auditory cortex. *Journal of the Acoustical Society of America*, 123, 899–909.

Metherate, R., & Cruikshank, S. J. (1999). Thalamocortical inputs trigger a propagating envelope of gamma-band activity in auditory cortex *in vitro*. *Experimental Brain Research*, 126, 160–174.

Middlebrooks, J. C. (2008). Auditory cortex phase locking to amplitude-modulated cochlear implant pulse trains. *Journal of Neurophysiology*, 100, 76–91.

Miller, J. L., & Eimas, P. D. (1995). Speech perception: From signal to word. *Annual Review of Psychology*, 46, 467–492.

Moore, B. C. J. (2008a). Basic auditory processes involved in the analysis of speech sounds. *Philosophical Transactions of the Royal Society B: Biological Sciences*, 363, 947–963.

Moore, B. C. J. (2008b). The role of temporal fine structure processing in pitch perception, masking, and speech perception for normal-hearing and hearing-impaired people. *Journal of the Association for Research in Otolaryngology*, 9, 399–406.

Morel, A., Garraghty, P. E., & Kaas, J. H. (1993). Tonotopic organization, architectonic fields, and connections of auditory cortex in macaque monkeys. *Journal of Comparative Neurology*, 335, 437–459.

Mukamel, R. et al., Mukamel, R., Nir, Y., Harel, M., Arieli, A., Malach, R., & Fried, I. (2010). Invariance of firing rate and field potential dynamics to stimulus modulation rate in human auditory cortex. *Human Brain Mapping*, July [epub ahead of print], doi:10.1002/hbm.21100

Nelken, I. (2008). Processing of complex sounds in the auditory system. *Current Opinions in Neurobiology*, 18, 413–417.

Nourski, K. V., Reale, R. A., Oya, H., Kawasaki, H., Kovach, C.K., Chen, H., et al. (2009). Temporal envelope of time-compressed speech represented in the human auditory cortex. *Journal of Neuroscience*, 29, 15564–15574.

Obleser, J., & Eisner, F. (2008). Pre-lexical abstraction of speech in the auditory cortex. *Trends in Cognitive Sciences*, 13, 14–19.
Obleser, J., Scott, S. K., & Eulitz, C. (2006). Now you hear it, now you don't: Transient traces of consonants and their nonspeech analogues in the human brain. *Cerebral Cortex*, 16, 1069–1076.
Obleser, J., Zimmermann, J., Van Meter, J., & Rauschecker, J. P. (2007). Multiple stages of auditory speech perception reflected in event-related fMRI. *Cerebral Cortex*, 17, 2251–2257.
Obleser, J., Eisner, F., & Kotz, S. A. (2008). Bilateral speech comprehension reflects differential sensitivity to spectral and temporal features. *Journal of Neuroscience*, 28, 8116–8124.
Panzeri, S., Brunel N., Logothetis, N. K., & Kayser, C. (2010). Sensory neural codes using multiplexed temporal scales. *Trends in Neuroscience*, 33, 111–120.
Parker, E. M. (1988). Auditory constraints on the perception of voice-onset time: The influence of lower tone frequency on judgments of tone-onset simultaneity. *Journal of the Acoustical Society of America*, 83, 1597–1607.
Petkov, C. I., Kayser, C., Steudel, T., Whittingstall, K., Augath, M., & Logothetis, N. K. (2008). A voice region in the monkey brain. *Nature Neuroscience*, 11, 367–374.
Pisoni, D. B. (1977). Identification and discrimination of the relative onset time of two component tones: Implications for voicing perception in stops. *Journal of the Acoustical Society of America*, 61, 1352–1361.
Pisoni, D. B., & Luce, P. A. (1987). Acoustic-phonetic representations in word recognition. *Cognition*, 25, 21–52.
Poeppel, D., Idsardi, W. J., & van Wassenhove, V. (2008). Speech perception at the interface of neurobiology and linguistics. *Philosophical Transactions of the Royal Society B: Biological Sciences*, 363, 1071–1086.
Poremba, A., & Mishkin, M. (2007). Exploring the extent and function of higher order auditory cortex in rhesus monkeys. *Hearing Research*, 229, 14–23.
Prather, J. F., Nowicki, S., Anderson, R. C., Peters, S., & Mooney, R. (2009). Neural correlates of categorical perception in learned vocal communication. *Nature Neuroscience*, 12, 221–228.
Raizada, R. D. S., & Poldrack, R. A. (2007). Selective amplification of stimulus differences during categorical processing of speech. *Neuron*, 56, 726–740.
Raizada, R. D. S., Tsao, F.-M., Liu, H.-M. & Kuhl, P. K. (2010). Quantifying the adequacy of neural representations for a cross-language phonetic discrimination task: Prediction of individual differences. *Cerebral Cortex*, 20, 1–12.
Rauschecker, J. P., & Scott, S. K. (2009). Maps and streams in the auditory cortex: Nonhuman primates illuminate human speech processing. *Nature Neuroscience*, 12, 718–724.
Rauschecker, J. P., & Tian, B. (2004). Processing of band-pass noise in the lateral auditory belt cortex of the rhesus monkey. *Journal of Neurophysiology*, 91, 2578–2589.
Reale, R. A., Calvert, G. A., Thesen, T., Jenison, R. L., Kawasaki, H., Oya, H., et al. (2007). Auditory-visual processing represented in the human superior temporal gyrus. Neuroscience, 145, 162–184.
Recio, A., Rhode, W. S., Kiefte, M., & Kluender, K. R. (2002). Responses to cochlear normalized speech stimuli in the auditory nerve of cat. *Journal of the Acoustical Society of America*, 111, 2213–2218.
Repp, B. H. (1982). Phonetic trading relations and context effects: New experimental evidence for a speech mode of perception. *Psychological Bulletin*, 92, 81–110.
Roman, S., Canévet, G., Lorenzi, C., Triglia, J.-M., & Liégeois-Chauvel, C. (2004). Voice onset time encoding in patients with left and right cochlear implants. *NeuroReport*, 15, 601–605.
Romanski, L. M., & Averbeck, B. B. (2009). The primate cortical auditory system and neural representation of conspecific vocalizations. *Annual Review of Neuroscience*, 32, 315–346.
Russ, B. E., Ackelson, A. L., Baker, A. E., & Cohen, Y. E. (2008a). Coding of auditory stimulus identity in the auditory non-spatial processing stream. *Journal of Neurophysiology*, 99, 87–95.
Russ, B. E., Orr, L. E., & Cohen, Y. E. (2008b). Prefrontal neurons predict choices during an auditory same-different task. *Current Biology*, 18, 1483–1488.

Sakai, M., Chimoto, S., Qin. L., & Sato, Y. (2009). Neural mechanisms of interstimulus interval-dependent responses in the primary auditory cortex of awake cats. *BMC Neuroscience*, 10, 10.
Schnupp, J. W. H., Hall, T. M., Kokelaar, R. F., & Ahmed, B. (2006). Plasticity of temporal pattern codes for vocalization stimuli in primary auditory cortex. *Journal of Neuroscience*, 26, 4785–4795.
Schreiner, C. E. (1998). Spatial distribution of responses to simple and complex sounds in the primary auditory cortex. *Audiology & Neurootology*, 3, 104–122.
Scott, S. K. (2005). Auditory processing—speech, space, and auditory objects. *Current Opinion in Neurobiology*, 15, 197–201.
Scott, B. H., Malone, B. J., & Semple, M. N. (2011). Transformation of temporal processing across auditory cortex of awake macaques. *Journal of Neurophysiology*, 105, 712–730.
Shamma, S. A. (1985). Speech processing in the auditory system I: The representation of speech sounds in the responses of the auditory nerve. *Journal of the Acoustical Society of America*, 78, 1612–1621.
Shannon, R. V., Zeng, F.-G., Kamath, V., Wygonski, J., & Ekelid, M. (1995). Speech recognition with primarily temporal cues. *Science*, 270, 303–304.
Shapleske, J., Rossell, S. L., Woodruff, P. W. R., & David, A. S. (1999). The planum temporale: A systematic, quantitative review of its structural, functional, and clinical significance. *Brain Research Reviews*, 29, 26–49.
Sharma, A., & Dorman, M. F. (2000). Neurophysiologic correlates of cross-language phonetic perception. *Journal of the Acoustical Society of America*, 107, 2697–2703.
Sinex, D. G. (1993). Auditory nerve fiber representations of cues to voicing in syllable-final stop consonants. *Journal of the Acoustical Society of America*, 94, 1351–1362.
Sinex, D. G., & McDonald, L. P. (1988). Average discharge rate representation of voice onset time in the chinchilla auditory nerve. *Journal of the Acoustical Society of America*, 83, 1817–1827.
Sinex, D. G., & Narayan, S. S. (1994). Auditory-nerve fiber representation of temporal cues in word-medial stop consonants. *Journal of the Acoustical Society of America*, 95, 897–903.
Sinnott, J. M., & Adams, F. S. (1987). Differences in human and monkey sensitivity to acoustic cues underlying voicing contrasts. *Journal of the Acoustical Society of America*, 82, 1539–1547.
Soli, S. D. (1983). The role of spectral cues in discrimination of voice onset time differences. *Journal of the Acoustical Society of America*, 73, 2150–2165.
Spitsyna, G., Warren, J. E., Scott, S. K., Turkheimer, F. E., & Wise, R. J. S. (2006). Converging language streams in the human temporal lobe. *Journal of Neuroscience*, 26, 7328–7336.
Steinschneider, M., & Fishman, Y. I. (2011). Enhanced physiologic discriminability of stop consonants with prolonged formant transitions in awake monkeys based on the tonotopic organization of primary auditory cortex. *Hearing Research*, 271, 103–114.
Steinschneider, M., Tenke, C., Schroeder, C., Javitt, D., Simpson, G. V., Arezzo, J. C., & Vaughan, H. G., Jr. (1992). Cellular generators of the cortical auditory evoked potential initial component. *Electroencephalography and Clinical Neurophysiology*, 84, 196–200.
Steinschneider, M., Reser, D., Schroeder, C. E., & Arezzo, J. C. (1995). Tonotopic organization of responses reflecting stop consonant place of articulation in primary auditory cortex (A1) of the monkey. *Brain Research*, 674, 147–152.
Steinschneider, M., Volkov, I. O., Noh, M. D., Garell, P. C., & Howard, M. A. III. (1999). Temporal encoding of the voice onset time phonetic parameter by field potentials recorded directly from human auditory cortex. *Journal of Neurophysiology*, 82, 2346–2357.
Steinschneider, M., Fishman, Y. I., & Arezzo, J. C. (2003). Representation of the voice onset time (VOT) speech parameter in population responses within primary auditory cortex of the awake monkey. *Journal of the Acoustical Society of America*, 114, 307–321.
Steinschneider, M., Volkov, I. O., Fishman, Y. I., Oya, H., Arezzo, J. C., & Howard, M. A. III. (2005). Intracortical responses in human and monkey primary auditory cortex support a temporal processing mechanism for encoding of the voice onset time phonetic parameter. *Cerebral Cortex*, 15, 170–186.

Steinschneider, M., Fishman, Y. I., & Arezzo, J. C. (2008). Spectrotemporal analysis of evoked and induced electroencephalographic responses in primary auditory cortex (A1) of the awake monkey. *Cerebral Cortex*, 18, 610–625.
Steinschneider, M., Nourski, K. V., Kawasaki, H., Oya, H., Brugge, J. F., & Howard, M. A. III. (2011). Intracranial study of speech-elicited activity on the human posterolateral superior temporal gyrus. *Cerebral Cortex*, 21, 2332–2347.
Stevens, K. N. (1981). Constraints imposed by the auditory system on the properties used to classify speech sounds: Data from phonology, acoustics, and psychoacoustics. In T. Myers, J. Laver, & J. Anderson (Eds.), *The cognitive representation of speech* (pp. 61–74). Amsterdam: North-Holland.
Stevens, K. N. (2002). Toward a model for lexical access based on acoustic landmarks and distinctive features. *Journal of the Acoustical Society of America*, 111, 1872–1891.
Stevens, K. N., & Blumstein, S. E. (1978). Invariant cues for place of articulation in stop consonants. *Journal of the Acoustical Society of America*, 64, 1358–1368.
Summerfield, Q., & Haggard, M. (1977). On the dissociation of spectral and temporal cues to the voicing distinction in initial stop consonants. *Journal of the Acoustical Society of America*, 62, 435–448.
Sussman, H. M., Bessell, N., Dalston, E., & Majors, T. (1997). An investigation of stop place of articulation as a function of syllable position: A locus equation perspective. *Journal of the Acoustical Society of America*, 101, 2826–2838.
Tallal, P. (2004). Improving language and literacy is a matter of time. *Nature Reviews Neuroscience*, 5, 721–728.
Tallal, P., Miller, S., & Fitch, R. H. (1993). Neurobiological basis of speech: A case for the preeminence of temporal processing. *Annals of the New York Academy of Sciences*, 682, 27–47.
Tavabi, K., Obleser, J., Dobel, C., & Pantev, C. (2007). Auditory evoked fields differentially encode speech features: An MEG investigation of the P50m and N100m time courses during syllable processing. *European Journal of Neuroscience*, 25, 3155–3162.
Tian, B., Reser, D., Durham, A., Kustov, A., & Rauschecker, J. P. (2001). Functional specialization in rhesus monkey auditory cortex. *Science*, 292, 290–293.
Trébuchon-Da Fonseca, A., Giraud, K., Badier, J.-M., Chauvel, P., & Liégeois-Chauvel, C. (2005). Hemispheric lateralization of voice onset time (VOT) comparison between depth and scalp EEG recordings. *NeuroImage*, 27, 1–14.
Tsunada, J., Lee, J. H., & Cohen, Y. E. (2011). Representation of speech categories in the primate auditory cortex. *Journal of Neurophysiology*, 105, 2634–2646.
Turkeltaub, P. E., & Coslett, H. B. (2010). Localization of sublexical speech perception components. *Brain & Language*, 114, 1–15.
Ulanovsky, N., Las, L., & Nelken, I. (2003). Processing of low-probability sounds by cortical neurons. *Nature Neuroscience*, 6, 391–398.
Ulanovsky, N., Las, L., Farkas, D., & Nelken, I. (2004). Multiple time scales of adaptation in auditory cortex neurons. *Journal of Neuroscience*, 24, 10440–10453.
Upadhyay, J., Silver, A., Knaus, T. A., Lindgren, K. A., Ducros, M., Kim, D.-S., & Tager-Flusberg, H. (2008). Effective and structural connectivity in the human auditory cortex. *Journal of Neuroscience*, 28, 3341–3349.
Vandermosten, M., Boets, B., Luts, H., Poelmans, H., Golestani, N., Wouters, J., & Ghesquière, P. (2010). Adults with dyslexia are impaired in categorizing speech and nonspeech sounds on the basis of temporal cues. *Proceedings of the National Academy of Sciences of the USA*, 177, 10389–10394.
Wallace, A. B., & Blumstein, S. E. (2009). Temporal integration in vowel perception. *Journal of the Acoustical Society of America*, 125, 1704–1711.
Wang, X., Merzenich, M. M., Beitel, R., & Schreiner, C. E. (1995). Representation of a species-specific vocalization in the primary auditory cortex of the common marmoset: Temporal and spectral characteristics. *Journal of Neurophysiology*, 74, 2685–2706.
Wang, X., Lu, T., Snider, R. K., & Liang, L. (2005). Sustained firing in auditory cortex by preferred stimuli. *Nature*, 435, 341–346.

Wang, X., Lu, T., Bendor, D., & Bartlett, E. (2008). Neural coding of temporal information in auditory thalamus and cortex. *Neuroscience*, 154, 294–303.

Werker, J. F., & Tees, R. C. (1999). Influences on infant speech processing: Toward a new synthesis. *Annual Review of Psychology*, 50, 509–535.

Werker, J. F., & Yeung, H. H. (2005). Infant speech perception bootstraps word learning. *Trends in Cognitive Sciences*, 9, 519–527.

Woolley, Fremouw, T. E., Hsu, A., & Theunissen, F. E. (2005). Tuning for spectro-temporal modulations as a mechanism for auditory discrimination of natural sounds. *Nature Neuroscience*, 8, 1371–1379.

Wright, B. A., Lombardino, L. J., King, W. M., Puranik, C. S., Leonard, C. M., & Merzenich, M. M. (1997). Deficits in auditory temporal and spectral resolution in language-impaired children. *Nature*, 387, 176–178.

Young, E. D. (2008). Neural representation of spectral and temporal information in speech. *Philosophical Transactions of the Royal Society B: Biological Sciences*, 363, 923–945.

Yrttiaho, S., Tiitinen, H., Alku, P., Miettinen, I., & May, P. J. (2010). Temporal integration of vowel periodicity in the auditory cortex. *Journal of the Acoustical Society of America*, 128, 224–234.

Zatorre, R. J., & Gandour, J. T. (2008). Neural specializations for speech and pitch: Moving beyond the dichotomies. *Philosophical Transactions of the Royal Society B: Biological Sciences*, 363, 1087–1104.

Zatorre, R. J., Belin, P., & Penhune, V. B. (2002). Structure and function of auditory cortex: music and speech. *Trends in Cognitive Sciences*, 6, 37–46.

Zeng, F.-G., & Liu, S. (2006). Speech perception in individuals with auditory neuropathy. *Journal of Speech, Language, and Hearing Research*, 49, 367–380.

Zevin, J. D., Yang, J., Skipper, J. I., & McCandliss, B. D. (2010). Domain general change detection accounts for "dishabituation" effects in temporal–parietal regions in functional magnetic resonance imaging studies of speech perception. *Journal of Neuroscience*, 30, 1110–1117.

Chapter 7
The Influence of Vision on Auditory Communication in Primates

Asif A. Ghazanfar and Chandramouli Chandrasekaran

Keywords Audiovisual speech • Auditory cortex • Crossmodal; multimodal • Monkey calls • Multisensory integration • Natural statistics • Neuronal oscillation • Phase resetting • Primate vocalizations • Superior temporal sulcus

7.1 Introduction: The Structure of Primate Communication Signals

All primate vocal signals are produced through the coordinated movements of the lungs, larynx (vocal folds), and the supralaryngeal vocal tract (Fitch & Hauser, 1995; Ghazanfar & Rendall, 2008). As a result, humans and other primates share a remarkable number of similarities in their vocal signaling, ranging from production mechanisms and signal structure to the inextricable link between vision and audition. In human speech, the signal—across all languages and contexts—is amplitude modulated, consisting of a rhythm that ranges from 2 to 7 Hz (Drullman, 1995; Greenberg et al., 2003; Chandrasekaran et al., 2009), roughly matching the timescale for syllable production. Such temporal modulation in similar frequency ranges also seems to be a common feature of several nonhuman primate vocalizations. For example, common marmoset (*Callithrix jacchus*) twitter calls are modulated in the 5–9 Hz range (Wang et al., 1995); squirrel monkey (*Saimiri sciureus*) vocalizations in the 6–10 Hz range (Godey et al., 2005); and finally cotton top tamarins (*Saguinus*

A.A. Ghazanfar (✉)
Departments of Psychology and Ecology & Evolutionary Biology,
Neuroscience Institute, Princeton University, Princeton, NJ 08540, USA
e-mail: asifg@princeton.edu

C. Chandrasekaran
Department of Psychology, Neuroscience Institute, Princeton University,
Princeton, NJ 08540, USA
e-mail: chandram@princeton.edu

Y.E. Cohen et al. (eds.), *Neural Correlates of Auditory Cognition*, Springer Handbook of Auditory Research 45, DOI 10.1007/978-1-4614-2350-8_7,
© Springer Science+Business Media New York 2013

oedipus), macaques (*Macaca* spp.), and chimpanzees (*Pan troglodytes*) all seem to have vocalizations modulated in the 3–8 Hz range (Cohen et al., 2007).

In addition to commonalities in temporal structure, vocal production in primates results in the deformation of the face around the oral aperture and other parts of the face (Hauser et al., 1993; Yehia et al., 2002; Chandrasekaran et al., 2009). In humans, this suggests that visual cues in speech should highly correlate with the spectral structure of the auditory component and modulated in a similar frequency range—which is precisely what is observed (Chandrasekaran et al., 2009). Nonhuman primate vocalizations also seem to share a similar link between acoustic output and facial dynamics. Different macaque monkey vocalizations are produced with unique lip configurations and mandibular positions, and the motion of such articulators influences the acoustics of the signal (Hauser et al., 1993; Hauser & Ybarra, 1994). Coo calls, like /u/ in speech, are produced with the lips protruded, while screams, like the /i/ in speech, are produced with the lips retracted (Fig. 7.1a).

In speech, the mouth opens and closes with a rhythm that matches the temporal structure of the acoustic output (Chandrasekaran et al., 2009). Oddly, however, for macaque vocalizations, the correlation of mouth opening to the temporal structure of the vocalizations is not apparent. In other words, though macaque vocalizations have an acoustic component that is modulated at a frequency that is between 2 and 7 Hz, their facial movements during vocal production do not have such a rhythm. Macaques produce a single ballistic mouth movement when producing vocalizations; an individual opens its mouth during an utterance and then closes it. Thus, the vocal tract origin of the temporal modulations embedded in the acoustics is not known. This suggests that although the acoustic rhythm in humans, apes, and monkeys is likely produced by the same mechanism, in humans the visual component was added at a later time during evolution. One idea is perhaps that the rhythmic visual component of audiovisual speech evolved in humans by the modification and co-opting of cyclical jaw movements in an ancestor (MacNeilage, 1998). For example, although rhythmic jaw movements are relatively rare during vocal production by monkeys, they are extremely common as facial communicative gestures. For example, lipsmacks, an affiliative signal produced during face-to-face contact, and teeth-grinds, a signal produced when a monkey is anxious, both involve cyclical movements of the mouth and are not accompanied by any vocalizations (Redican, 1975); notably, they have a temporal structure that matches audiovisual speech (2–7 Hz) (Ghazanfar et al., 2010).

In addition to facial deformations, movements of other parts of the head and the body not directly associated with the articulation process also accompany speech. These are potentially informative. In humans, eyebrow raises, nods, and other head movements, manual gestures, and blinks usually accompany speech (Munhall & Vatikiotis-Bateson, 2004). Similarly, in macaques, visual cues not part of the articulatory process are also present. For example, bark vocalizations are usually accompanied by retracted ears and a lowered head, whereas pant threats occur with ears forward and a raised head (Partan, 2002).

There is, naturally, a robust temporal correlation between auditory and visual components of vocalizations—if not the rhythm, then the duration. However, the relative timing of the two modalities is such that, in humans and other primates,

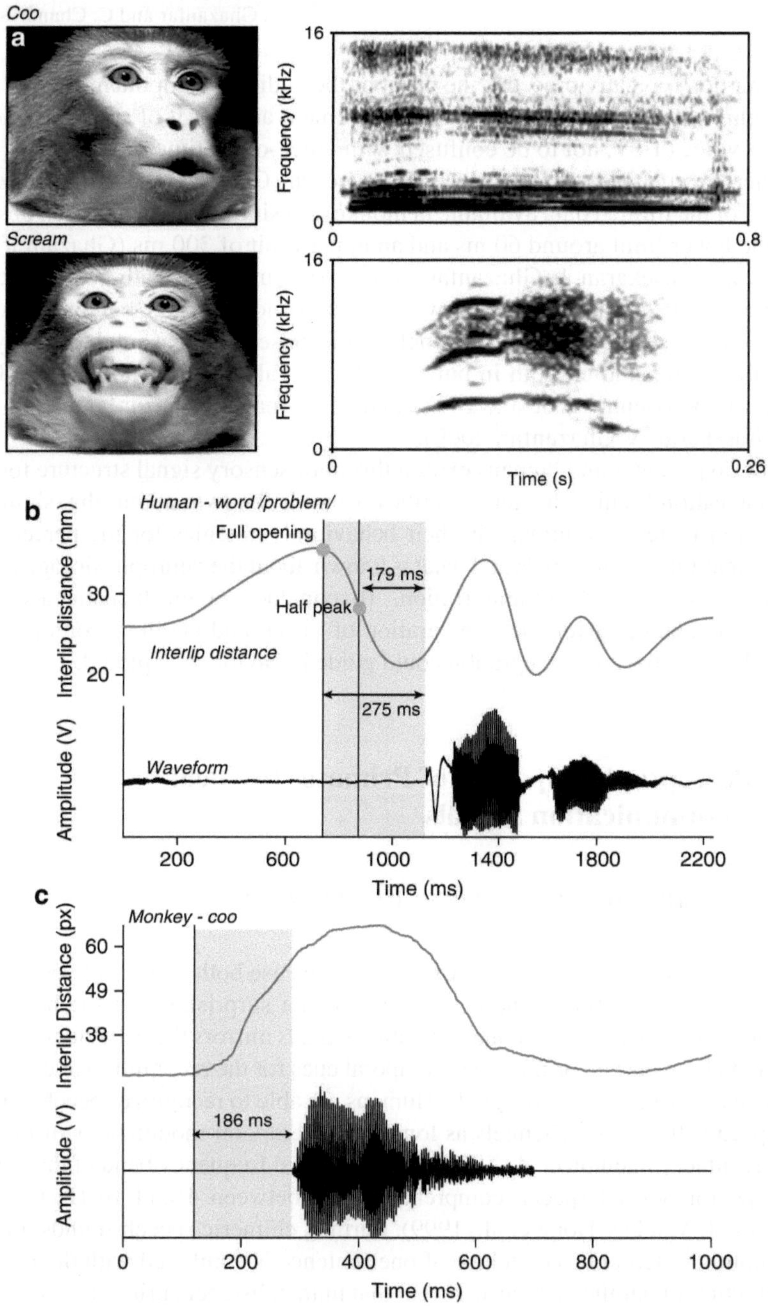

Fig. 7.1 Structure of communication signals. (**a**) Examples of the facial expressions produced concomitantly with vocalizations. Rhesus monkey coo and scream were taken at the midpoint of the expressions with their corresponding spectrograms. (**b**) Visual and auditory dynamics during the production of the word "PROBLEM" by a single speaker. Upper panel denotes the interlip distance as a function of time. The lower panel shows waveform of the sound. Dots denote the full and half opening points of the mouth. Note that the mouth dynamics seem to start nearly 275 ms before the onset of the sound. (**c**) Visual and auditory dynamics during a coo vocalization. Figure conventions as in **b**. Note again how mouth opening precedes onset of the sound even for nonhuman primate vocalizations

facial motion typically precedes the onset of the auditory component (Fig. 7.1b, c). This delay between the onset of mouth movement and onset of sound, termed the time-to-voice (TTV; not to be confused with voice-onset time or VOT), can range anywhere from 100 to 300 ms for human speech (Chandrasekaran & Ghazanfar, 2009). For the limited data available in monkeys, a similar range has been observed with the lower limit around 60 ms and an upper limit of 300 ms (Ghazanfar et al., 2005; Chandrasekaran & Ghazanfar, 2009). Although this feature of speech and primate vocalizations is well known, the physiological and behavioral consequences of onset of facial motion before the onset of the sound are only now beginning to be studied both in humans (Abry et al., 1996; Munhall & Tohkura, 1998; van Wassenhove et al., 2007) and in primates (Ghazanfar et al., 2005; Chandrasekaran & Ghazanfar, 2009).

How do primates and humans exploit this multisensory signal structure for vocal communication? This chapter describes what is known about the similarities between primates and humans in their behavioral strategies for the perception of vocal communication signals and what is known about the neurophysiological bases of multisensory vocal communication. To conclude, a mechanistic account to explain the neurophysiological integration of visual and auditory communication channels, and how such integration could guide behavior, is explored.

7.2 Perceptual Responses of Primates to Vocal Communication Signals

7.2.1 Temporal Cues in Vocal Recognition

Given that both humans and other extant primates use both facial and vocal expressions as communication signals, it is perhaps not surprising that the behavior of primates in response to such communication signals mirrors the behavior of humans. First, both humans and primates use temporal cues for the recognition of conspecific (i.e., from the same species) signals. Humans are able to recognize speech with very few spectral frequency channels as long as the temporal modulations of the envelope are intact (Shannon et al., 1995) and the critical frequency band of the temporal envelope for normal speech comprehension is between 4 and 16 Hz (Drullman et al., 1994; Van Der Horst et al., 1999). Further, chimeric speech sounds, in which the amplitude (temporal) envelope of one sentence is combined with the fine spectral structure of another sentence, reveal that human listeners primarily use the temporal envelope in speech perception (Smith et al., 2002).

Few studies have investigated the role of different acoustic features in primate vocal recognition, but the available data strongly suggest that the temporal structure of vocalizations is critical. For example, at the population level, wild rhesus monkeys tend to orient their heads to the right when they hear conspecific sounds (Hauser & Andersson, 1994; cf. Teufel et al., 2010 for a critique of this approach). When the amplitude envelopes of grunts and shrill barks are stretched beyond the species-typical range, they no longer orient to the right for these calls (Hauser et al., 1998). Similarly, when tested with time-reversed versions of harmonic arches or shrill barks—a manipulation that changes the direction of the envelope but preserves the overall spectral content—rhesus monkeys again orient toward the left as though they are not hearing the calls as conspecific signals (Ghazanfar et al., 2001). These studies suggest that the amplitude envelope is critical for call recognition in rhesus monkeys. Along the same lines (but using a different methodology), tamarin monkeys will continue to respond vocally to conspecific "long calls" that have the species-typical amplitude envelope even when the fine spectral structure is replaced with white noise (Ghazanfar et al., 2002). They, however, respond significantly less to calls *without* the species-typical envelope.

7.2.2 Multisensory Matching of Faces and Voices

The link between facial motion and vocalizations presents an obvious opportunity to exploit the concordance of both channels. Thus, it is not surprising that many primates other than humans recognize the correspondence between the visual and auditory components of vocal signals. Rhesus and Japanese macaques (*Macaca mulatta* and *Macaca fuscata*), capuchins (*Cebus apella*), and chimpanzees (*Pan troglodytes*), the only non-human primates tested thus far, all recognize auditory–visual correspondences between their various vocalizations (Ghazanfar & Logothetis, 2003; Parr, 2004; Evans et al., 2005). For example, rhesus monkeys readily match the facial expressions of "coo" and "threat" calls with their associated vocal components (Ghazanfar & Logothetis, 2003). Perhaps more pertinent, rhesus monkeys can also segregate competing voices in a chorus of coos, much as humans might with speech in a cocktail party scenario, and match them to the correct number of individuals seen cooing on a video screen (Jordan et al., 2005). Finally, macaque monkeys use formants (i.e., vocal tract resonances) as acoustic cues to assess age-related body size differences among conspecifics (Ghazanfar et al., 2007). They do so by linking across modalities the body size information embedded in the formant spacing of vocalizations (Fitch, 1997) with the visual size of animals who are likely to produce such vocalizations (Ghazanfar et al., 2007). Taken together these data suggest that humans are not at all unique in their ability to receive communication information across modalities.

7.2.3 Eye Movement Strategies

Although both humans and other primates readily link facial expressions with appropriate, congruent vocal expressions, the cues and behavioral strategies they use to make such matches are not known. One method for investigating behavioral strategies is the measurement of eye movement patterns. For example, when human participants are given no task or instruction regarding what acoustic cues to attend to, they will consistently look at the eye region more than the mouth when viewing

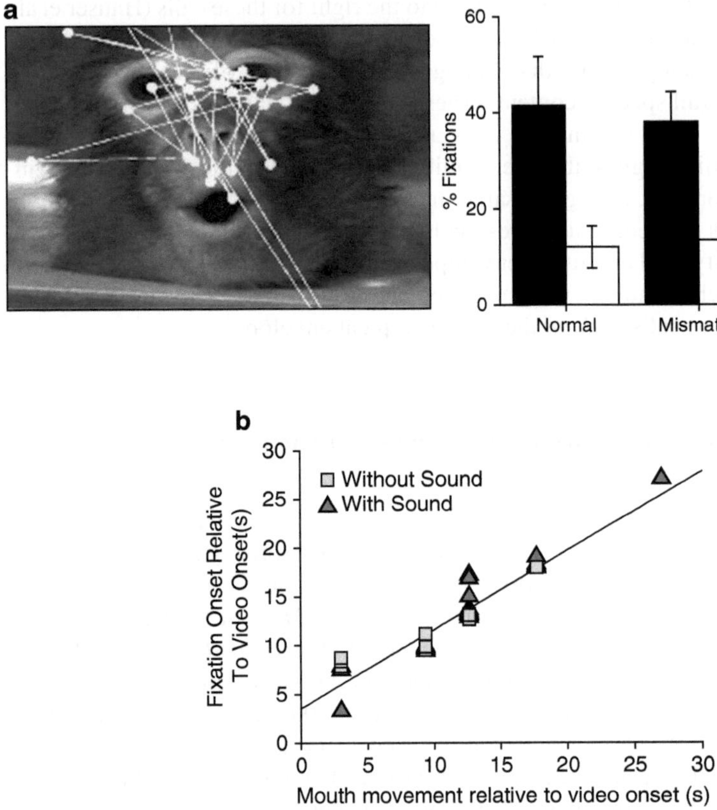

Fig. 7.2 Monkey eye movement strategy for audiovisual communication signals. (**a**) The average fixation on the eye region versus the mouth region across three subjects while viewing a 30-s video of a vocalizing conspecific. The audio track had no influence on the proportion of fixations falling onto the mouth or the eye region. Error bars represent SEM. (**b**) When monkeys do saccade to the mouth region, it is tightly correlated with the onset of mouth movements ($r = 0.997, p < 0.00001$)

videos of human speakers (Vatikiotis-Bateson et al., 1998; Klin et al., 2002). Macaque monkeys exhibit a very similar strategy. The eye movement patterns of monkeys viewing conspecifics producing vocalizations reveal that monkeys spend most of their time inspecting the eye region relative to the mouth (Ghazanfar et al., 2006) (Fig. 7.2a). When they did fixate on the mouth, it correlated highly with the onset of mouth movements (Fig. 7.2b). Indeed, such a strategy is probably adaptive, as suggested by evidence from human paradigms. For example, human subjects asked to identify words fixated more on the mouth region at the onset of facial motion (Lansing & McConkie, 2003).

7.2.4 Behavioral Integration of Faces and Voices

All of the above experiments with monkeys and apes test whether or not they can *match* faces to voices. None provide direct evidence as to whether multisensory communication signals provide a behavioral advantage. To bridge this gap, monkeys were trained to detect auditory, visual, or audiovisual vocalizations embedded in noise as fast and as accurately as possible (Chandrasekaran et al., 2011). A free-response task was designed to approximate a natural face-to-face vocal communication event. In such settings, the vocal components of the communication signals are degraded by environmental noise. The face and its motion, on the other hand, are usually perceived clearly. In the task, monkeys responded to "coo" calls that are affiliative vocalizations commonly produced by macaque monkeys in a variety of contexts (Fig. 7.3). All vocalizations had five different levels of sound intensity and were embedded in a constant background noise. For dynamic faces, computer-generated monkey avatars were used (Fig. 7.3a). The use of avatars allowed the restriction of facial motion to the mouth region and ensured constant lighting and background and the parameterization of the size of the mouth opening while keeping eye and head positions constant. The degree of mouth opening was in accordance with the intensity of the associated vocalization: Greater sound intensity was coupled to larger mouth openings by the dynamic face. Two coos were paired with two monkey avatars, respectively, and this pairing was kept constant (Fig. 7.3a).

During the task, one avatar face would be continuously on the screen for a block of trials; the background noise was also continuous (Fig. 7.3b). In the "visual only (V)" condition, this avatar would move its mouth without any corresponding auditory component; that is, it silently produced a coo. In the "auditory-only (A)" condition, the vocalization normally paired with the *other* avatar (which is not on the screen) is presented with the *static* face of the avatar. Finally, in the "audiovisual (AV)" condition, the avatar moves its mouth accompanied by the corresponding vocalization and in accordance (degree of mouth opening) with its intensity. Each condition (V, A, or AV) was presented after a variable (drawn from a uniform distribution) interval between 1 and 3 s, and subjects indicated the detection of an event (visible mouth motion, auditory signal, or both) by pressing a lever. Under these task conditions, monkeys (like humans in similar experiments) integrated faces and voices, allowing them to decrease their reaction times significantly relative to unimodal conditions (Chandrasekaran et al., 2011)(Fig. 7.3c). Currently, this is the only evidence for a behavioral advantage for combining faces and voices in a nonhuman primate species.

Taken together, these behavioral data suggest that humans are not at all unique among primates in their ability to perceive and integrate communication signals across modalities. As described in Sections 7.3 and 7.4, the neural mechanisms by which monkeys process these signals are also similar to those used by humans.

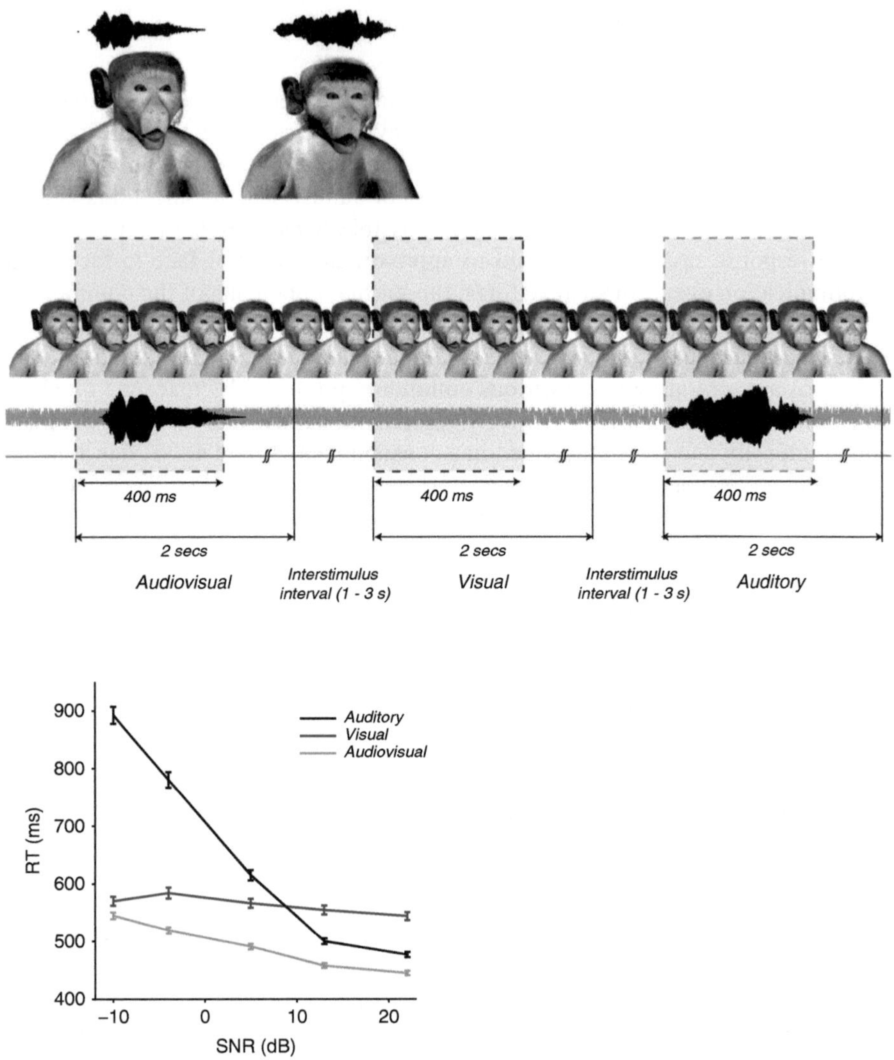

Fig. 7.3 Audiovisual integration of coo calls in a free-response paradigm. (**a**) Waveform of coo vocalizations detected by the monkeys and their respective avatars below. (**b**) Free-response paradigm task structure. An avatar face was always on the screen. Visual, auditory, and audiovisual stimuli were randomly presented with an interstimulus interval of 1–3 s drawn from a uniform distribution. Responses within a 2-s window after stimulus onset were considered to be hits. Responses in the interstimulus interval are considered to be false alarms and led to timeouts. (**c**) Mean RTs obtained by pooling across all sessions as a function of SNR for the unisensory and multisensory conditions for one monkey. Error bars denote standard error of the mean estimated using bootstrapping. The *x*-axes denote SNR in dB. The *y*-axes depict RT in milliseconds

7.3 Neocortical Processing of Face/Voice Signals in Monkeys

Although it is generally recognized that humans and other animals use the different senses in an integrated fashion, it is assumed that, at the neural level, these senses are, for the most part, processed independently but then converge at critical nodes in the neocortex. This idea extends as far back as Leonardo da Vinci's (1452–1519) research into the neuroanatomy of the human brain. He suggested that there was an area above the pituitary fossa where the five senses converged (the "sensu comune") (Pevsner, 2002). Until recently, this basic tenet of neocortical organization had not changed to a large degree since da Vinci's time, as it had long been argued that different regions of the cortex have different functions segregated according to sense modality. Some regions receive visual sensations, others auditory sensations, and still others tactile sensations (so and so forth, for olfaction and gustation). Each of these sensory regions is thought to send projections that converge on an "association area" that then enables the association of between the different senses and between the senses and movement.

Thus, according to this traditional line of thinking, the linking of vision with audition in the multisensory vocal perception described in the preceding text would be attributed to the functions of association areas such as the superior temporal sulcus in the temporal lobe or the principal and intraparietal sulci located in the frontal and parietal lobes, respectively. Although these regions may certainly play important roles, they are certainly not necessary for all types of multisensory behaviors (Ettlinger & Wilson, 1990), nor are they the sole regions for multisensory convergence (Ghazanfar & Schroeder, 2006; Driver & Noesselt, 2008). The auditory cortex, in particular, has many potential sources of visual inputs (Ghazanfar & Schroeder, 2006), and this is borne out in the increasing number of studies demonstrating visual modulation of auditory cortical activity (Ghazanfar et al., 2005, 2008; Kayser et al., 2008). The focus here is on those auditory cortical studies investigating face/voice integration specifically.

7.3.1 The Auditory Cortical Node in Audiovisual Communication

Two types of neural signals in auditory cortex have been investigated: spiking activity and local field potential activity. Spiking activity represents the output responses (in the form of action potentials) of one or more neurons in a given area. Local field potentials (LFPs), on the other hand, largely reflect input processes that occur on a slower timescale, including synaptic potentials. Local field potentials can be further broken down into distinct rhythms. In fact, the neocortex in general produces stereotypical rhythms or *oscillations* (Buzsaki & Draguhn, 2004). These neocortical oscillations are regular amplitude modulations, representing fluctuations between low and high excitability states. These modulations range in frequency from very slow (e.g., 0.05 Hz) to very fast (e.g., 500 Hz). The different frequencies are thought to represent the size of the underlying neural networks that produce them, with higher

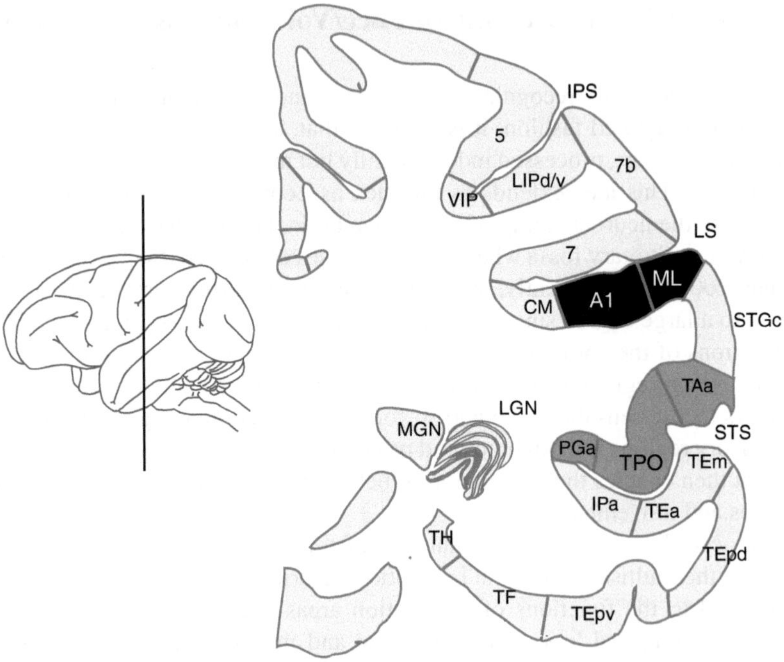

Fig. 7.4 Structures in the temporal lobe sensitive to multisensory vocal communication signals. Coronal slice through the macaque brain showing the regions in the temporal lobe highlighting auditory areas that have been shown to integrate visual and auditory components of communication signals (in dark gray) and the upper bank of the STS (lighter gray) which is sensitive to dynamic faces and is source of visual input into auditory cortex

frequency oscillations mediated by smaller networks of neurons than lower frequency oscillations. The set of oscillations that the most is known about are labeled with Greek letters: δ (1–3 Hz), θ (3–8 Hz), α (8–14 Hz), β (14–30 Hz), and γ (>30 Hz). These different oscillations are associated with different brain states and, under certain conditions, multiple rhythms can occur at the same time in a particular neocortical area or even across different areas (Ghazanfar et al., 2008; Chandrasekaran & Ghazanfar, 2009).

Recordings from both primary (A1) and lateral belt (middle lateral belt area, ML) auditory cortex in the monkey reveal that responses to the voice are influenced by the presence of a dynamic face (Fig. 7.4). This was consistent with evoked potential studies in humans (Besle et al., 2004; van Wassenhove et al., 2005). Monkey subjects viewing unimodal and bimodal versions of two different species-typical vocalizations ("coos" and "grunts") show both enhanced and suppressed LFP responses in the bimodal condition relative to the unimodal auditory condition (Ghazanfar et al., 2005). These modulations of neural signal strength are considered "integrative" in the sense that the change in response magnitude is significantly greater than the strongest unimodal response, and in most cases, greater than the sum of the two unimodal responses. In monkeys, the combination of faces and voices led to integrative responses in the vast majority of auditory cortical sites—both in primary auditory cortex and the lateral belt auditory cortex. The data demonstrated that LFP signals in the auditory cortex are capable of multisensory integration of facial and vocal signals in monkeys (Ghazanfar et al., 2005), and such a hypothesis has subsequently been shown to be true also at the level of spiking activity from single neurons in the lateral belt cortex (Ghazanfar et al., 2008).

One possibility is that *any* arbitrary visual signal paired with a vocalization could lead to audiovisual integration in the auditory cortex, perhaps through arousal effects. Replacing the dynamic faces with dynamic discs that mimicked the aperture and displacement of the mouth tested this possibility. In human psychophysical experiments, such artificial dynamic stimuli can still lead to enhanced speech detection, but not to the same degree as a real face (Bernstein et al., 2004; Schwartz et al., 2004). Thus, these dynamics discs were high contrast, biological motion stimuli. When LFP or spiking activity were investigated after presentations with dynamic discs, far less integration was seen when compared to when real monkey faces were presented (Ghazanfar et al., 2005, 2008) (Fig. 7.5a, b). This was true primarily for the lateral belt auditory cortex and was observed to a lesser extent in the primary auditory cortex.

7.3.2 The Superior Temporal Sulcus Is a Source for Visual Modulation of Auditory Cortical Activity

What is the source of face-specific visual input into auditory cortex? Although there are several visually sensitive regions that project to auditory cortex (Cappe & Barone, 2005; Ghazanfar & Schroeder, 2006), one region ideally suited to modulate activity in auditory cortex is the upper bank of the superior temporal sulcus (STS) (Bruce et al., 1981; Baylis et al., 1987; Hikosaka et al., 1988)(see Fig. 7.3). The STS is an excellent candidate for visual input into auditory cortex because neurons in this region are predominantly visual (Bruce et al., 1981), highly sensitive to biological motion such as dynamic faces (Ghazanfar et al., 2010), and can integrate faces and voices (Barraclough et al., 2005; Chandrasekaran & Ghazanfar, 2009) (see Fig. 7.4b, c). Reciprocal anatomical connections are known to be present between parts of the STS and the belt region of auditory cortex (Barnes & Pandya, 1992; Seltzer & Pandya, 1994).

The functional relationships between STS and the lateral belt region of auditory cortex during audiovisual vocalization processing was tested in the following manner: (1) recording LFPs from the lateral belt region of auditory cortex and STS concurrently; (2) breaking the LFP signal into the different frequency bands reflecting neural oscillations; and (3) measuring the correlations between these oscillations as a function of stimulus condition using a "cross-spectrum" analysis. The focus of the analyses were on the high-frequency γ band (>30Hz) oscillations and revealed the activity between the auditory cortex and the STS was more strongly correlated during the presentation of faces and voices together relative to the unimodal conditions

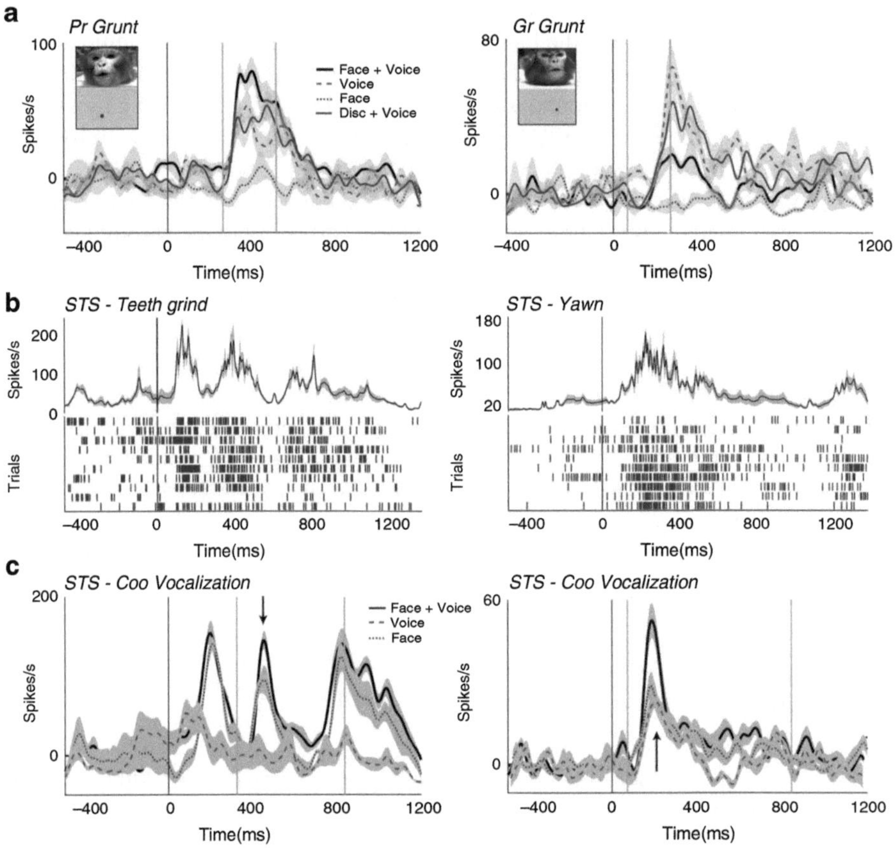

Fig. 7.5 Integrating vision and audition in auditory cortex and the upper bank of STS. (**a**) Single-neuron examples of multisensory integration of Face+Voice stimuli compared with Disk+Voice stimuli in the lateral belt area of auditory cortex. The left panel shows an enhanced response when voices are coupled with faces, but no similar modulation when coupled with dynamic disks. The right panel shows similar effects for a suppressed response. The *x*-axes show time aligned to onset of the face (black vertical line). Gray vertical lines indicate the onset and offset of the voice signal. The *y*-axes depict the firing rate of the neuron in spikes per second. Shaded regions denote the SEM. (**b**) Neurons in the upper bank of the STS respond to dynamic faces. The left panel shows a neuron in the upper bank of the STS that responded to a teeth-grind. The right panel shows another neuron that responded well to a yawn. The *x*-axes show time aligned to onset of the face (black vertical line) in milliseconds. The *y*-axes depict the firing rate of the neuron in spikes per second. Shaded regions denote the SEM. (**c**) Single-neuron examples of multisensory integration of communication signals in the upper bank of the STS. The left panel shows an enhanced response when voices are coupled with faces. The right panel shows similar effects for another coo vocalization. Conventions as in **A**

(Ghazanfar et al., 2008)(Fig. 7.6a). Because the cross-spectrum analysis conflates coordinated changes in strength with those of timing, a separate analysis called "phase coherence" was used to examine only changes in timing. This analysis revealed that correlation strength changes between the two structures were also

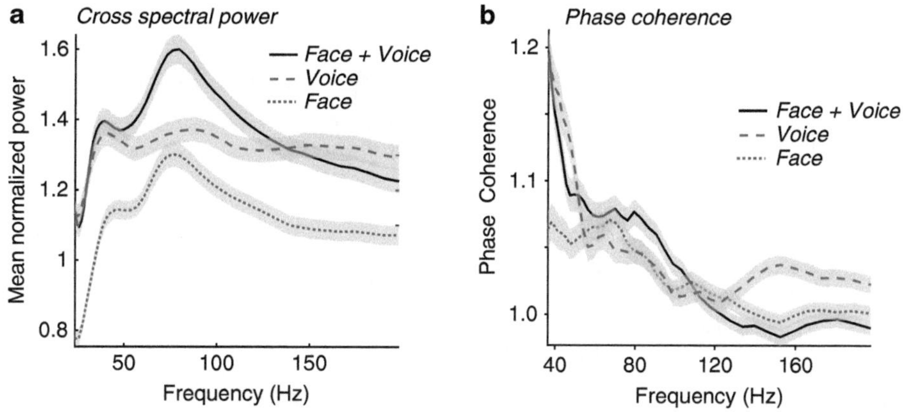

Fig. 7.6 STS is a source of visual input into auditory cortex. **(a)** Cross-spectral power between the LFPs in auditory cortex and the upper bank of the STS from 0 to 300 ms. The *x*-axis depicts neural frequency band in Hz. The *y*-axis depicts mean normalized cross-spectral power. Shaded regions denote the SEM across all electrode pairs and calls. **(b)** Population phase concentration between STS and auditory cortex from 0 to 300 ms after voice onset. The *x*-axis depicts neural frequency band in Hz. The *y*-axis depicts the average normalized phase concentration. Shaded regions denote the SEM across all electrode pairs and calls

driven by tight temporal coordination of their respective γ oscillations (Fig. 7.6b). In sum, faces and voices generate stronger functional interactions between the auditory cortex and the STS.

These data suggest that the influence of vision via the STS does not drive auditory cortical neurons, but rather *modulates* their excitability, leading to enhancement or suppression. Further, the influence of the STS on auditory cortex was not merely restricted to γ oscillations. Spiking activity also seems to be modulated, but not "driven," by ongoing activity arising from the STS. That is, the spiking activity in auditory cortex seemed to be influenced by the activity in the STS that preceded it. Two lines of evidence suggest this "modulation" scenario. First, visual influences on single neurons were most robust when in the form of dynamic faces and were apparent only when neurons had a significant response to a vocalization (i.e., there were no overt responses to faces alone). Second, these integrative responses were often "face-specific" and had a wide distribution of latencies, which suggested that the face signal was an ongoing signal that influenced auditory responses (Ghazanfar et al., 2008).

Local field potential signals from both the auditory cortex and the STS have multiple bands of oscillatory activity generated in responses to stimuli that might mediate different functions (Lakatos et al., 2005; Chandrasekaran & Ghazanfar, 2009). In the STS, these different bands of oscillatory activity seem to integrate faces and voices differently (Chandrasekaran & Ghazanfar, 2009), and such integration was dependent on the time-to-voice (Fig. 7.7). Because different oscillation frequencies imply different cortical network sizes, these differences in integration may reflect different underlying multisensory computations using networks with

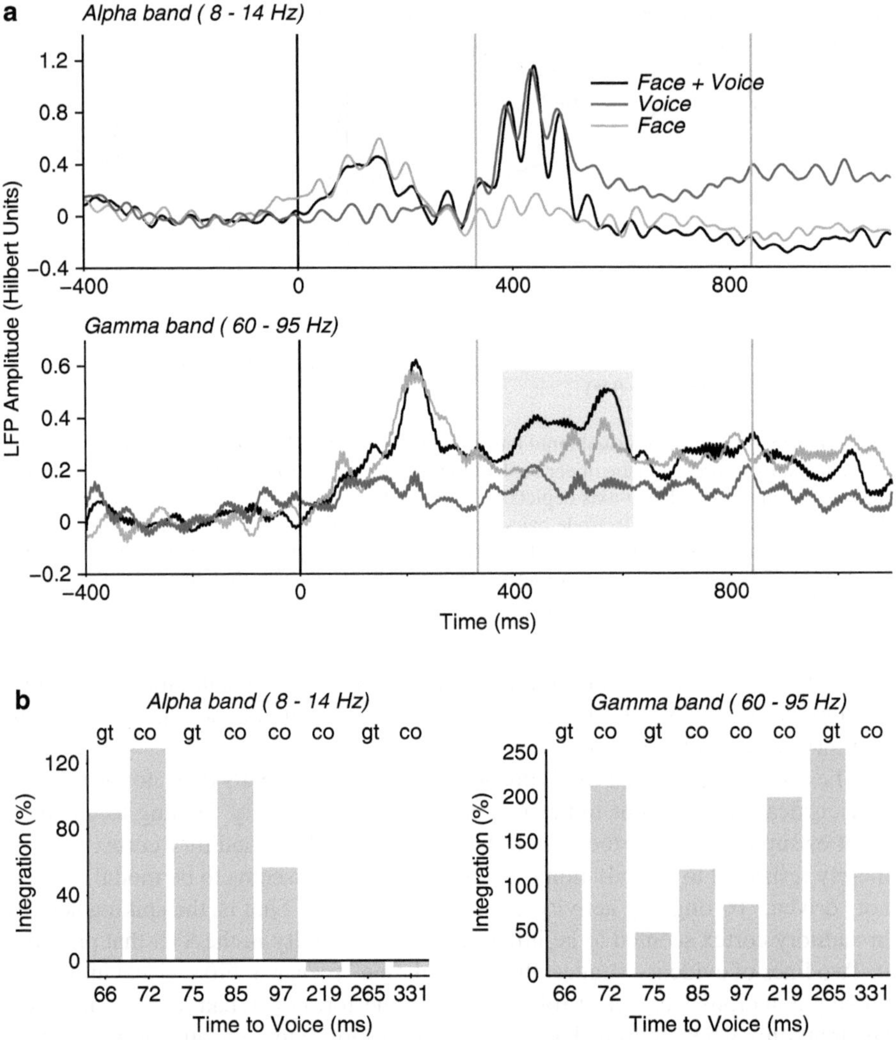

Fig. 7.7 Different neural frequency bands integrate faces and voices differently. (**a**) Top, baseline corrected alpha band activity (8–14 Hz) for the three conditions, face + voice, face alone, and voice alone for a coo call with a 331-ms time-to-voice. Bottom panel, baseline corrected gamma (60–95 Hz) band activity for the same coo call. The *x*-axes depict time in milliseconds. The *y*-axes depicts baseline corrected power in Hilbert units. The black line denotes onset of the face; solid gray lines, the onset and offset of the voice. (**b**) Left, percent integration of the peak face + voice alpha response relative to the voice alone response as a function of the time-to-voice. The *x*-axes depict time to voice in milliseconds; the *y*-axes depict integration in percent. Labels co and gt denote coos and grunts respectively. Right, percent integration of the peak face + voice gamma response relative to the voice alone response as a function of the time to voice. Conventions as in left panel

different spatial scales (Senkowski et al., 2008). Below 20 Hz, and in response to naturalistic audiovisual stimuli, there are directed interactions from auditory cortex to STS, whereas above 20 Hz, there are directed interactions from STS to auditory cortex (Kayser & Logothetis, 2009). Given that mouth movements are in the range of 2–7 Hz, at least in humans, (Chandrasekaran et al., 2009), and eye movements such as saccades and microsaccades are in the range of 3–4 Hz (Otero-Millan et al., 2008; Shepherd et al., 2010), it is possible that these lower frequency interactions between the STS and auditory cortex also represent distinct multisensory processing channels.

The source of visual influences on auditory cortex is not likely to be STS alone. Other areas, consistent with their connectivity and response properties (e.g., sensitivity to faces and voices), could also (and very likely) have a visual influence on auditory cortex. These include the ventrolateral prefrontal cortex (Romanski et al., 2005; Sugihara et al., 2006) and the amygdala (Gothard et al., 2007; Kuraoka & Nakamura, 2007).

7.4 Mechanism for Visual Modulation of Auditory Cortical Signals: Phase Resetting

How do visual inputs modulate, but not drive, auditory cortical responses? One candidate mechanism for this modulation is "phase resetting" (Lakatos et al., 2007). Phase resetting occurs when an ongoing neural oscillation, in essence, starts over after the presentation of a stimulus (Fig. 7.8a). Schroeder et al. (2008), based on results from somatosensory–auditory integration (Lakatos et al., 2007), hypothesized that, during audiovisual speech, the onset of mouth motion before the voice could lead to a phase reset of ongoing oscillations in auditory cortex. Subsequent auditory inputs falling on high-excitability peaks of this reset oscillation will be amplified whereas auditory inputs falling on the low-excitability peaks of this oscillation would be suppressed (Fig. 7.8b). Schroeder and colleagues originally proposed it for LFPs and multiunit activity, but this idea could be extended to single neurons quite readily. For example, if LFPs are thought to reflect synaptic input and thereby membrane potential dynamics (Logothetis, 2002; Okun et al., 2010), then one hypothesis is that the onset of mouth motion depolarizes or hyperpolarizes the *membrane potential* of auditory neurons, and the time of arrival of auditory inputs would lead to multisensory responses being either enhanced or suppressed relative to the auditory alone responses. If one assumes that membrane potential dynamics track the LFP (Logothetis, 2002; Okun et al., 2010), and that visual inputs are weak and modulatory, this change in the membrane potential would in principle be observed as a phase reset in the LFP.

The original phase resetting hypothesis claim was made based on physiological data generated by using punctate (stimuli were <10 ms in duration) auditory and somatosensory stimuli. Therefore, the prediction is not straightforward with regard to how continuous mouth motion seen during the production of vocalizations would

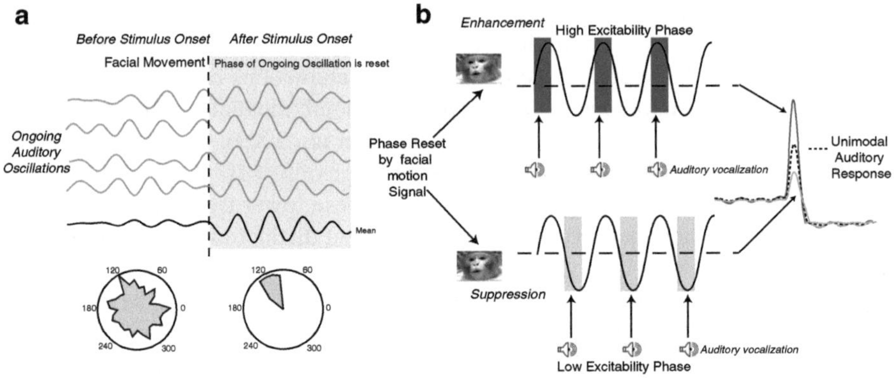

Fig. 7.8 Phase resetting as a mechanism for enhancement and suppression. (**a**) Before the onset of the visual stimulus, the phase of the auditory cortical oscillation is random and thus their mean across trials is flat and the phase distribution in the circular plot is widely dispersed. However, once the face comes on, the oscillations are reset and thus aligned with each other, leading to a consistent phase distribution over trials. Numbers around circular plots represent degree of phase angle. (**b**) In the overall mechanism of multisensory integration, the visual signal resets the phase of the ongoing oscillation (as in **a**) and when the auditory input arrives in a high-excitability phase then the response is enhanced. In contrast, when the auditory input arrives in the low-excitability region of the reset oscillation, then the response is suppressed

modify such processing in primary auditory cortex, or how shifting eye movements of observers (Vatikiotis-Bateson et al., 1998; Lansing & McConkie, 2003; Ghazanfar et al., 2006) would modify the phase resetting of ongoing oscillations (Ghazanfar & Chandrasekaran, 2007). A dynamic, vocalizing face is a complex sequence of sensory events, but one that elicits fairly stereotypical eye movements: humans and other primates fixate on the eyes but then saccade to mouth when it moves before saccading back to the eyes (Lansing & McConkie, 2003; Ghazanfar et al., 2006). Eye position influences single neuron and local field potential activity in multiple regions of auditory cortex (Werner-Reiss et al., 2003; Fu et al., 2004). Therefore, one possibility is that the *eye fixations* at the onset of mouth movements (Fig. 7.2b) send a signal to auditory cortex, which resets the phase of an ongoing oscillation. Such effects have been, for example, already seen in the primary visual cortex (Rajkai et al., 2008). This eye movement signal thus primes auditory cortex to amplify or suppress (depending on the timing) the neural response to a subsequent auditory signal originating from the mouth. Given that mouth movements precede the voiced components of both human (Abry et al., 1996; Chandrasekaran & Ghazanfar, 2009; Chandrasekaran et al., 2009) and monkey vocalizations (Ghazanfar et al., 2005; Chandrasekaran & Ghazanfar, 2009), the temporal order of visual to auditory signals is consistent with this idea. This hypothesis is also supported (though indirectly) by the finding that suppressed versus enhanced responses to face/voice signals in auditory cortex and the STS are influenced by the timing of mouth movements relative to the onset of the voice (Ghazanfar et al., 2005; Chandrasekaran & Ghazanfar, 2009). Finally, as STS neurons seem to be involved in saccades to visual targets (Scalaidhe et al., 1995) and respond to mouth movements and biological motion (Puce & Perrett, 2003; Barraclough et al., 2005; Ghazanfar et al., 2010), they would be a perfect candidate again to modulate auditory cortex during this process (Ghazanfar et al., 2008).

7.5 Conclusions

The temporal structure of vocalizations and their link to orofacial movements are critical for both human and nonhuman primate vocal recognition. Facial movements assist in improved perception, especially in noisy and unstructured auditory environments, helping the receiver better segment auditory cues from the cacophony of signals around them (Sumby & Pollack, 1954; Helfer & Freyman, 2005; Ross et al., 2007; Chandrasekaran et al., 2011). Although several association cortical areas in the temporal, parietal, and frontal lobes integrate visual and auditory cues, it is now increasingly recognized that much if not all of the neocortex is multisensory (Ghazanfar & Schroeder, 2006). For vocal processing in primates, auditory cortical neurons and networks are strongly modulated by visual signals coming from the face region. One source of this visual input is the upper bank of the superior temporal sulcus. The mechanisms by which this visual input modulates auditory cortical neurons is unknown, but "phase resetting" may be a plausible mechanism.

Acknowledgments The authors gratefully acknowledge the scientific contributions and numerous discussions with Ipek Kulahci, Joost Maier, Darshana Narayanan, Stephen Shepherd, and Daniel Takahashi. This work was supported by NIH R01NS054898, NSF BCS-0547760 CAREER Award and the James S. McDonnell Scholar Award.

References

Abry, C., Lallouache, M. T., & Cathiard, M. A. (1996). How can coarticulation models account for speech sensitivity in audio-visual desynchronization? In D. Stork & M. Henneke (Eds.), *Speechreading by humans and machines: Models, systems and applications* (pp. 247–255). Berlin: Springer-Verlag.

Barnes, C. L., & Pandya, P. K. (1992). Efferent cortical connections of multimodal cortex of the superior temporal sulcus in the rhesus monkey. *Journal of Comparative Neurology*, 318(2), 222–244.

Barraclough, N. E., Xiao, D., Baker, C. I., Oram, M. W., & Perrett, D. I. (2005). Integration of visual and auditory information by superior temporal sulcus neurons responsive to the sight of actions. *Journal of Cognitive Neuroscience*, 17(3), 377–391.

Baylis, G. C., Rolls, E. T., & Leonard, C. M. (1987). Functional subdivisions of the temporal lobe neocortex. *The Journal of Neuroscience*, 7(2), 330–342.

Bernstein, L. E., Auer, E. T., & Takayanagi, S. (2004). Auditory speech detection in noise enhanced by lipreading. *Speech Communication*, 44(1–4), 5–18.

Besle, J., Fort, A., Delpuech, C., & Giard, M.-H. (2004). Bimodal speech: Early suppressive visual effects in human auditory cortex. *European Journal of Neuroscience*, 20(8), 2225–2234.

Bruce, C., Desimone, R., & Gross, C. G. (1981). Visual properties of neurons in a polysensory area in superior temporal sulcus of the macaque. *Journal Of Neurophysiology*, 46(2), 369–384.

Buzsaki, G., & Draguhn, A. (2004). Neuronal oscillations in cortical networks. *Science*, 304, 1926–1929.

Cappe, C., & Barone, P. (2005). Heteromodal connections supporting multisensory integration at low levels of cortical processing in the monkey. *European Journal of Neuroscience*, 22(11), 2886–2902.

Chandrasekaran, C., & Ghazanfar, A. A. (2009). Different neural frequency bands integrate faces and voices differently in the superior temporal sulcus. *Journal of Neurophysiology*, 101(2), 773–788.

Chandrasekaran, C., Trubanova, A., Stillittano, S., Caplier, A., & Ghazanfar, A. A. (2009). The natural statistics of audiovisual speech. *PLoS Computational Biology*, 5(7), e1000436.

Chandrasekaran, C., Lemus, L., Trubanova, A., Gondan, M., & Ghazanfar, A. A. (2011). Monkeys and humans share a common computation for face/voice integration. *PLoS Computational Biology*, 7, e1002165.

Cohen, Y. E., Theunissen, F., Russ, B. E., & Gill, P. (2007). Acoustic features of rhesus vocalizations and their representation in the ventrolateral prefrontal cortex. *Journal of Neurophysiology*, 97(2), 1470–1484.

Driver, J., & Noesselt, T. (2008). Multisensory interplay reveals crossmodal influences on 'sensory-specific' brain regions, neural responses, and judgments. *Neuron*, 57(1), 11–23.

Drullman, R. (1995). Temporal envelope and fine structure cues for speech intelligibility. *The Journal of the Acoustical Society of America*, 97(1), 585–592.

Drullman, R., Festen, J. M., & Plomp, R. (1994). Effect of reducing slow temporal modulations on speech reception. *The Journal of the Acoustical Society of America*, 95(5 Pt 1), 2670–2680.

Ettlinger, G., & Wilson, W. A. (1990). Cross-modal performance: Behavioural processes, phylogenetic considerations and neural mechanisms. *Behavioural Brain Research*, 40, 169–192.

Evans, T. A., Howell, S., & Westergaard, G. C. (2005). Auditory-visual cross-modal perception of communicative stimuli in tufted capuchin monkeys (*Cebus apella*). *Journal of Experimental Psychology-Animal Behavior Processes*, 31, 399–406.

Fitch, W. T. (1997). Vocal tract length and formant frequency dispersion correlate with body size in rhesus macaques. *The Journal of the Acoustical Society of America*, 102(2), 1213–1222.

Fitch, W. T., & Hauser, M. (1995). Vocal production in nonhuman primates: Acoustics, physiology, and functional constraints on ldquohonestrdquo advertisement. *American Journal of Primatology*, 37(3), 191–219.

Fu, K. M. G., Shah, A. S., O'Connell, M. N., McGinnis, T., Eckholdt, H., Lakatos, P., et al. (2004). Timing and laminar profile of eye-position effects on auditory responses in primate auditory cortex. *Journal Of Neurophysiology*, 92(6), 3522–3531.

Ghazanfar, A. A., & Chandrasekaran, C. F. (2007). Paving the way forward: Integrating the senses through phase-resetting of cortical oscillations. *Neuron*, 53(2), 162–164.

Ghazanfar, A. A., & Logothetis, N. K. (2003). Facial expressions linked to monkey calls. *Nature*, 423(6943), 937–938.

Ghazanfar, A. A., & Rendall, D. (2008). Evolution of human vocal production. *Current Biology*, 18(11), R457–460.

Ghazanfar, A. A., & Schroeder, C. E. (2006). Is neocortex essentially multisensory? *Trends in Cognitive Sciences*, 10(6), 278–285.

Ghazanfar, A. A., Smith-Rohrberg, D., & Hauser, M. D. (2001). The role of temporal cues in rhesus monkey vocal recognition: Orienting asymmetries to reversed calls. *Brain Behavior And Evolution*, 58(3), 163–172.

Ghazanfar, A. A., Smith-Rohrberg, D., Pollen, A. A., & Hauser, M. D. (2002). Temporal cues in the perception of long calls by tamarins. *Animal Behaviour*, 64, 427–438.

Ghazanfar, A. A., Maier, J. X., Hoffman, K. L., & Logothetis, N. K. (2005). Multisensory integration of dynamic faces and voices in rhesus monkey auditory cortex. *The Journal of Neuroscience*, 25(20), 5004–5012.

Ghazanfar, A. A., Nielsen, K., & Logothetis, N. K. (2006). Eye movements of monkey observers viewing vocalizing conspecifics. *Cognition*, 101(3), 515–529.

Ghazanfar, A. A., Turesson, H. K., Maier, J. X., van Dinther, R., Patterson, R. D., & Logothetis, N. K. (2007). Vocal-tract resonances as indexical cues in rhesus monkeys. *Current Biology*, 17(5), 425–430.

Ghazanfar, A. A., Chandrasekaran, C., & Logothetis, N. K. (2008). Interactions between the superior temporal sulcus and auditory cortex mediate dynamic face/voice integration in rhesus monkeys. *The Journal of Neuroscience*, 28(17), 4457–4469.

Ghazanfar, A. A., Chandrasekaran, C., & Morrill, R. J. (2010). Dynamic, rhythmic facial expressions and the superior temporal sulcus of macaque monkeys: Implications for the evolution of audiovisual speech. *European Journal of Neuroscience*, 31(10), 1807–1817.

Godey, B. T., Atencio, C. A., Bonham, B. H., Schreiner, C. E., & Cheung, S. W. (2005). Functional organization of squirrel monkey primary auditory cortex: Responses to frequency-modulation sweeps. *Journal of Neurophysiology*, 94(2), 1299–1311.

Gothard, K. M., Battaglia, F. P., Erickson, C. A., Spitler, K. M., & Amaral, D. G. (2007). Neural responses to facial expression and face identity in the monkey amygdala. *Journal of Neurophysiology*, 97(2), 1671–1683.

Greenberg, S., Carvey, H., Hitchcock, L., & Chang, S. (2003). Temporal properties of spontaneous speech—a syllable-centric perspective. *Journal of Phonetics*, 31(3–4), 465–485.

Hauser, M. D., & Andersson, K. (1994). Left hemisphere dominance for processing vocalizations in adult, but not infant, rhesus monkeys: Field experiments. *Proceedings of the National Academy of Sciences of the USA*, 91(9), 3946–3948.

Hauser, M. D., & Ybarra, M. S. (1994). The role of lip configuration in monkey vocalizations—experiments using xylocaine as a nerve block. *Brain and Language*, 46(2), 232–244.

Hauser, M. D., Evans, C. S., & Marler, P. (1993). The role of articulation in the production of rhesus-monkey, *Macaca mulatta*, vocalizations. *Animal Behaviour*, 45(3), 423–433.

Hauser, M. D., Agnetta, B., & Perez, C. (1998). Orienting asymmetries in rhesus monkeys: The effect of time-domain changes on acoustic perception. *Animal Behaviour*, 56(1), 41–47.

Helfer, K. S., & Freyman, R. L. (2005). The role of visual speech cues in reducing energetic and informational masking. *The Journal of the Acoustical Society of America*, 117(2), 842–849.

Hikosaka, K., Iwai, E., Saito, H., & Tanaka, K. (1988). Polysensory properties of neurons in the anterior bank of the caudal superior temporal sulcus of the macaque monkey. *Journal of Neurophysiology*, 60(5), 1615–1637.

Jordan, K. E., Brannon, E. M., Logothetis, N. K., & Ghazanfar, A. A. (2005). Monkeys match the number of voices they hear to the number of faces they see. *Current Biology*, 15(11), 1034–1038.

Kayser, C., & Logothetis, N. K. (2009). Directed interactions between auditory and superior temporal cortices and their role in sensory integration. *Frontiers in Integrative Neuroscience*, 3, 7. doi: 10.3389/neuro.07.007.2009.

Kayser, C., Petkov, C. I., & Logothetis, N. K. (2008). Visual modulation of neurons in auditory cortex. *Cerebral Cortex*, 18, 1560–1574.

Klin, A., Jones, W., Schultz, R., Volkmar, F., & Cohen, D. (2002). Visual fixation patterns during viewing of naturalistic social situations as predictors of social competence in individuals with autism. *Archives of General Psychiatry*, 59, 809–816.

Kuraoka, K., & Nakamura, K. (2007). Responses of single neurons in monkey amygdala to facial and vocal emotions. *Journal of Neurophysiology*, 97(2), 1379–1387.

Lakatos, P., Shah, A. S., Knuth, K. H., Ulbert, I., Karmos, G., & Schroeder, C. E. (2005). An oscillatory hierarchy controlling neuronal excitability and stimulus processing in the auditory cortex. *Journal of Neurophysiology*, 94(3), 1904–1911.

Lakatos, P., Chen, C.-M., O'Connell, M. N., Mills, A., & Schroeder, C. E. (2007). Neuronal oscillations and multisensory interaction in primary auditory cortex. *Neuron*, 53(2), 279–292.

Lansing, C. R., & McConkie, G. W. (2003). Word identification and eye fixation locations in visual and visual-plus-auditory presentations of spoken sentences. *Perception & Psychophysics*, 65(4), 536–552.

Logothetis, N. K. (2002). The neural basis of the blood-oxygen-level-dependent functional magnetic resonance imaging signal. *Philosophical Transactions of the Royal Society B: Biological Sciences*, 357(1424), 1003–1037.

MacNeilage, P. F. (1998). The frame/content theory of evolution of speech production. *Behavioral and Brain Sciences*, 21(4), 499–511.

Munhall, K. G., & Tohkura, Y. (1998). Audiovisual gating and the time course of speech perception. *The Journal of the Acoustical Society of America*, 104(1), 530–539.

Munhall, K. G., & Vatikiotis-Bateson, E. (2004). Spatial and temporal constraints on audiovisual speech perception In G. A. Calvert, C. Spence, & B. E. Stein (Eds.), *The handbook of multisensory processes* (pp. 177–188). Cambridge, MA: MIT Press.

Okun, M., Naim, A., & Lampl, I. (2010). The subthreshold relation between cortical local field potential and neuronal firing unveiled by intracellular recordings in awake rats. *The Journal of Neuroscience*, 30(12), 4440–4448.

Otero-Millan, J., Troncoso, X. G., Macknik, S. L., Serrano-Pedraza, I., & Martinez-Conde, S. (2008). Saccades and microsaccades during visual fixation, exploration, and search: Foundations for a common saccadic generator. *Journal of Vision*, 8(14), 1–18.

Parr, L. A. (2004). Perceptual biases for multimodal cues in chimpanzee (*Pan troglodytes*) affect recognition. *Animal Cognition*, 7(3), 171–178.

Partan, S. R. (2002). Single and multichannel facial composition: Facial expressions and vocalizations of rhesus macaques (*Macaca mulatta*). *Behaviour*, 139, 993–1027.

Pevsner, J. (2002). Leonardo da Vinci's contributions to neuroscience. *Trends in Neurosciences*, 25, 217–220.

Puce, A., & Perrett, D. I. (2003). Electrophysiology and brain imaging of biological motion. *Philosophical Transactions of the Royal Society B: Biological Sciences*, 358(1431), 435–445.

Rajkai, C., Lakatos, P., & Schroeder, C. E. (2008). Visual fixation-related neuronal activity in the auditory cortex of monkeys. Paper presented at the Society for Neuroscience, Washington, DC, November 15–19. Abstract 770.2

Redican, W. K. (1975). Facial expressions in nonhuman primates. In L. A. Rosenblum (Ed.), *Primate behavior: Developments in field and laboratory research*. (pp. 103–194). New York: Academic Press.

Romanski, L. M., Averbeck, B. B., & Diltz, M. (2005). Neural representation of vocalizations in the primate ventrolateral prefrontal cortex. *Journal of Neurophysiology*, 93(2), 734–747.

Ross, L. A., Saint-Amour, D., Leavitt, V. M., Javitt, D. C., & Foxe, J. J. (2007). Do you see what I am saying? Exploring visual enhancement of speech comprehension in noisy environments. *Cerebral Cortex*, 17(5), 1147–1153.

Scalaidhe, S. P., Albright, T. D., Rodman, H. R., & Gross, C. G. (1995). Effects of superior temporal polysensory area lesions on eye movements in the macaque monkey. *Journal of Neurophysiology*, 73(1), 1–19.

Schroeder, C. E., Lakatos, P., Kajikawa, Y., Partan, S., & Puce, A. (2008). Neuronal oscillations and visual amplification of speech. *Trends in Cognitive Sciences*, 12(3), 106–113.

Schwartz, J.-L., Berthommier, F., & Savariaux, C. (2004). Seeing to hear better: evidence for early audio-visual interactions in speech identification. *Cognition*, 93(2), B69–B78.

Seltzer, B., & Pandya, D. N. (1994). Parietal, temporal, and occipita projections to cortex of the superior temporal sulcus in the rhesus monkey: A retrograde tracer study. *Journal of Comparative Neurology*, 343(3), 445–463.

Senkowski, D., Schneider, T. R., Foxe, J. J., & Engel, A. K. (2008). Crossmodal binding through neural coherence: Implications for multisensory processing. *Trends in Neurosciences*, 31(8), 401–409.

Shannon, R. V., Zeng, F.-G., Kamath, V., Wygonski, J., & Ekelid, M. (1995). Speech recognition with primarily temporal cues. *Science*, 270(5234), 303–304.

Shepherd, S. V., Steckenfinger, S. A., Hasson, U., & Ghazanfar, A. A. (2010). Human-monkey gaze correlations reveal convergent and divergent patterns of movie viewing. *Current Biology*, 20, 649–656.

Smith, Z. M., Delgutte, B., & Oxenham, A. J. (2002). Chimaeric sounds reveal dichotomies in auditory perception. *Nature*, 416(6876), 87–90.

Sugihara, T., Diltz, M. D., Averbeck, B. B., & Romanski, L. M. (2006). Integration of auditory and visual communication information in the primate ventrolateral prefrontal cortex. *The Journal of Neuroscience*, 26(43), 11138–11147.

Sumby, W. H., & Pollack, I. (1954). Visual contribution to speech intelligibility in noise. *Journal of the Acoustical Society of America*, 26(2), 212–215.

Teufel, C., Ghazanfar, A. A., & Fischer, J. (2010). On the relationship between lateralized brain function and orienting asymmetries. *Behavioral Neuroscience*, 124, 437–445.

Van Der Horst, R., Leeuw, A. R., & Wouter, A. D. (1999). Importance of temporal-envelope cues in consonant recognition. *Journal of the Acoustical Society of America*, 105(3), 1801–1809.

van Wassenhove, V., Grant, K. W., & Poeppel, D. (2005). Visual speech speeds up the neural processing of auditory speech. *Proceedings of the National Academy of Sciences of the USA*, 102(4), 1181–1186.

van Wassenhove, V., Grant, K. W., & Poeppel, D. (2007). Temporal window of integration in auditory-visual speech perception. *Neuropsychologia*, 45(3), 598–607.

Vatikiotis-Bateson, E., Eigsti, I. M., Yano, S., & Munhall, K. G. (1998). Eye movement of perceivers during audiovisual speech perception. *Perception & Psychophysics*, 60(6), 926–940.

Wang, X., Merzenich, M. M., Beitel, R., & Schreiner, C. E. (1995). Representation of a species-specific vocalization in the primary auditory cortex of the common marmoset: Temporal and spectral characteristics. *Journal of Neurophysiology*, 74(6), 2685–2706.

Werner-Reiss, U., Kelly, K. A., Trause, A. S., Underhill, A. M., & Groh, J. M. (2003). Eye position affects activity in primary auditory cortex of primates. *Current Biology*, 13(7), 554–562.

Yehia, H. C., Kuratate, T., & Vatikiotis-Bateson, E. (2002). Linking facial animation, head motion and speech acoustics. *Journal of Phonetics*, 30(3), 555–568.

Chapter 8
Neurophysiology of Attention and Memory Processing

Amy Poremba and James Bigelow

Keywords A1 • Auditory cortex • Delayed matching-to-sample • Habituation • Lateral intraparietal area • Medial geniculate body • Monkey • Neural plasticity • Non-human primate • Prefrontal cortex • Short-term memory • Stimulus-specific adaptation • Superior colliculus • Superior temporal gyrus • Working memory

Abbreviations

AC	auditory cortex
BF	best frequency
CF	characteristic frequency
CS	conditioned stimulus
DMTS	delayed matching-to-sample
dTP	dorsal temporal pole
ERP	event-related potential
ISI	interstimulus interval
LED	light-emitting diode
ITC	inferior temporal cortex
LFP	local field potential
LIP	lateral intraparietal
MGB	medial geniculate body
PETH	perievent time histogram
PFC	prefrontal cortex
PST	peristimulus time

A. Poremba (✉) • J. Bigelow
Department of Psychology, University of Iowa, E11 SSH, Iowa City, IA 52242, USA
e-mail: amy-poremba@uiowa.edu; james-bigelow@uiowa.edu

Y.E. Cohen et al. (eds.), *Neural Correlates of Auditory Cognition*, Springer Handbook of Auditory Research 45, DOI 10.1007/978-1-4614-2350-8_8,

© Springer Science+Business Media New York 2013

SC	superior colliculus
SOA	stimulus onset asynchrony
STG	superior temporal gyrus
STRF	spectrotemporal receptive field
TORCS	temporally orthogonal ripple combinations
TRN	thalamic reticular nucleus
vPFC	ventrolateral PFC
WM	working memory

8.1 Introduction

The "standard cocktail party" description of hearing voices in a crowded environment and being able to tune in to a conversation in which the person quickly identifies his or her name being mentioned may seem simple on the surface because people and animals can be conditioned to respond to important signals. However, Cherry's 1953 cocktail party example is compelling and often used because it invokes many processes ubiquitous to auditory processing, such as frequency encoding and stream segregation to attention and memory. One must remember the sound of his or her own name and be able to separate that specific sound from the other sounds that are translated into background noise in reference to the attended source. Responses to well-known signals, such as names, are the product of experience-induced plasticity located many places along the auditory pathway. This translation into the unique processing of a specific stimulus is an example of long-term memory in the auditory system. This evoked response to names is not only an example of long-term memory in the auditory system, but it also engenders attention to the conversation where it originated, bringing in working and short-term memory as well. The following sections discuss what is known of the cortical processing related to auditory attention and memory.

Neural processing of attention, that is, the act of focusing on a particular stimulus for processing, and memory, that is, holding a stimulus representation over time, are most likely complementary and synergistic in many ways but may vary in their neural bases. Attentional focus may increase the memory for particular stimuli. Even though it may not be required for all types of memory formation, the corollary is also true, that memories may modulate attentional focus on particular stimuli. This chapter provides an overview of these neural processes focusing on neurophysiological studies of nonhuman primates primarily at the cortical level, with some discussion of relevant findings in other animals. The study of the neural correlates of memory and attention is still fairly new, although there is still much to ascertain about auditory stimulus processing in general. Understanding neuronal encoding of auditory stimuli across a wide number of attentional and memory demands within and across modalities and various contexts will lead to a unified view of the neural circuitry and mechanisms underlying these critical processes.

8.2 Neural Correlates of Auditory Attention

Auditory attention is a broad term and many of its proposed component processes are similar to those across other modalities. The basic premise of attention is that a particular stimulus in the context will draw a listener's interest either because of the internal or external salience of the information. This focus process may or may not accompany a suppression of other incoming information. Two major organizing principles of attention are bottom-up and top-down processing capabilities. Briefly stated, bottom-up processing occurs when attention is paid automatically to a particular stimulus because it is salient in the moment (e.g., siren, loud noise, scream), or perhaps because it was very familiar or, conversely, infrequent (i.e., stimulus-driven). Top-down processing is typically the influence of cortical processing pertaining to goals, expectations, and/or memory on what and how information will be voluntarily processed from the available incoming sensory system. There are many excellent reviews on the different forms of attentional processes (e.g., Knudsen, 2007). As there is yet no established neural circuit for auditory attention, this chapter discusses attention broadly as any stimulus when heard in different contexts that brings about some change in the neural processing pattern.

8.2.1 The Role of Primary Auditory Cortex in Unimodal Attention Tasks

For many years the prevailing hypothesis concerning the primary functions of the core auditory cortical regions (A1 and R) indicated that these regions faithfully encode and represent incoming auditory stimuli (Yin et al., 2008). Many of the studies reviewed in this chapter show that even these primary areas may be modulated by attentional processing such as behavioral context and stimulus salience. For example, there were early indications that activity in the primary cortex could be modulated by factors other than the stimulus itself (Hubel et al., 1959; Miller et al., 1972; Hocherman et al., 1976; Benson & Hienz, 1978) despite the heavy early focus on anesthetized recording preparations.

Early studies of the cat primary auditory cortex (AC) casually mentioned neurons that seemed to respond to auditory stimuli when the subject was "paying attention" (Hubel et al., 1959; Evans & Whitfield, 1964). The percentage of neurons estimated to have evoked activity under probable conditions of auditory attention, for either arousal to a particular stimulus or a generally attentive state, was 10%–30%. Hubel et al. (1959) suggest that enhanced neural activity is likely brought about by novelty, stimulus meaning, or matching stimulation between auditory and visual modalities (see Fritz et al., 2007 for review). Although these early studies were not well controlled, they certainly foreshadowed the types of attentional activity found in primary auditory cortices as reviewed in this chapter.

In the 1970s the first controlled attentional study in primate AC was established when Miller et al. (1972) demonstrated differentiated activity between a simple auditory discrimination behavioral task and passive listening. Neuron responsiveness levels varied across three different conditions or contexts. The neurons were more responsive in well-trained monkeys performing the task than in trained animals that were not performing the task, compared with naïve monkeys in which responsiveness of the neurons were maintained only when the stimuli were not repetitive. This study suggested that although experience plays a role in focusing future attention, concurrent attentional mechanisms are also modulating the primary AC during task performance.

There were also some very early studies of awake monkeys performing selective auditory attention tasks. One such study conducted by Benson and Hienz (1978; see Fig. 8.1) showed recording from the primary AC in rhesus monkeys (*Macaca mulatta*). The monkeys were trained to perform a selective attention task in which they pressed a lighted response key on the side of the attended ear but made no response to stimuli presented to the unattended ear. Firing rate to the stimuli was significantly increased when presented to the attended ear in nearly 20% of the recorded units. There were no observed changes in the discharge pattern, or the spontaneous activity, between attended and nonattended ear presentations. These obvious attention-related changes in firing occurred as early as 20 ms, suggesting no top-down influence but rather local or bottom-up processing of the attention-related activity.

Several studies demonstrate that behavioral context has a strong influence over which auditory stimuli should be given attention within a conditioning task. Durif et al. (2003) reported that 33% of neurons within A1 and the middle lateral auditory belt (ML) of rhesus monkeys exhibited different responses to the exact same auditory stimulus when presented as either instruction or a confirmation cue in the same conditional acousticomotor task. The authors concluded that although the physical characteristics of a sound such as frequency and intensity are important in forming the frequency receptive field heavily represented in A1, the behavioral context created a strong modulation of this neuronal activity. This further confirms the role of attention in primary AC as single-unit responses vary with behavioral context.

Several studies in recent years have demonstrated attention-related rapid changes in the neuronal spectrotemporal receptive fields (STRFs) in primary auditory cortices of ferrets (*Mustela putorius furo*; Fritz et al., 2003, 2005a, c). A basic conditioning paradigm of conditioned avoidance is used wherein the ferret is trained to lick a spout for water delivery while one to seven reference sounds are delivered and the animal learns to avoid drinking during a target stimulus to avoid a mild shock. Thus the animal learns to attend to the target and to ignore the reference sounds. Typically, the reference sounds are ripple stimuli called TORCS (temporally orthogonal ripple combinations), that is, broadband stimuli that change spectrally over time, and help to characterize the STRF of individual neurons. The target stimuli vary across study but are spectrally and temporally distinct from the reference sounds. Across many variations of this task, there is generally an enhancement of the STRFs at the frequency or frequencies of the target stimuli and a suppression of the STRFs at and near the reference frequencies. These changes in overall amplitude and spectral selectivity of the STRF changes are similar whether there is enhancement at the target or suppression at the reference. These STRF differential changes do not occur

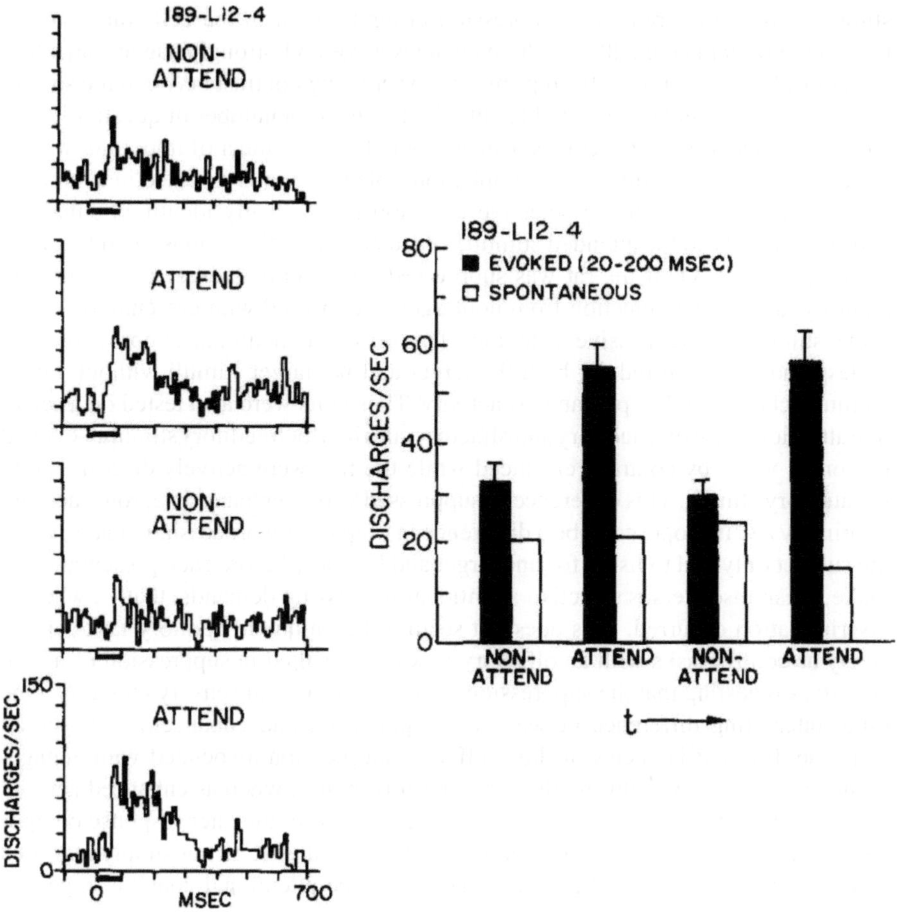

Fig. 8.1 PST histograms of unit 189-L12-4 for a sequence of Non-Attend and Attend conditions. Bar graphs represent the mean-evoked activity and standard error of the mean in each of the PST histograms over a period of 20–200 ms after stimulus onset and the spontaneous activity for the 100-ms period preceding stimulus onset. The behavioral conditions are arranged in chronological sequence beginning at the top for the histograms and at the left for the bar graphs. Stimulus: 8.0 kHz, 65 dB SPL, 100 ms duration. Stimulus duration indicated by bar below time axis of PST histograms. (Reproduced from Benson & Hienz; 1978; Figure 3, with permission)

during passive listening before the task, but may linger for various time periods during passive listening after the task. Fritz et al. (2007) suggest a contrast-matched filter hypothesis for these dynamic changes in STRF that occur during the behavioral tasks. This filter, through the changes in receptive field properties of A1, will enhance the contrast between reference and target stimuli. When this conditioned avoidance task is made more difficult by embedding the target stimulus in noise, and presumably increased attention is needed, high stimulus-to-noise ratio stimuli cause a greater enhancement at the target stimulus whereas low stimulus-to-noise ratio stimuli cause an overall gain suppression brought about in the cells far from the target tone (Atiani et al., 2009). These changes across A1 strongly suggest an attentional mechanism that is utilized primarily for creating contrast between the stimuli that may or may not be generated locally. It also leaves a number of questions to be answered, such as what the consequences would be if a number of important stimuli were used at the same time, for example, multiple key words repeated in a talk.

Whereas studies utilizing selective attention have usually identified enhanced neuronal activity to the attended stimuli, Otazu et al. (2009) demonstrated that neuronal activity in A1 of the rat was suppressed while animals were performing an auditory task (target detection from nontargets) contrasted with presentation of the same stimuli during passive listening. The observed neuronal suppression was widespread and occurred for both the target and nontarget stimuli without a concomitant change in the spontaneous activity. These rats were also tested on a selective attention task using auditory and olfactory cues in which auditory stimulus-evoked responses were, by contrast, enhanced while the rats were actively discriminating the auditory stimuli. This difference in suppressed versus enhanced neuronal activity in primary AC is proposed to be a difference in preparatory processing, that is, when the subject only had to listen for one target and be ready before each presentation to make a response, versus selective attentional processing demands, that is, wherein discrimination occurred. This does not seem to be unique to auditory tasks exclusively as control tasks with an olfactory task also resulted in suppression of neural activity, suggesting that the suppression occurs regardless of sensory modality. One other interesting difference between the suppression and enhancement observed were the different latencies of those effects. Suppression associated with engagement was evident even during the early-onset response, whereas enhanced activity observed during selective attention was rather limited to the later response components. The authors conclude that neuronal activity in the AC is not simply a limited resource to be increased when necessary, as unique attentional contexts engender distinct neural processing mechanisms. Overall these findings suggest a great deal of consideration should be paid to exactly what type of task the subject is being asked to accomplish and in what context said activity is occurring.

8.2.2 Primary AC and Attention Across Multiple Sensory Modalities

An early study investigating selective auditory attention by Hocherman et al. (1976) brings into focus current findings that show the primary AC to be directly affected by attentional tasks, even though it is a primary sensory area and the neurons located there may be linked to other modalities as well, particularly vision. In this study, single-unit responses in primary AC, including surrounding areas, showed increased firing both when the sound was relevant and also when some increases with the light was relevant (Fig. 8.2). Rhesus monkeys were trained to push a lever to the left or right depending on light or sound after presentation of a combined stimulus, and they determined relevance by comparing whether or not responses were made to the same

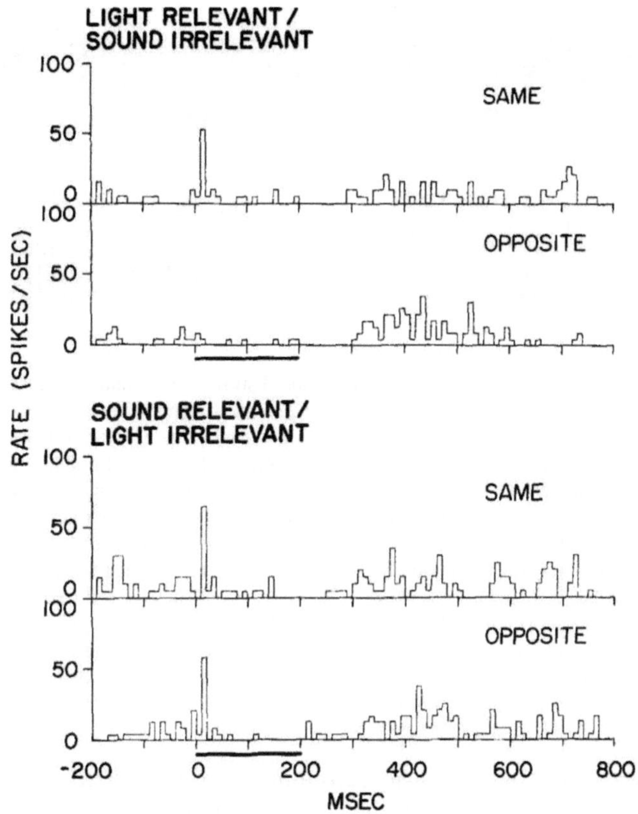

Fig. 8.2 PST histograms for correct responses in conditions III, Sound Relevant and IV, Light Relevant for unit 261-12-2R. The acoustic stimulus for all four histograms was identical: 2.6-kHz tone bursts, 200 ms duration. (Reproduced from Hocherman et al., 1976; Figure 6, with permission)

light or sound stimulus. Significant changes were typically identified in response strength instead of pattern and could arise as early as 15–20 ms after stimulus onset. Two-thirds of the neurons showed differences in response strength dependent on whether sound or light was the relevant cue. Also, neuronal responses were generally stronger when both the light and the sound stimulus indicated the lever to be pushed in the same direction, independent of which modality was relevant. Because latency of the initial response is so short, it suggests that the interaction may be subcortical.

In another selective-attention task across modalities, Lakatos et al. (2009) have shown a possible mechanism for supramodal control of neurophysiological context by attention: oscillatory phase resetting. Monkeys were required to attend to either a visual stream of flashing lights or an auditory stream of tone cues to identify a target. Both streams were simultaneously occurring but offset temporally, and the monkeys had to attend to one modality while ignoring the other. Alternating blocks of which modality was attended to were presented continuously. The primary sensory cortices of rhesus monkeys showed phase locking in the gamma and theta

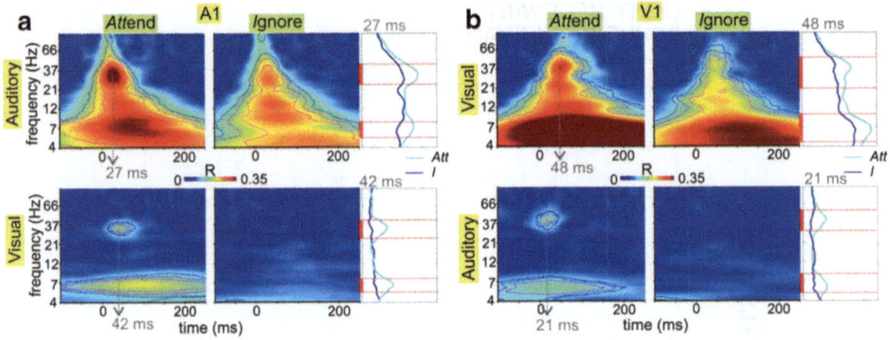

Fig. 8.3 Pooled ITC related to attended versus ignored stimuli in A1 and V1. (**a**) Time-frequency plots show ITC values from selected supragranular electrode channels averaged across all A1 sites ($n = 19$) related to attended and ignored auditory (upper) and visual (lower) stimuli. Contour plots demarcate regions of pooled ITC values that are significantly larger (at 95%, 99%, and 99.9% confidence levels) than the ITC value above which a significantly nonrandom phase distribution can be assumed (calculated using the Rayleigh statistic, $p < 0.01$). Traces to the right display ITC values for frequencies between 4 and 100 Hz related to attended (light blue) and nonattended (dark blue) stimuli at the time of the maximal gamma ITC peak to attended stimuli marked by gray arrows (the exact latency is shown for each ITC peak). Red lines on the frequency axis denote frequency ranges where ITC related to attended stimuli is significantly greater than ITC related to ignored ones (dependent t-test, $p < 0.01$, $n = 19$). (**b**) Same as in **a**, but pooled ITC values displayed are from supragranular V1 sites ($n = 25$) related to attended and ignored visual (upper) and auditory (lower) stimuli. (Reproduced from Lakatos et al., 2009; Figure 2, with permission)

bands through current source density recordings along the different layers of cortex, and were contingent upon whether the nonpreferred or the preferred stimuli within the auditory or visual modalities were given attention (Fig. 8.3). Target stimuli, by contrast, enhanced evoked activity to the specific stimulus in addition to the phase resetting. Phase setting of the neuronal oscillations in primary cortical areas was concentrated in the supragranular layers. This type of phase setting is modulatory and seems to be a form of top-down attentional control, perhaps mediated by input from the thalamic reticular formation (TRN) that may be receiving its direction from the prefrontal cortex (PFC; Zikopoulos & Barbas, 2006).

8.2.3 Areas Outside of Primary AC and Auditory Attention

Recent work has demonstrated that the PFC is involved in attentional auditory tasks. Single-unit recordings from the frontal cortex of ferrets performing a tone detection task showed rapid attention-modulated changes of STRFs similar to their observed changes in primary auditory cortex (A1) during the same task (Fritz et al., 2010; Fig. 8.4). Local field potentials (LFPs) from the frontal cortex showed large event-related potentials (ERPs) during passive listening to all future task sounds. During the actual behavioral auditory task, ERPs continued to be enhanced to the target

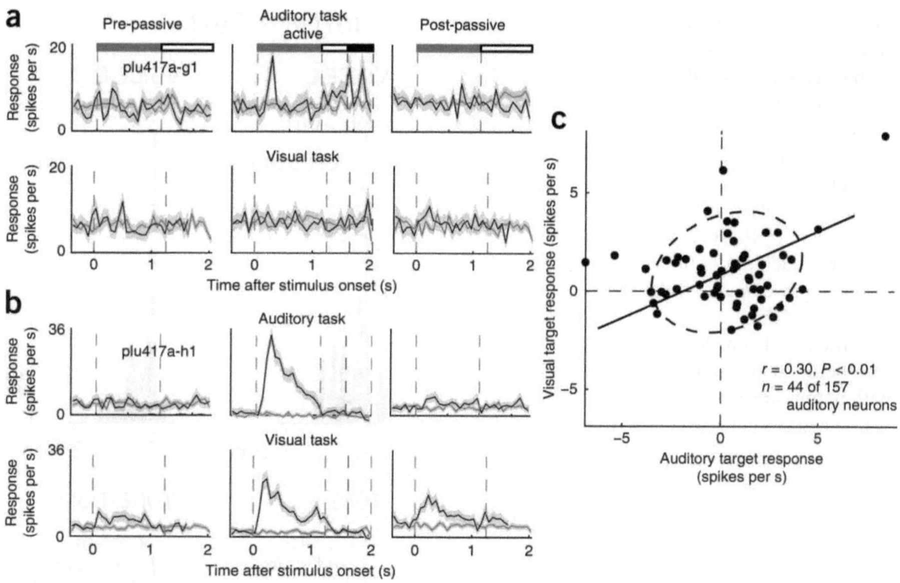

Fig. 8.4 Bimodal and unimodal sensory responses in frontal cortex. (**a**) Example of single-unit responses to successive auditory and visual tasks. This cell showed modulation during behavior only to the auditory target. (**b**) A second neuron showed target modulation by both the visual and the auditory stimuli. This cell also responded persistently after behavior, particularly to the visual target. (**c**) Scatterplot of auditory (horizontal axis) and visual (vertical axis) target responses showing that cells responded similarly if they were bimodal (i.e., responded to both auditory and visual stimuli, $n = 93$, $r = 0.30$, $p < 0.01$). (Reproduced from Fritz et al., 2010; Figure 7, with permission)

stimuli but selectively suppressed for the reference stimuli. Fritz et al. (2010) suggested that this differential processing is a form of selective gating and a possible mechanism for auditory attention similar to that observed for visual attention. This prefrontal selective gating may be the driving force for projections to the thalamic reticular nucleus (TRN) for attentional modulation of the primary AC.

Neurons in the TRN region of the rat have been shown by Yu et al. (2009) to respond more to deviant pure-tone stimuli than high probability pure tones, suggesting this area plays a role in auditory attention. The target output of the auditory portion of the TRN is the medial geniculate body (MGB). MGB neurons in this same study also exhibited more change detection, albeit to a smaller extent, than the thalamic reticular nucleus, with the TRN neurons either enhancing or (more often) suppressing the MGB neuronal responses. This is another example of a top-down central nervous system strategy for inducing attentional shifts.

Other areas involved in the prefrontal region in auditory attention include area 8b of the rhesus monkey. This area, labeled by Lucchetti et al. (2008) as the premotor ear-eye field (PEEF), can be considered part of the premotor cortex or PFC, and is involved during the interaction of auditory attention and visual gaze control. Area 8b is connected reciprocally with regions of the AC (Barbas & Mesulam, 1981; Barbas & Pandya, 1989). More than 20% of the neurons recorded in this area

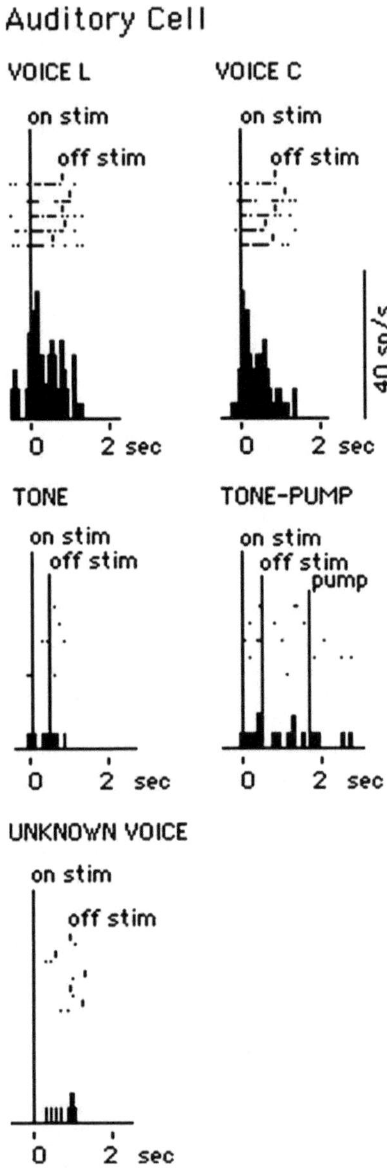

Fig. 8.5 Example of auditory cell active for the voices of both researchers (VOICE L and VOICE C). The discharge was weak and not synchronized when other auditory stimuli were produced (TONE, TONE-PUMP, and UNKNOWN VOICE). Each dot in the rasters represents one spike. Bin width: 50 ms. (Reproduced from Bon & Lucchetti, 2006; Figure 2, with permission)

responded to complex auditory environmental stimuli, but not to simple auditory stimuli such as pure tones (Bon & Lucchetti, 2006; see Fig. 8.5). The outcome of their findings suggests that PEEF may be involved in integrating auditory input with motor output related to orienting ear and eye movements. In the study of Lucchetti et al. (2008), most auditory-responsive neurons were evoked to stimuli from the contralateral side, and the firing of these auditory-responsive cells and the auditory–motor neurons was frequently inhibited when the animal was visually fixated. The results suggest that this area, much like the frontal eye fields, is important for directing attention and integrating these actions with eye movements for incoming visual information. These findings (Bon & Lucchetti, 2006; see Fig. 8.5) may result from the large visual component associated with the auditory stimulus, wherein neurons seem typically to respond to complex stimuli. For instance, the researcher's voice may stimulate the monkey to envision the image of the researcher, therefore indicating that it is less about "complex" versus "simple" and perhaps more related to a combination of auditory and visual meaning.

Another region in which attention modulated by behavioral context and salience plays a role in neuronal responsiveness to auditory stimuli is the parietal cortex. Neurons in the rhesus monkey parietal cortex, in particular the lateral intraparietal cortex (LIP), will respond to both simple band-pass noise bursts and complex monkey vocalizations when more salient visual cues are absent from the behavioral context and also when auditory stimulus location is important to task performance (Gifford & Cohen, 2004). As the auditory stimuli become more salient during various predictive cueing tasks with or without memory requirements through task requirements or experience, the parietal neurons increase their responses to auditory stimuli (Cohen et al., 2004). This flexibility in the primary spatial/visual parietal encoding has been observed in several instances, and as suggested by Cohen (2009) may be strongly influenced by auditory processing for shifts of attention. Parietal neurons may be another part of the attentional system where their role is not simply in auditory attention, but rather the intersection of modalities interacting, competing, or synergistically affecting one another.

Other studies also demonstrate an interaction between the visual and auditory systems modulating attentional processing. Winkowski and Knudsen (2006) used their barn owl model to demonstrate sharpened spatial selectivity and enhanced midbrain responses to the same auditory stimuli encoded by the forebrain site. Responsiveness to auditory stimuli represented at other midbrain locations was also suppressed by forebrain stimulation. The results suggest that auditory responses are regulated by top-down influences of the gaze control circuitry. Regulation of sensory gain by another processing system may be a common central nervous system strategy for attentional control.

Background firing rate of single- and multiunit recordings from locus coeruleus neurons, considered to be from the noradrenergic population in awake moneys, was higher during periods of alertness or agitation than at other times (Grant et al., 1988). Significant evoked-unit responses were observed when the intensity of the sounds was greater or when the sharp rise times simulated a more abrupt onset of sounds; in general, the more complex but naturalistic sounds were the best predictors of which auditory stimuli increased the firing rate (sometimes in a phasic manner). Grant et al. (1988) propose that sensory stimuli leading to significant evoked activity in the locus coeruleus are those "conspicuous to that species," meaning these stimuli tended to attract attention because they were unexpected. This pattern of evoked activity fits with the theory that locus coeruleus is involved when vigilance and/or alarm states are established, thus evoking a change in the focus of attention.

Other neural structures within the auditory input pathway and tangentially-related areas also demonstrate attention-related changes. In addition to recordings in the AC of the rat (Section 8.2.1), the Otazu et al. (2009) study also assessed neuronal firing in the MGB during an auditory task in which a target tone was presented among nontarget tones to the right or left ear. The MGB had enhanced spontaneous firing rates compared to a passive listening condition, but no change in evoked activity during auditory task engagement. The superior colliculus is another structure that receives auditory input and may contribute to interactions between auditory/visual attentional and/or orienting processing systems. Deactivations including both upper and intermediate layers in the cat lead to contralateral neglect of auditory stimuli (Lomber et al., 2001). The superior colliculus may be part of both the bottom-up and top-down processing systems.

Attentional processing in the auditory system is a critical part of normal hearing function because there is a need to distinguish between background and repetitive sounds. This type of discrimination may be important for ignoring some sounds unimportant for behavior, or for attending to novel sounds, which may be critical to assess and perhaps process for action. Novelty detection has been demonstrated in a number of regions primarily by the mechanism of habituation or feedback from the higher auditory system at many lower auditory system structures, and perhaps by active suppression at higher auditory system structures. Pérez-González et al. (2005) found in anesthetized rats that neurons in the dorsal and external cortex of the inferior colliculus did not respond when a sound was repeated multiple times, but did show recovered excitability when a new sound or the stimulus parameters were changed. The authors suggest that this pattern of novelty detection is inherent to the inferior colliculus as the latency of the neurons that habituate to repeated stimuli does not differ from those that are nonhabituating (latencies of 18.2 ms for novelty units and 20.2 ms for nonhabituating units). This phenomenon appears to be preattentive because this feature of novelty detection has been observed in many species at lower levels of the auditory system (Kraus et al., 1994; King et al., 1995), suggesting that these subcortical pathways are used in directing attention by producing orienting responses to novel sounds.

There are several studies pointing to the role of suppression in the auditory system processing regions, and in particular across the corpus callosum. One early study of this phenomenon in ferret AC, albeit in an anesthetized preparation, demonstrated that suppression was the most prevalent influence across the corpus callosum. This was in response to monopolar stimulation of the right hemisphere on responses of single units in the left AC (Kitzes & Doherty, 1994). There were also some excitation responses and a mixed class of influence with both suppression and excitation. Suppression duration was often more than 100 ms with latency of 2–4 ms. This suppression of the opposite hemisphere in response to stimulation of its homotopic region of connection across the corpus callosum may be a mechanism for attentional processing by limiting the responses to a particular auditory stimulus to one hemisphere. This may lead to specialized processing or lateralization of processing within the auditory system. A study using rhesus monkeys demonstrated a suppression of the right hemisphere presumably by the left hemisphere via the corpus callosum when processing species-specific communication signals (Poremba et al., 2004). This suppression induced lateralization of processing to the left temporal pole region of the superior temporal gyrus (STG) specifically for that sound type.

Different mechanisms likely exist within the primary processing hemisphere for a particular sound presentation and contralaterally for attentional processing.

Overall, Section 8.2.2 reviews many regions outside of primary cortical regions, both cortically and subcortically, that are involved in modulating the processing of auditory information. This is typically because of interaction with other neural systems, which may require focusing on particular pieces of auditory information, and/or the integration of auditory information with other components of sensory information presented through other modalities. Attention then to auditory or auditory information related to other sensory modalities appears to be spread across a large number of structures that are not necessarily a unified neural circuit but rather widespread because of the number of systems that may need to utilize auditory information.

8.2.4 Attentional Processing Mechanisms

In the visual system, increases in neuronal firing are usually noted during attention-requiring tasks (Desimone & Duncan, 1995); however, firing rates have been shown to both increase and decrease during tasks requiring auditory attention (Hocherman et al., 1976; Miller et al., 1980). Observing both increases and decreases in firing rates during auditory attention, as Hromádka & Zador (2007) suggest, is in agreement with James's 1890 work that essentially posited attention as a limited resource requiring some information be ignored or withdrawn from so other information can come to the forefront. Increasing activation of some inputs and inhibiting yet others could bring about implementation of this concept at the neuronal level. This is apparent even in the omega auditory neuron of crickets, in which selective attention is encoded as excitation or inhibition depending on the side of stimulation. When an intense sound is presented representing an ipsilateral stimulus, there is selective encoding due to intensity-dependent inhibition. This intensity-dependent inhibition is in competition with the neuron's normally excitatory response to a sound, for example, a lower intensity stimulus from the contralateral side evokes a less intense response, making it a subthreshold signal while the high-intensity stimulus (representing the ipsilateral side) remains above threshold (Pollack, 1988). These patterns of excitation and inhibition observed in a number of auditory attention tasks could certainly be modified or driven by experience and therefore have a direct relationship to memory processing.

Another issue is related to how preattentive processing should be viewed. Is this level of processing completely independent, or is there some modulation of early stimulus processing by task requirements, experience, and/or attention? Even the outer hair cells have been suggested to play a role in attentional mechanisms (Guinan, 2010) and one can find many studies suggesting that areas below the cortex are involved in attention (e.g., see Perez-Gonzalez et al., 2005, discussed in Section 8.2.3). A recent set of hypotheses put forth by Shamma et al. (2010) proposed that one hearing process humans and animals perform is to attend to one sound source culled from many and to follow that particular sound source over time. Shamma et al. (2010) propose that this forces "stream formation" dependent on the cortical neural correlates encoding various features of the stimulus with tight temporal coherence to be "bound" together. This binding and stream formation may be selected by the attentional system and modulated by items such as the context of the task or the experience of the listener. This proposed theoretical framework is being used to investigate a number of interesting problems within the auditory processing field such as auditory scene analysis and spectrotemporally complex auditory signals.

8.2.5 Future Questions Regarding Auditory Attention and Neural Activity

Certainly a strong theme to be explored, also applying to components of memory processing, is the importance of context in attentional processing. Context may be internal to the task, internal to the subject, or externally mediated, and it appears to be the guiding factor in attentional processing. Contextual differences mediating attention need to be examined during all experiments and control conditions added as necessary across single-unit, multiunit, and LFP studies. In addition, although many neural structures show attention-related changes in firing patterns or background activity, there are many other regions not yet examined, including cortical areas downstream to primary AC. Exploring these additional regions will inform the search for the complete circuits underlying a number of diverse attentional processing types.

8.3 Auditory Memory

Herein, auditory memory is broadly defined as any sequence of change in neural activity patterns reflecting a past history with a physical stimulus. Auditory researchers have been interested in a few basic forms of memory, which include neurophysiological adaptation, working memory, and long-term retention and plasticity. These rudimentary divisions offer a preliminary framework from which to describe the current literature across the limited number of brain regions explored thus far.

8.3.1 Stimulus-Specific Adaptation

Sections 8.1 and 8.2 show that observed neuronal responses are rarely attributable solely to the acoustic properties of the stimulus, but also depend on attention and context. Another factor that influences evoked neural activity is the history of previous stimulus presentations. For example, an AC cell might be effectively driven by a tone when presented for the first time, but if the same tone is repeated in rapid succession, the elicited response will be significantly attenuated. This basic form of physiological memory is known as stimulus-specific adaptation (SSA; Jääskeläinen et al., 2007).

Although adaptation processes have typically been studied using brief interstimulus presentation intervals (e.g., Hocherman & Gilat, 1981; Brosch & Schreiner,

1997), Werner-Reiss et al. (2006) provided evidence that adaptation can occur after longer intervals. In this study, awake rhesus monkeys were presented with pairs of noise stimuli separated by intervals ranging from 200 to 5000 ms. Although responses to the second sound were most attenuated after intervals of 1 s or less, significant attenuation was observed at the primary AC population level after intervals up to 5 s, the longest interval tested. An important manipulation in this study showed these effects were not different between passively listening monkeys and monkeys that were actively engaged in a fixation task. This suggests the observed adaptation processes are relatively automatic and are not dependent on behavioral state.

A complementary form of physiological facilitation has been described by Brosch et al. (1999) in the primary AC and parts of caudomedial belt cortex of cynomologous monkeys (*Macaca fascicularis*; see also Bartlett & Wang, 2005). In this study, the response elicited by a second sound was evaluated while the stimulus onset asynchrony (SOA) and frequency of a first tone were manipulated systematically. Many single cells (42%) and the majority of multiunit groups (67%) showed enhanced responses when the second tone was preceded by a nonidentical first tone. The authors reported that maximal response facilitation is produced by frequency differences of approximately one octave and SOA of approximately 120 ms.

SSA has been studied in many species (e.g., Werner-Reiss et al., 2006), and using a wide range of acoustic stimuli, ranging from simple tones to complex vocalizations (e.g., Brosch et al., 1999). Recent studies have suggested that SSA may be closely related to the mismatch negativity (e.g., Ulanovsky et al., 2003; Jääskeläinen et al., 2004), which is a human scalp-recorded potential thought to index auditory sensory memory. Together, neurophysiological response adaptation and facilitation provide basic mechanisms for monitoring change in a dynamic auditory environment.

8.3.2 Working Memory

Behaviorally-relevant events such as conspecific vocalizations are often available in the sensory environment for only a brief period of time. Thus, animals must frequently maintain neural representations of events in the absence of direct stimulation to guide adaptive behavior (Goldman-Rakic, 1995). Many researchers have adopted the term "working memory" (WM) to describe the temporary retention of information that is needed to accomplish a prospective action or goal (Constantinidis and Procyk, 2004). According to Fuster (2008), it is this goal-oriented, future perspective that differentiates WM from passive forms of short-term memory such as SSA.

For many decades, investigators have successfully employed WM tasks such as delayed response and delayed matching-to-sample (DMTS) to study visual WM and its neural substrates in monkeys (e.g., Fuster & Alexander, 1971; Mishkin, 1978). Monkeys readily acquire above-chance performance on these tasks after several weeks of training. By contrast, investigators that have attempted to train monkeys on WM tasks in the auditory modality have met with difficulty (Cohen et al., 2005). For instance, Colombo and D'Amato (1986) reported that only four of eight capuchin monkeys (*Cebus apella*) were able to learn an auditory DMTS task after years of failure.

Recently, Fritz et al. (2005b) reported that naïve rhesus monkeys required approximately 15,000 trials to learn the auditory DMTS rule, whereas only approximately 500 trials were required to learn a visual version of the task (Mishkin & Delacour, 1975). Further, researchers who have successfully trained monkeys on auditory DMTS tasks have been able to maintain criterion-level performance only at relatively short delays. For example, Kojima (1985) reported that Japanese macaques (*Macaca fuscata*) required many trials to advance from a 7-s to 9-s interstimulus interval (ISI) by small increments, and that performance fell to chance levels at 16 s. Fritz et al. (2005b) reported that monkeys were able to maintain 85% performance with delays up to approximately 10 s, but performance fell below the 75% "forgetting threshold" at delays beyond approximately 35 s. By contrast, monkeys seem to have little difficulty with visual DMTS delays of several minutes (e.g., Murray & Mishkin, 1998); indeed, Overman and Doty (1980) reported that pig-tailed macaques (*Macaca nemestrina*) were able to make accurate matches after delays of 24–96 hours. Importantly, cynomologus monkeys have also performed tactile DMTS at delays extending beyond several minutes (e.g., Buffalo et al., 1999), which is substantially greater than the longest achievable auditory delay that has been reported in the literature. This suggests that monkeys seem to have a unique difficulty performing auditory versions of WM tasks originally used to investigate visual WM.

Several authors have offered hypotheses to explain why monkeys might have difficulty with auditory memory tasks (e.g., Kojima, 1985; Colombo & D'Amato, 1986; Sakurai, 1990). A particularly interesting possibility emerges from lesion experiments performed by Fritz et al. (2005b). In this study, auditory recognition memory in rhesus monkeys was impaired by lesions of the rostral STG or the medial temporal lobe. However, lesions of the entorhinal and perirhinal cortices, which disrupt visual and tactile recognition memory, failed to produce these deficits. Notably, the auditory forgetting thresholds of intact rhesus monkeys were remarkably similar to the visual forgetting thresholds in subjects with rhinal lesions. Based on these results, the authors suggest that memory for auditory stimuli is not significantly supported by the rhinal cortices in monkeys, leading to qualitative differences in memory processing for auditory and visual stimuli.

Given the inherent difficulty of studying auditory memory in monkeys, it is perhaps not surprising that far fewer studies have examined auditory compared to visual memory processes. Notwithstanding these challenges, a number of laboratories have successfully trained monkeys to perform auditory WM tasks to study the neural substrates of auditory WM. Many of the experimental results thus far are consistent with the much larger visual WM literature, which suggests that the PFC interacts with sensory areas in the temporal and parietal lobes to maintain representations of behaviorally-relevant stimuli.

8.3.2.1 Neural Correlates of Auditory Working Memory in PFC

Neurophysiological recordings, lesion and inactivation studies, and neuroimaging data uniformly point to a critical role for the PFC in retaining and integrating sensory information for future use in goal-directed behavior (Miller & Cohen, 2001;

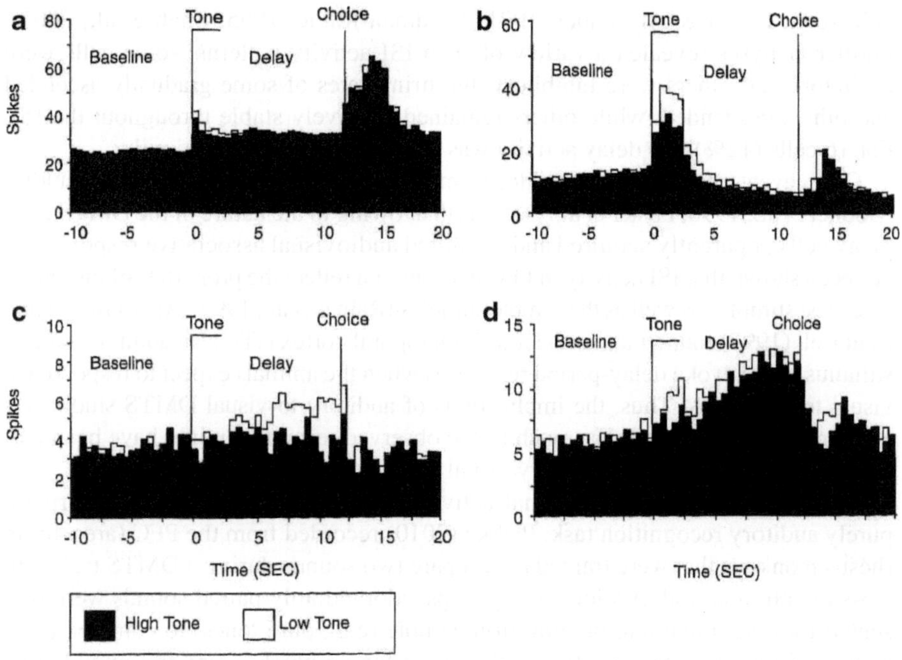

Fig. 8.6 Average frequency histograms (0.5-s bin size) from four cells showing different types of delay period activity. (**a**) A cell excited by the presentation of the memorandum and persistently activated (nondifferentially) during the delay period. (**b**) A cell showing differential excitation ($p < 0.01$) during presentation of the memorandum as well as during the delay; activity decreased in the course of the delay. (**c**) A cell showing differential activity ($p < 0.01$) during the delay, firing after one of the stimuli gradually increased in the delay period. Note the sharp drop of firing back to baseline after the choices. (**d**) A cell similar to that in **c**, but not firing differentially. Cells **b** and **d** show a slight acceleration before the beginning of each trial, as previously observed in a visual task (Fuster et al., 1982); it probably reflects preparation for the tone in a paradigm in which trails occurred at approximately equal intervals (about 50 s). (Reproduced from Bodner et al., 1996; Figure 2, with permission)

Fuster, 2008). The literature describing these functions of the PFC largely comprises visual experiments. Because there is now abundant evidence for auditory regions in the rhesus monkey PFC (Poremba et al., 2003; Romanski & Averbeck, 2009), these memory functions presumably extend to the auditory modality. However, researchers have only begun the experimental work to demonstrate and describe these functions.

Fuster and colleagues (Bodner et al., 1996; see also Fuster et al., 2000) provided some of the earliest neurophysiological evidence that the PFC supports WM for auditory stimuli. In their experiment, rhesus monkeys were trained to remember tones for a 10-s to 17-s delay period. Depending on whether the sample stimulus was a high or low tone, the monkey was rewarded for pressing a green or red button, respectively. Of 138 neurons recorded around the area of the principal sulcus, 68 (49%) showed discharges during the ISI that differed significantly from baseline (Fig. 8.6). Persistent changes in firing rate during the ISI have traditionally been interpreted as a "memory trace" or sustained representation of a stimulus during

WM (e.g., Fuster & Alexander, 1971; Goldman-Rakic, 1995; Shafi et al., 2007). Further analyses revealed a variety of such ISI activity patterns: some cells were excitatory and others were inhibitory, the firing rates of some gradually ascended and others descended, while others remained relatively stable throughout the ISI. For 16 cells (12%), the delay activity was specific to the sample stimulus.

One caveat regarding the auditory "memory cells" described in these studies (Bodner et al., 1996; Fuster et al., 2000) is that, owing to the nature of the DMTS task, many cells apparently acquired and exhibited audiovisual associative responses. It has been shown that ISI activity in PFC neurons can reflect the properties of the anticipated test stimulus as well as the sample stimulus (Rainer et al., 1999). Also, Gibson and Maunsell (1997) found that, in the inferior temporal cortex (ITC), an auditory sample stimulus could evoke delay-period responses when the animals expect to respond to a visual test stimulus. Thus, the implications of auditory-to-visual DMTS studies for auditory WM are limited, inasmuch as the observed responses might have been partially driven by the learned auditory/visual associations.

Only recently has PFC neuronal activity in primates been examined during a purely auditory recognition task. Plakke (2010) recorded from the PFC (area 46) in rhesus monkeys that were trained to compare two sounds during a DMTS task with a 5-s retention period. A wide variety of pseudorandomly paired sounds were presented for discrimination, ranging from simple (e.g., pure tones) to complex (e.g., conspecific monkey vocalizations). Passive listening blocks were presented before and after the task for comparison. The results from this study are in general agreement with the tone/light DMTS experiments conducted by Fuster and colleagues (Bodner et al., 1996; Fuster et al., 2000). Many cells showed delay-period responses, although a smaller proportion exhibited sustained activity throughout the entire ISI. In addition to cue-elicited and ISI-related activity, a variety of cells were observed that encoded other task-related events such as the behavioral response and reward delivery. Cells that encoded multiple task events (e.g., cue-elicited responses plus reward-related activity) were more common than cells that only encoded one event. An interesting result from this experiment was that the sound selectivity of individual neurons was frequently observed to change between passive and task conditions. This suggests PFC cells can be selective for different acoustic features depending on the nature of the task. Population multiunit analyses revealed that matching test stimuli typically elicited enhanced responses compared to nonmatching stimuli. These results extend and strengthen the evidence that the PFC is instrumental in mediating auditory WM.

Cohen and colleagues (Russ et al., 2008; Lee et al., 2009) have studied decision-making processes in the ventrolateral PFC (vPFC) of rhesus monkeys. The behavioral task in these experiments required the monkeys to discriminate the human-vocalized words *bad* and *dad* and intermediate morphs. Each trial began with two to four presentations of a sample stimulus (1600-ms ISI on average), which was always a 100% morph prototype. This was followed by either a matching (60%–100% morphs) or nonmatching (0%–40% morphs) test stimulus, and the monkeys were rewarded for correct categorizations. On trials with 50% morphs, rewards were given randomly based on the monkey's cumulative accuracy for the session. One of the primary conclusions from these experiments was that, on average, vPFC neurons coded the monkeys' decisions better than the acoustic properties of the stimuli. In other words, the neural activity more often reflected what the monkey *actually chose* than what the monkey *should have chosen* based on the relationship between the sample and test stimuli. These results identify a relationship between individual vPFC neurons and perceptual choices. This relationship is further strengthened by Orr et al. (2008), who reported that transcranial magnetic stimulation of the PFC disrupts decision making during this task (described in Lee et al., 2009).

Russ et al. (2008) also indicated that matching test stimuli tended to elicit a relatively high firing rate in some vPFC neurons compared to nonmatching stimuli. Such "match enhancement" has been observed in PFC neurons during visual WM (e.g., Miller et al., 1996), indicating that visual and auditory WM may share some common neural mechanisms. It has been suggested that match enhancement in PFC cells might act as a feedback mechanism to influence visual responses in ITC (Miller & Desimone, 1994; Miller et al., 1996; Woloszyn & Sheinberg, 2009). Further studies are needed to understand fully the functional significance of enhancement and suppression in the auditory system and how it relates to auditory recognition memory.

In addition to the investigations of nonspatial WM reviewed in the preceding text, several experiments have investigated the neurophysiological correlates of auditory spatial WM in the PFC. One of the first investigations of the neural basis of auditory memory in the primate cortex was conducted by Joseph and Barone (1987), who recorded single-unit activity in the dorsolateral PFC of rhesus monkeys. Subjects were required to fixate on a central point while an auditory stimulus consisting of 20 clicks per second was presented either on the right or the left side for one second. After a delay, left and right key lights were illuminated and the monkeys were rewarded for making a saccade to the light position cued by the sound. Many cells showed an increased firing rate during the delay period between the auditory cue and the choice of lights. For some neurons, this delay-related activity was selective for either the right or left cue position, whereas other neurons did not show spatial preference. These results are consistent with previous visual experiments indicating that the dorsolateral PFC is instrumental in retaining spatial information cued by either visual or auditory stimuli.

Kikuchi-Yorioka and Sawaguchi (2000) extended these results by examining the spatial mnemonic activity of neurons between the arcuate sulcus and the caudal principal sulcus. Further, they directly compared spatial WM between sensory modalities by employing an oculomotor delayed-response task that alternately presented visual and auditory trials. One rhesus monkey and one Japanese macaque were trained to fixate on a central light-emitting diode (LED) and were cued by either a speaker (500-Hz tone) or a green LED in one of four locations. After a 3-s retention period, the subject was rewarded for making a saccade to the correct cue location. The authors focused their analyses on cells with delay-period activity that was specific to the location of the cue. Of 190 such neurons, 53% retained only the visually cued locations, 17% were auditory-specific, and 57% responded to either modality.

It should be noted that the mnemonic function of neurons exhibiting delay-specific activity during delayed-response tasks could conceivably be confounded with preparatory motor activity, because the animal knows which response to make as soon as the first stimulus is presented. To circumvent this possible explanation, Artchakov et al. (2007) studied audiospatial WM in rhesus monkeys using a spatial DMTS task. Under this contingency, the animal does not know which response will be made until the test stimulus is presented, thus freeing the retention delay of preparatory motor activity. For each trial in this experiment, the subject was presented with either an auditory (500-Hz tone) or visual (green LED) sample stimulus on the left or right side of a central fixation LED. After a 2-s or 3-s retention period, the monkeys were rewarded for correctly identifying whether or not the test stimulus was presented in the same location. Each trial was unimodal (i.e., both stimuli were either auditory or visual), and trial types were presented in random order. The largest proportion of task-related neurons was activated by visual cues only (67%). A smaller subset of neurons responded only to tones (19%), and another subset was responsive to either cue modality (14%). Analyses of the auditory-responsive neurons indicate that single cells encode and retain auditory spatial information throughout the ISI period, and show ISI activity to be location-specific for some neurons. As in previous PFC experiments, activity correlated with behavioral responses and reward delivery was observed. The results from this experiment reinforce previous evidence suggesting that PFC cells subserve audiospatial memory.

In an interesting follow-up experiment, Artchakov et al. (2009) investigated the effects of distracters on neuronal activity during audiospatial and visuospatial memory. The spatial DMTS task was similar to that of Artchakov et al. (2007) except that distracter stimuli (100 or 300 ms) interrupted the ISI for a randomized portion of the trials. The auditory distracter was a tone alternating between the left and right speakers at a rate of 50 Hz. The visual distracter was the simultaneous presentation of an additional eight LEDs surrounding the central fixation LED. The effects of auditory and visual distracters were examined for both auditory and visual DMTS trials. Some neurons showed spatially-selective activity during the ISI that was impaired by the presentation of distracters (i.e., the degree of spatial-selectivity was reduced, in some cases to nonsignificance). For other cells, the spatial tuning of the delay-period activity was maintained in spite of distracters. Interestingly, some cells showed no spatially-tuned ISI activity during control trials, but developed spatially-tuned ISI activity after a distracter was presented. These effects were found in the data from both auditory and visual trials. Based on this result, the authors suggest cells that are latent during relatively easy conditions might be recruited to perform more demanding cognitive tasks. Collectively, the results from this experiment indicate that auditory PFC regions have mechanisms for managing interference to accomplish goal-directed behavior.

Two of the previously discussed experiments (Kikuchi-Yorioka & Sawaguchi, 2000; Artchakov et al., 2007) employed tasks in which the activity of single PFC neurons was recorded while both auditory and visual trials were performed by the subjects. Both of these studies noted that fewer PFC cells responded during auditory rather than visual trials. These findings replicate the results of an earlier experiment by Watanabe (1992), who recorded PFC neuronal activity while Japanese macaques performed an auditory and visual association task. In this experiment, one of two auditory (tones) or visual (simple patterns) stimuli was presented as a cue. After a 1-s delay period, a red or green light was then presented for 1 s. If the monkey released the lever during the light presentation, a juice reward was either delivered or not depending on which cue was presented. A sample of 289 neurons was recorded around the arcuate sulcus and caudal principal sulcus. Cue-elicited, delay-period, and reward-related responses were observed. The responses of 13 cells (4.5%) were specific to auditory stimuli, 136 cells (47%) were visual-specific, and 140 cells (48.5%) were bimodal. Thus, all three studies employed tasks in which auditory and visual stimuli were presented, and all three reported samples of PFC neurons with substantially smaller proportions of auditory-responsive than visually-responsive cells. Interestingly, in all three studies the monkeys had more difficulty with the auditory trials: Artchakov et al. (2007) reported 70% accuracy for auditory and 98% for visual trials, Kikuchi-Yorioka and Sawaguchi (2000) reported 80% accuracy for auditory and 90% for visual trials, and Watanabe (1992) reported that it took twice as long for the monkeys to learn the auditory task as compared to the visual task, and also that response latencies were greater for auditory trials. These results raise the possibility that the relatively small number of available auditory PFC neurons might be related to the difficulty monkeys have with auditory WM tasks. It should be noted that the auditory cues in all three of these studies were pure tones, which may have negatively biased the cellular responses (Wollberg & Sela, 1980; Romanski & Goldman-Rakic, 2002). Also, many of the cells reported by Kikuchi-Yorioka and Sawaguchi (2000) and Watanabe (1992) were recorded in the prearcuate cortex, where visual anatomical inputs and physiological responses are dominant. With these caveats in mind, future experiments comparing neuronal responses in PFC areas of substantial visual and auditory overlap (Poremba et al., 2003) during WM for a variety of auditory and visual stimuli may shed further light on this issue.

The experiments reviewed in this section demonstrate that the PFC is indeed instrumental in retaining and discriminating auditory stimuli and coordinating these events with motor output for goal-directed behavior. These functions extend to both the spatial and nonspatial attributes of the stimuli. Although many of these functions replicate findings reported in the visual literature, there are some indications that auditory mnemonic resources may be less extensive in monkey PFC. Thus the extent to which auditory and visual memory processing are equivalent within the PFC remains in question. As auditory WM research within the PFC has been relatively limited, more studies are needed to adequately describe auditory mnemonic representations in the PFC.

8.3.2.2 Neural Correlates of Auditory WM in the STG and Parietal Lobe

The STG has been directly implicated in auditory memory by several lesion studies (Colombo et al., 1990, 1996; Fritz et al., 2005b). The rostral end of STG forms the dorsal temporal pole (dTP), which is a higher-order auditory region (Poremba et al., 2004; Poremba & Mishkin, 2007) with reciprocal connections to PFC (Markowitsch

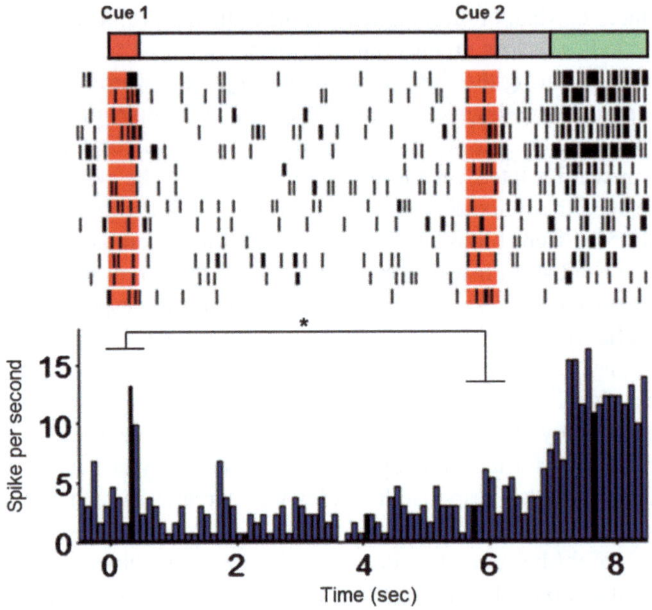

Fig. 8.7 An example of a unit showing match suppression when the monkey correctly identified two matching sounds. Asterisks and brackets denote a significant activity difference between cue 1 and cue 2 (paired sample *t*-test, $*p < 0.05$). Each bin is 100 ms. Colored boxes represent discrete events during an auditory delayed matching-to-sample task: red, sound presentations; white, 5-s memory delays; gray, 1-s wait period; green, 1.5-s response period. Black ticks represent action potentials in a raster plot of spike activity. Match suppression occurred when the two matching sounds were a sound clip of natural phenomena ("fire roaring"; cf. Ng, 2011)

et al., 1985; Kondo et al., 2003). Ng (2011) recorded neuronal activity in dTP while rhesus monkeys performed an auditory DMTS task involving simple and complex sounds. During the 5-s ISI, some cells showed either transient inhibitory or excitatory activity. Although these responses were not enduring, recent evidence suggests that sustained ISI activity may not be a necessary characteristic of cells involved in WM (e.g., Woloszyn & Sheinberg, 2009). Similar to AC (Gottlieb et al., 1989; see Section 8.3.2.3), population analyses revealed that matching test stimuli tended to suppress neuronal firing in dTP neurons (Fig. 8.7). These studies indicate that retention and recognition of auditory stimuli involves the STG, including dTP.

In the parietal lobe, area LIP has traditionally been understood as a center for visuospatial processing. Yet under certain conditions, or following the appropriate behavioral training, LIP has been shown to respond to auditory stimuli (Cohen et al., 2005). Of interest to this section, LIP neurons have been recorded during auditory memory-guided saccade tasks. For instance, in an experiment by Mazzoni et al. (1996), monkeys were presented with either an auditory (white-noise burst) or visual (light) cue on the left or right of a central fixation point. After maintaining fixation throughout a 1250-ms memory delay, the animal was rewarded for making a saccade to the location of the cue. Of 80 neurons recorded from LIP, 44 showed cue and/or sustained memory period responses during auditory trials. The majority of the cue and delay responses were spatially selective. Most of these spatially-tuned cells shared direction preference during auditory and visual trials. This activity likely represents retention of the stimulus location or the location of the next intended saccade.

8.3.2.3 Neural Correlates of Auditory WM in the AC

Gottlieb et al. (1989) recorded single-cell activity from core and belt regions of AC of a baboon (*Papio anubis*) during an auditory WM task and passive listening blocks presented before and after the task. During both the DMTS task and the passive listening blocks, the monkey was exposed to two tones separated by a 1-s ISI. During task performance, the monkey was rewarded for correctly identifying whether the tones were identical (right button press) or different (left button press). During the passive listening conditions, no behavioral responses were made and no rewards were given. As in the PFC, neurophysiological analyses revealed a variety of sustained responses throughout the ISI. These included excitatory and suppressive responses, which could gradually increase or decrease, or remain relatively constant throughout the ISI. A limitation of this experiment is that the ISI was only 1 s, and indeed, some ISI responses were observed during the passive listening conditions. However, a comparison between the behavioral and passive conditions revealed that, for the majority of neurons (65%), the firing rate during the ISI was greater during the DMTS task (Fig. 8.8). Further, the proportion of neurons that exhibited stimulus-specific delay-period activity (i.e., the firing rate depended on the frequency of the first tone) was greater for the DMTS task (50% of cells) than during passive listening (32% of cells). This suggests that the ISI-related activity observed in many cells was indeed related to maintaining the sample tone. In addition, some cells (23%) showed firing rates during the test stimulus presentation that were dependent on whether it matched the first. The firing rates of these cells were lower for matching stimuli than nonmatching stimuli, and were either similar to or lower than the baseline firing rate. The presence of suppressed responses (i.e., below baseline firing) suggests processes that are distinct from stimulus-specific adaptation. These so-called "match-dependent" responses were never observed during passive listening. Together, these results provide evidence that AC is part of the circuitry underlying the short-term retention and discrimination of tones.

Sakurai (1990) also reported AC cells with mnemonic properties in rats performing a tone-discrimination task. A portion of the neurons in this study exhibited increased activity that was sustained throughout the duration of the 3-s ISI period (20% of cells). Further experiments revealed some individual AC cells participating in two different forms memory, whereas other cells participate in one form but not another (Sakurai, 1994). These results extend the evidence that AC has a role in auditory WM. This notion is further strengthened by studies showing auditory WM deficits produced by lesions in cats (Neff et al., 1975) and temporary inactivations of the AC in gerbils (Wang et al., 2006).

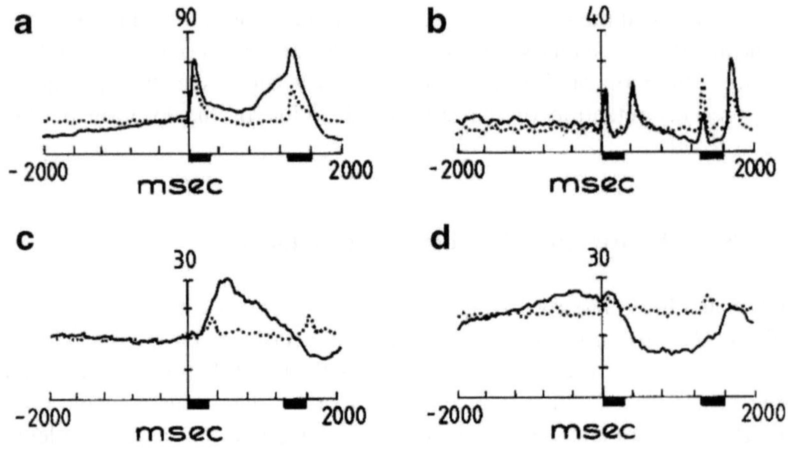

Fig. 8.8 PETH describing activity of four units around the first tone in the performance (solid line) and control (dotted line) states. The first and second tones are marked by black bars under the abscissa. The full scale of the ordinates is given in spikes/s. The PETHs in the performance state present different types of ISI responses. To obtain smoother histograms the data were pooled from trials of various frequencies of the first tone. Because of the similar ISI responses to all the frequencies, this pooling was possible. (Reproduced from Gottlieb et al., 1989; Figure 4A–D, with permission)

The putative role of AC in auditory WM is further supported by several human studies. For example, Lu et al. (1992) reported that physiological measures (MEG) of memory for tone amplitude localized to AC closely correlated with behavioral measures. In another study, Brechmann et al. (2007) examined AC activity using functional magnetic resonance imaging in human subjects performing a 2-back memory task for frequency-modulated tones. They found that activation of the left planum temporale was highly correlated with behavioral accuracy. This indicates that auditory information being used in WM finds enduring representation in AC that is maintained in spite of intervening stimuli.

In addition to sensory, discriminative, and stimulus-retention functions, a number of studies have reported AC cells (A1 and R) that apparently encode task-critical but non-auditory events (see Scheich and Brosch, Chapter 3). For example, in an experiment reported by Yin et al. (2008), rhesus monkeys were rewarded for releasing a touch bar every time a target melody was presented. A subset of the neurons recorded in this study exhibited activity related to the behavioral response and/or reward anticipation. These findings corroborate an earlier report from Brosch et al. (2005), who found single- and multiunit responses in primary AC and posterior auditory fields to response-bar touches and releases as well as activity elicited by a cue light. Some of these neurons were active only during these nonauditory events, whereas others were modulated by both acoustic and nonacoustic stimuli. It should be noted that these responses were not observed when the monkey was engaged in an analogous visual task. This suggests that motor, somatosensory, and visual inputs interact with AC neurons primarily within contexts that engage the AC.

Decision-related processing in AC has only recently been investigated, and inconsistent conclusions have been reported. Lemus et al. (2009a) trained rhesus monkeys to discriminate acoustic flutter stimuli separated by a variable 1-s to 3-s delay. They compared psychometric and neurometric discrimination functions and concluded that the role of the AC (A1) was sensory only; they found no evidence of decision-related processes. These higher-order processes were detectable upstream in the ventral premotor cortex (Lemus et al., 2009b). However, using a similar paradigm, Niwa et al. (2009) were able to detect evidence for decision-related processing in the A1 region of AC (described in Sutter & Shamma, 2010). Because such processing has also been reported in areas that are both upstream (PFC: Russ et al., 2008; Lee et al., 2009; see Section 8.3.2.1) and downstream (MGB: Komura et al., 2005) to various regions within the STG, it is not unexpected that decision-related activity might be observed in the AC. Further research and alternative approaches to evaluating decision-making processes may resolve these apparent conflicts.

Although relatively few neurophysiological investigations of AC during auditory working memory have been conducted, the aforementioned results agree with a principle of cerebral cortical function that has emerged over the past several decades: sensory cortex that mediates sensation and perception also forms a part of the circuitry that supports memory for that modality (Pasternak & Greenlee, 2005).

8.3.3 *Plasticity and Long-Term Retention*

Changes in neuronal response properties that accompany long-term behavioral training have been reported in a variety of contexts. For example, Beitel et al. (2003) reported that in the A1 region of AC, cellular response properties were altered according to the reward probability of a stimulus after tone discrimination training. In this study, owl monkeys (*Aotus trivirgatus*) detected increases in the envelope frequency of sinusoidally amplitude modulated (SAM) tones. Population neuronal responses showed relatively enhanced responses to envelope frequencies frequently paired with reward compared to frequencies that were infrequently or never rewarded. A recent study by Yin et al. (2008) described the neurons of A1 and R regions of AC in rhesus monkeys that were trained to detect target tone sequences (melodies). Melody-specific enhanced and suppressed responses were observed more frequently in trained versus naïve monkeys, and, on average, the latency of these responses was reduced in trained animals. In addition, component tones of the target melody came to elicit significant responses in some neurons unresponsive to that tone when it was presented in isolation.

An important study by Recanzone et al. (1993) described global changes in the tonotopic organization of primary AC in owl monkeys occurring after several weeks of tone discrimination training. Experimental monkeys were rewarded for detecting tone changes within specific frequency ranges, and these animals were compared to naïve monkeys and control monkeys exposed to tones that were not behaviorally relevant. Representation in the AC of behaviorally relevant frequencies was expanded in the experimental animals, and the cortical area responsive to these frequencies correlated with behavioral performance (Fig. 8.9). These and other results have led to the notion that the response characteristics of AC neurons are shaped by task contingencies to optimize behavioral performance and maximize reward.

Neuronal response plasticity in the adult mammalian auditory system that accompanies specific experience has been the subject of extensive inquiry (Suga & Ma, 2003; Weinberger, 2004a, b). Specific, learning-induced neuronal plasticity can be induced very rapidly (e.g., within a few minutes; Edeline et al., 1993) and undergo post-training consolidation (Galván & Weinberger, 2002), and it has been shown to persevere across physiological states of arousal (e.g., awake vs. anesthetized). In addition, Weinberger et al. (1993) have shown that these changes can endure for extensive periods of time. In this study, the best frequency of AC neurons was first determined in naïve guinea pigs. These animals were then exposed to 20 to 30 pairings of a non–best-frequency tone conditioned stimulus with a mild foot shock. One hour after training, the cells showed typical shifts in tuning toward the CS. The important result of this experiment was that this change in tuning was still observable in many cells when tested 1, 2, 4, and 8 weeks (the longest period tested) after conditioning (Fig. 8.10). This establishes a role for AC in the long-term retention of behaviorally relevant auditory cues.

Learning-induced physiological plasticity appears to be a widely encompassing phenomenon. It is induced by a variety of learning paradigms, including classical (Bakin & Weinberger, 1990) and operant conditioning (e.g., Bakin et al., 1996; Blake et al., 2002), and it has been observed in primary (Bakin & Weinberger, 1990) and secondary (e.g., Diamond & Weinberger, 1986) auditory cortical areas as well as auditory thalamus (e.g., Edeline & Weinberger, 1991a, b). A variety of species have been used to demonstrate this neural plasticity, including guinea pigs, rats, cats, bats, monkeys, and humans (Weinberger, 2004b). In addition to electrophysiological recordings, these changes have also been detected via animal and human neuroimaging methods (e.g., Gonzalez-Lima & Scheich, 1986; Morris et al., 1998).

Given the specificity and perseverance of learning-induced plasticity, AC has been proposed as a site of long-term storage for auditory associative information. Experimental data implicates both top-down and bottom-up mechanisms in the production of these memory traces (e.g., Edeline, 1999; Weinberger, 2004a; Polley et al., 2006). These conclusions are consonant with auditory WM literature inasmuch as they expand the role of primary AC to include learning and memory processes.

8.3.4 Future Questions Pertaining to Auditory Memory

Several subcortical structures have been shown to play a part in primate visual WM (Constantinidis & Procyk, 2004). For example, Fuster and Alexander (1973) observed a variety of DMTS-related responses in the mediodorsal nucleus of the thalamus, including sustained firing during the ISI. Another study by Kawagoe et al. (1998) demonstrated sustained delay activity and reward anticipation in the caudate

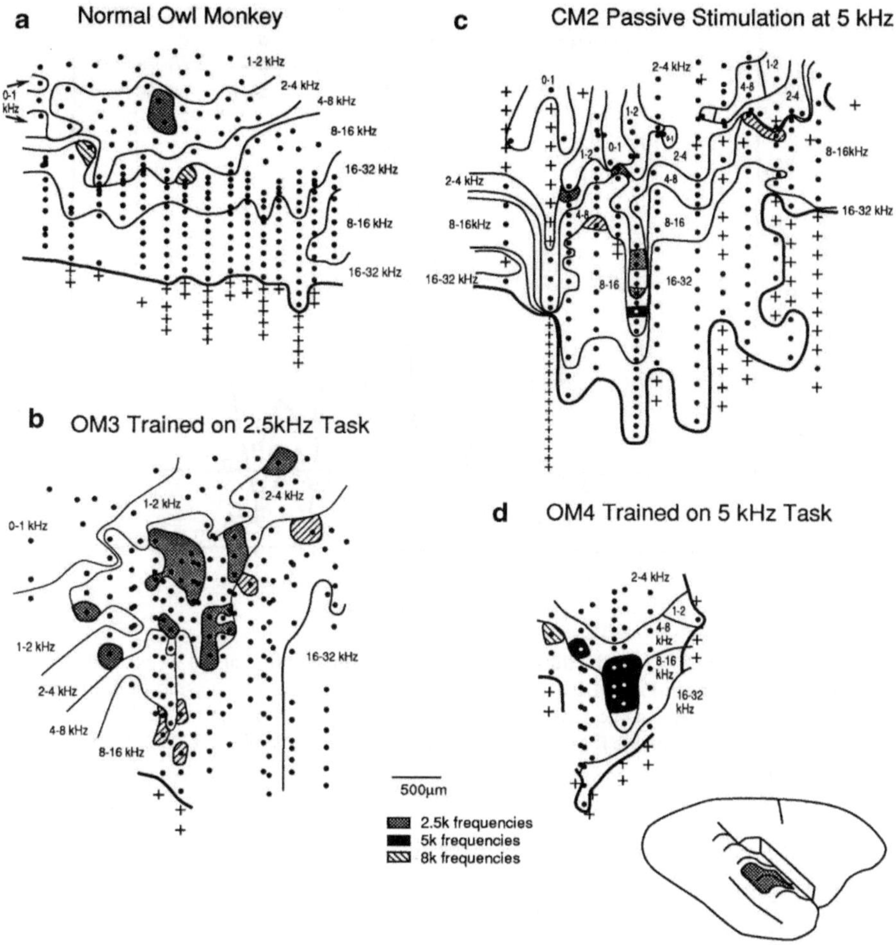

Fig. 8.9 Cortical representations of CF in A1 of four adult owl monkeys. Thin lines define boundaries of cortical locations with CFs within one octave. Stippled regions encircle cortical locations where neurons were recorded with CFs in the frequency range used in the 2.5-kHz task, solid regions represent octave. Stippled regions encircle cortical locations where neurons were recorded with CFs in the frequency range used in the 2.5-kHz task, solid regions represent frequencies used in the 5-kHz task, and hatched regions represent the frequency range used in the 8-kHz task. The cortical areas representing a given frequency range were approximated by connecting the 50% distance values to the neighboring recording sites with CFs outside the given frequency range. Pluses denote recording sites with neuronal responses not consistent with properties of A1 neurons. (**a**) is from a representative normal owl monkey (N2), (**b**) is from a monkey trained at 2.5 kHz (OM3), (**c**) shows the monkey passively stimulated with the frequencies used in the 5-kHz task (CM2), and (**d**) shows the representation of the monkey trained at the 5-kHz task (OM4). (Reproduced from Recanzone et al., 1993; Figure 10, with permission)

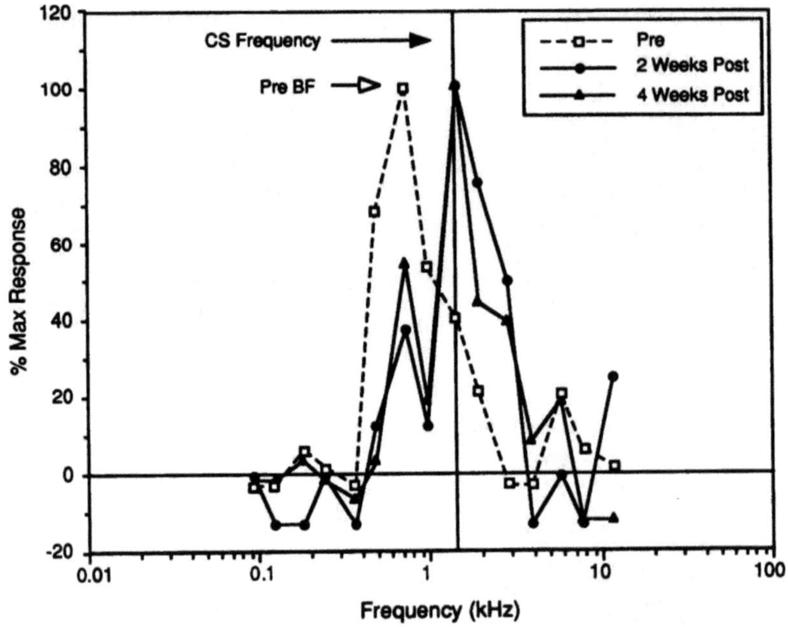

Fig. 8.10 Frequency RFs (60 dB) illustrating very long-term retention of > 4 weeks and complete shift of tuning. For pretraining, BF = 0.75 kHz and CS = 1.5 kHz. Note the shift of tuning so that the CS frequency became the new BF, shown here 2 weeks and approximately 32 days after training, the last recording available for this subject. (Reproduced from Weinberger et al., 1993; Figure 2, with permission)

nucleus of Japanese macaques during a memory-guided saccade task. Recordings in rats (Sakurai, 1990) indicate that some MGB cells are active during WM delay periods. In light of this finding, and previous research in the primate visual WM, it is expected that nuclei in the thalamus and basal ganglia contribute to auditory WM in primates. A more complete understanding of the involvement of these subcortical structures in auditory WM awaits experimental investigation.

There is now considerable anatomical, functional, and physiological evidence from primates that dorsal and ventral auditory cortical pathways specialize in processing spatial and nonspatial stimulus attributes, respectively (e.g., Kaas & Hackett, 1999; Romanski et al., 1999; Rauschecker & Tian, 2000). These specialized processing streams are comparable to the well-described "what" and "where" pathways in the visual system (Ungerleider & Mishkin, 1982; Ungerleider & Haxby, 1994). Data from human experiments indicate that maintenance of spatial and nonspatial information during WM differentially engages these pathways for the visual (e.g., Courtney et al., 1996) and auditory (Rämä et al., 2004; Rämä, 2008) modalities. Future neurophysiological recordings in primates should address these functional dissociations. Current evidence is consistent with a model wherein auditory WM is accomplished through a distributed network, which prominently includes the PFC, as well as the STG and AC. Future experiments will no doubt expand this model and elaborate on the functions of each of its components.

8.4 Overall Summary for Neural Correlates of Auditory Attention and Memory

There is no question that bottom-up and top-down processing are both major players in auditory attention, and for that matter, auditory memory as well. Both bottom-up processing and top-down processing pathways are well supported by the known anatomical connections (Winer & Lee, 2007). Context surrounding the auditory stimuli is another important factor in determining which sounds will be attended and remembered, and multiple modality processing figures prominently in focusing attention and associating meaning with important signals (Scheich et al., 2011).

Many of the mechanisms for focused or selective attention are closely linked to interactions between sensory systems when stimulus processing is occurring in multiple modalities, that is, part of the overall context. The influence of other sensory systems on the auditory system itself may occur at a very early stage of processing to help select the object of attention from the available context. The work of Schroeder and colleagues (Lakatos et al., 2009) suggests that the mechanisms for attention are similar among modalities, and the early latency of oscillatory behavior linking between primary sensory areas suggest that at least some forms of attention may be universal in their mechanism and under supramodal attentional control.

Interestingly, many of the same neural structures show firing pattern changes in both attention and memory-related auditory tasks (e.g., primary AC, PFC, parietal cortex), and how integrated or dependent these processes are at each level or as an integrated circuit is to be determined. It is clear that the primary AC is affected by attentional mechanisms and demonstrates both short- and long-term plasticity. Whether the primary AC directs attention and/or functions as the repository of auditory memories remain open questions.

There are many regions still to explore within the auditory core, belt, and parabelt regions, as well as other auditory-related cortices (Poremba et al., 2003; Hackett, 2008). Other animal models to be considered for information concerning the circuitry and mechanisms involved in auditory attention and memory include the songbird model system (Shea & Margoliash, 2010), wherein meaningful communication takes place by the performance, learning, and shaping of bird songs. In addition, different neurotransmitter systems, for example, acetylcholine, dopamine, and norephinephrine, also play a large role in neural activity patterns (Harley, 2004; Weinberger, 2004a; Plakke et al., 2008; Scheich et al., 2011) and should be considered carefully for their distributions within the neural structures identified as relating to auditory attention and memory.

In summary, the neurophysiological investigations of primate auditory attention and memory have just begun. Many of the findings resemble well-established visual attention and memory processes, but important differences have also been noted, for example, excitatory/inhibitory changes and memory differences. Current evidence is consistent with a model wherein context and experience are key modulators of auditory attention, learning, and memory.

Acknowledgments We thank Amy Opheim, Chi-wing Ng, and Ryan Opheim for assisting with the manuscript preparation. A. Poremba is supported by NIH, NIDCD, DC0007156.

References

Artchakov, D., Tikhonravov, D., Vuontela, V., Linnankoski, I., Korvenoia, A., & Carlson, S. (2007). Processing of auditory and visual location information in the monkey prefrontal cortex. *Experimental Brain Research*, 180(3), 469–479.

Artchakov, D., Tikhonravov, D., Ma, Y., Neuvonen, T., Linnankoski, I., & Carlson, S. (2009). Distracters impair and create working memory-related neuronal activity in the prefrontal cortex. *Cerebral Cortex*, 19(11), 2680–2689.

Atiani, S., Elhilali, M., David, S. V., Fritz, J. B., & Shamma, S. A. (2009). Task difficulty and performance induce diverse adaptive patterns in gain and shape of primary auditory cortical receptive fields. *Neuron*, 61(3), 467–480.

Bakin, J. S., & Weinberger, N. M. (1990). Classical conditioning induces CS-specific receptive field plasticity in the auditory cortex of the guinea pig. *Brain Research*, 536(1–2), 271–286.

Bakin, J. S., South, D. A., & Weinberger, N. M. (1996). Induction of receptive field plasticity in the auditory cortex of the guinea pig during instrumental avoidance conditioning. *Behavioral Neuroscience*, 110(5), 905–913.

Barbas, H., & Mesulam, M. (1981). Organization of afferent input to subdivisions of area 8 in the rhesus monkey. *Journal of Comparative Neurology*, 200(3), 407–431.

Barbas, H., & Pandya, D. N. (1989). Architecture and intrinsic connections of the prefrontal cortex in the rhesus monkey. *Journal of Comparative Neurology*, 286(3), 356–375.

Bartlett, E. L., & Wang, X. (2005). Long-lasting modulation by stimulus context in primate auditory cortex. *Journal of Neurophysiology*, 94(1), 83–104.

Beitel, R. E., Schreiner, C. E., Cheung, S. W., Wang, X., & Merzenich, M. M. (2003). Reward-dependent plasticity in the primary auditory cortex of adult monkeys trained to discriminate temporally modulated signals. *Proceedings of the National Academy of Sciences of the USA*, 100(19), 11070–11075.

Benson, D. A., & Hienz, R. D. (1978). Single-unit activity in the auditory cortex of monkeys selectively attending left vs. right ear stimuli. *Brain Research*, 159(2), 307–320.

Blake, D. T., Strata, F., Churchland, A. K., & Merzenich, M. M. (2002). Neural correlates of instrumental learning in primary auditory cortex. *Proceedings of the National Academy of Sciences of the USA*, 99(15), 10114–10119.

Bodner, M., Kroger, J., & Fuster, J. M. (1996). Auditory memory cells in dorsolateral prefrontal cortex. *NeuroReport*, 7(12), 1905–1908.

Bon, L., & Lucchetti, C. (2006). Auditory environmental cells and visual fixation effect in area 8B of macaque monkey. *Experimental Brain Research*, 168(3), 441–449.

Brechmann, A., Gaschler-Markefski, B., Sohr, M., Yoneda, K., Kaulisch, T., & Scheich, H. (2007). Working memory specific activity in auditory cortex: Potential correlates of sequential processing and maintenance. *Cerebral Cortex*, 17(11), 2544–2552.

Brosch, M., & Schreiner, C. E. (1997). Time course of forward masking tuning curves in cat primary auditory cortex. *Journal of Neurophysiology*, 77(2), 923–943.

Brosch, M., Schulz, A., & Scheich, H. (1999). Processing of sound sequences in macaque auditory cortex: Response enhancement. *Journal of Neurophysiology*, 82(3), 1542–1559.

Brosch, M., Selezneva, E., & Scheich, H. (2005). Nonauditory events of a behavioral procedure activate auditory cortex of highly trained monkeys. *Journal of Neuroscience*, 25(29), 6797–6806.

Buffalo, E. A., Ramus, S. J., Clark, R. E., Teng, E., Squire, L. R., & Zola, S. M. (1999). Dissociation between the effects of damage to perirhinal cortex and area TE. *Learning and Memory*, 6(6), 572–599.

Cherry, E. C. (1953). Some experiments on the recognition of speech, with one and with two ears. *Journal of the Acoustical Society of America*, 25(5), 975–979.

Cohen, Y. E. (2009). Multimodal activity in the parietal cortex. *Hearing Research*, 258(1–2), 100–105.

Cohen, Y. E., Cohen, I. S., & Gifford, G. W., III. (2004). Modulation of LIP activity by predictive auditory and visual cues. *Cerebral Cortex*, 14(12), 1287–1301.

Cohen, Y. E., Russ, B. E., & Gifford, G. W., III. (2005). Auditory processing in the posterior parietal cortex. *Behavioral and Cognitive Neuroscience Review*, 4(3), 218–231.

Colombo, M., & D'Amato, M. R. (1986). A comparison of visual and auditory short-term memory in monkeys (*Cebus apella*). *Quarterly Journal of Experimental Psychology B*, 38(4), 425–448.

Colombo, M., D'Amato, M. R., Rodman, H. R., & Gross, C. G. (1990). Auditory association cortex lesions impair auditory short-term memory in monkeys. *Science*, 247(4940), 336–338.

Colombo, M., Rodman, H. R., & Gross, C. G. (1996). The effects of superior temporal cortex lesions on the processing and retention of auditory information in monkeys (*Cebus apella*). *Journal of Neuroscience*, 16(14), 4501–4517.

Constantinidis, C., & Procyk, E. (2004). The primate working memory networks. *Cognitive, Affective, & Behavioral Neuroscience*, 4(4), 444–465.

Courtney, S. M., Ungerleider, L. G., Keil, K., & Haxby, J. V. (1996). Object and spatial visual working memory activate separate neural systems in human cortex. *Cerebral Cortex*, 6(1), 39–49.

Desimone, R., & Duncan, J. (1995). Neural mechanisms of selective visual attention. *Annual Review of Neuroscience*, 18, 193–222.

Diamond, D. M., & Weinberger, N. M. (1986). Classical conditioning rapidly induces specific changes in frequency receptive fields of single neurons in secondary and ventral ectosylvian auditory cortical fields. *Brain Research*, 372(2), 357–360.

Durif, C., Jouffrais, C., & Rouiller, E. M. (2003). Single-unit responses in the auditory cortex of monkeys performing a conditional acousticomotor task. *Experimental Brain Research*, 153(4), 614–627.

Edeline, J. M. (1999). Learning-induced physiological plasticity in the thalamo-cortical sensory systems: A critical evaluation of receptive field plasticity, map changes and their potential mechanisms. *Progress in Neurobiology*, 57(2), 165–224.

Edeline, J. M., & Weinberger, N. M. (1991a). Subcortical adaptive filtering in the auditory system: Associative receptive field plasticity in the dorsal medial geniculate body. *Behavioral Neuroscience*, 105(1), 154–175.

Edeline, J. M., & Weinberger, N. M. (1991b). Thalamic short-term plasticity in the auditory system: Associative retuning of receptive fields in the ventral medial geniculate body. *Behavioral Neuroscience*, 105(5), 618–639.

Edeline, J. M., Pham, P., & Weinberger, N. M. (1993). Rapid development of learning-induced receptive field plasticity in the auditory cortex. *Behavioral Neuroscience*, 107(4), 539–551.

Evans, E. F., & Whitfield, I. C. (1964). Classification of unit responses in the auditory cortex of the unanaesthetized and unrestrained cat. *Journal of Physiology*, 171(3), 476–493.

Fritz, J. B., Shamma, S., Elhilali, M., & Klein, D. (2003). Rapid task-related plasticity of spectrotemporal receptive fields in primary auditory cortex. *Nature Neuroscience*, 6(11), 1216–1223.

Fritz, J. B., Elhilali, M., & Shamma, S. A. (2005a). Differential dynamic plasticity of A1 receptive fields during multiple spectral tasks. *Journal of Neuroscience*, 25(33), 7623–7635.

Fritz, J. B., Mishkin, M., & Saunders, R. C. (2005b). In search of an auditory engram. *Proceedings of the National Academy of Sciences of the USA*, 102(26), 9359–9364.

Fritz, J. B., Shamma, S., & Elhilali, M. (2005c). One click, two clicks: The past shapes the future in auditory cortex. *Neuron*, 47(3), 325–327.

Fritz, J. B., Elhilali, M., David, S. V., & Shamma, S. A. (2007). Does attention play a role in dynamic receptive field adaptation to changing acoustic salience in A1? *Hearing Research*, 229(1–2), 186–203.

Fritz, J. B., David, S. V., Radtke-Schuller, S., Yin, P., & Shamma, S. A. (2010). Adaptive, behaviorally gated, persistent encoding of task-relevant auditory information in ferret frontal cortex. *Nature Neuroscience*, 13(8), 1011–1019.

Fuster, J. M. (2008). Overview of prefrontal functions: The temporal organization of behavior. In J. M. Fuster, *The prefrontal cortex* (pp. 333–385). Amsterdam; Boston: Academic Press/Elsevier.

Fuster, J. M., & Alexander, G. E. (1971). Neuron activity related to short-term memory. *Science*, 173(997), 652–654.

Fuster, J. M., & Alexander, G. E. (1973). Firing changes in cells of the nucleus medialis dorsalis associated with delayed response behavior. *Brain Research*, 61, 79–91.

Fuster, J. M., Bauer, R. H., & Jervey, J. P. (1982). Cellular discharge in the dorsolateral prefrontal cortex of the monkey in cognitive tasks. *Experimental Neurology*, 77(3), 679–694.
Fuster, J. M., Bodner, M., & Kroger, J. K. (2000). Cross-modal and cross-temporal association in neurons of frontal cortex. *Nature*, 405(6784), 347–351.
Galván, V. V., & Weinberger, N. M. (2002). Long-term consolidation and retention of learning-induced tuning plasticity in the auditory cortex of the guinea pig. *Neurobiology of Learning and Memory*, 77(1), 78–108.
Gibson, J. R., & Maunsell, J. H. (1997). Sensory modality specificity of neural activity related to memory in visual cortex. *Journal of Neurophysiology*, 78(3), 1263–1275.
Gifford, G. W., III, & Cohen, Y. E. (2004). Effect of a central fixation light on auditory spatial responses in area LIP. *Journal of Neurophysiology*, 91(6), 2929–2933.
Goldman-Rakic, P. S. (1995). Cellular basis of working memory. *Neuron*, 14(3), 477–485.
Gonzalez-Lima, F., & Scheich, H. (1986). Neural substrates for tone-conditioned bradycardia demonstrated with 2-deoxyglucose. II. Auditory cortex plasticity. *Behavioural Brain Research*, 20(3), 281–293.
Gottlieb, Y., Vaadia, E., & Abeles, M. (1989). Single unit activity in the auditory cortex of a monkey performing a short term memory task. *Experimental Brain Research*, 74(1), 139–148.
Grant, S. J., Aston-Jones, G., & Redmond, D. E., Jr. (1988). Responses of primate locus coeruleus neurons to simple and complex sensory stimuli. *Brain Research Bulletin*, 21(3), 401–410.
Guinan, J. J., Jr. (2010). Cochlear efferent innervation and function. *Current Opinion in Otolaryngology & Head and Neck Surgery*, 18(5), 447–453.
Hackett, T. A. (2008). Anatomical organization of the auditory cortex. *Journal of the American Academy of Audiology*, 19(10), 774–779.
Harley, C. W. (2004). Norepinephrine and dopamine as learning signals. *Neural Plasticity*, 11(3–4), 191–204.
Hocherman, S., & Gilat, E. (1981). Dependence of auditory cortex evoked unit activity on interstimulus interval in the cat. *Journal of Neurophysiology*, 45(6), 987–997.
Hocherman, S., Benson, D. A., Goldstein, M. H., Jr., Heffner, H. E., & Hienz, R. D. (1976). Evoked unit activity in auditory cortex of monkeys performing a selective attention task. *Brain Research*, 117(1), 51–68.
Hromádka, T., & Zador, A. M. (2007). Toward the mechanisms of auditory attention. *Hearing Research*, 229(1–2), 180–185.
Hubel, D. H. (1959). Single unit activity in striate cortex of unrestrained cats. *Journal of Physiology*, 147, 226–238.
Jääskeläinen, I. P., Ahveninen, J., Bonmassar, G., Dale, A. M., Ilmoniemi, R. J., Levänen, S., & Belliveau, J. W. (2004). Human posterior auditory cortex gates novel sounds to consciousness. *Proceedings of the National Academy of Sciences of the USA*, 101(17), 6809–6814.
Jääskeläinen, I. P., Ahveninen, J., Belliveau, J. W., Raij, T., & Sams, M. (2007). Short-term plasticity in auditory cognition. *Trends in Neurosciences*, 30(12), 653–961.
James, W. (1890). *Principles of psychology*. New York: Henry Holt and Co.
Joseph, J. P., & Barone, P. (1987). Prefrontal unit activity during a delayed oculomotor task in the monkey. *Experimental Brain Research*, 67(3), 460–468.
Kaas, J. H., & Hackett, T. A. (1999). 'What' and 'where' processing in auditory cortex. *Nature Neuroscience*, 2(12), 1045–1047.
Kawagoe, R., Takikawa, Y., & Hikosaka, O. (1998). Expectation of reward modulates cognitive signals in the basal ganglia. *Nature Neuroscience*, 1(5), 411–416.
Kikuchi-Yorioka, Y., & Sawaguchi, T. (2000). Parallel visuospatial and audiospatial working memory processes in the monkey dorsolateral prefrontal cortex. *Nature Neuroscience*, 3(11), 1075–1076.
King, C., McGee, T., Rubel, E. W., Nicol, T., & Kraus, N. (1995). Acoustic features and acoustic changes are represented by different central pathways. *Hearing Research*, 85(1–2), 45–52.
Kitzes, L. M., & Doherty, D. (1994). Influence of callosal activity on units in the auditory cortex of ferret (*Mustela putorius*). *Journal of Neurophysiology*, 71(5), 1740–1751.

Knudsen, E. I. (2007). Fundamental components of attention. *Annual Review of Neuroscience*, 30, 57–78.

Kojima, S. (1985). Auditory short-term memory in the Japanese monkey. *International Journal of Neuroscience*, 25(3–4), 255–262.

Komura, Y., Tamura, R., Uwano, T., Nishijo, H., & Ono, T. (2005). Auditory thalamus integrates visual inputs into behavioral gains. *Nature Neuroscience*, 8(9), 1203–1209.

Kondo, H., Saleem, K. S., & Price, J. L. (2003). Differential connections of the temporal pole with the orbital and medial prefrontal networks in macaque monkeys. *The Journal of Comparative Neurology*, 465(4), 499–523.

Kraus, N., McGee, T., Littman, T., Nicol, T., & King, C. (1994). Nonprimary auditory thalamic representation of acoustic change. *Journal of Neurophysiology*, 72(3), 1270–1277.

Lakatos, P., O'Connell, M. N., Barczak, A., Mills, A., Javitt, D. C., & Schroeder, C. E. (2009). The leading sense: Supramodal control of neurophysiological context by attention. *Neuron*, 64(3), 419–430.

Lee, J. H., Russ, B. E., Orr, L. E., & Cohen, Y. E. (2009). Prefrontal activity predicts monkeys' decisions during an auditory category task. *Frontiers in Integrative Neuroscience*, 3(16), 1–12.

Lemus, L., Hernández, A., & Romo, R. (2009a). Neural codes for perceptual discrimination of acoustic flutter in the primate auditory cortex. *Proceedings of the National Academy of Sciences of the USA*, 106(23), 9471–9476.

Lemus, L., Hernández, A., & Romo, R. (2009b). Neural encoding of auditory discrimination in ventral premotor cortex. *Proceedings of the National Academy of Sciences of the USA*, 106(34), 14640–14645.

Lomber, S. G., Payne, B. R., & Cornwell, P. (2001). Role of the superior colliculus in analyses of space: Superficial and intermediate layer contributions to visual orienting, auditory orienting, and visuospatial discriminations during unilateral and bilateral deactivations. *Journal of Comparative Neurology*, 441(1), 44–57.

Lu, Z. L., Williamson, S. J., & Kaufman, L. (1992). Behavioral lifetime of human auditory sensory memory predicted by physiological measures. *Science*, 258(5088), 1668–1670.

Lucchetti, C., Lanzilotto, M., & Bon, L. (2008). Auditory-motor and cognitive aspects in area 8B of macaque monkey's frontal cortex: A premotor ear-eye field (PEEF). *Experimental Brain Research*, 186(1), 131–141.

Markowitsch, H. J., Emmans, D., Irle, E., Streicher, M., & Preilowski, B. (1985). Cortical and subcortical afferent connections of the primate's temporal pole: A study of rhesus monkeys, squirrel monkeys, and marmosets. *Journal of Comparative Neurology*, 242(3), 425–458.

Mazzoni, P., Bracewell, R. M., Barash, S., & Andersen, R. A. (1996). Spatially tuned auditory responses in area LIP of macaques performing delayed memory saccades to acoustic targets. *Journal of Neurophysiology*, 75(3), 1233–1241.

Miller, E. K., & Cohen, J. D. (2001). An integrative theory of prefrontal cortex function. *Annual Reviews of Neuroscience*, 24, 167–202.

Miller, E. K., & Desimone, R. (1994). Parallel neuronal mechanisms for short-term memory. *Science*, 263(5146), 520–522.

Miller, E. K., Erickson, C. A., & Desimone, R. (1996). Neural mechanisms of visual working memory in prefrontal cortex of the macaque. *Journal of Neuroscience*, 16(16), 5154–5167.

Miller, J. M., Sutton, D., Pfingst, B., Ryan, A., Beaton, R., & Gourevitch, G. (1972). Single cell activity in the auditory cortex of Rhesus monkeys: Behavioral dependency. *Science*, 177(47), 449–451.

Miller, J. M., Dobie, R. A., Pfingst, B. E., & Hienz, R. D. (1980). Electrophysiologic studies of the auditory cortex in the awake monkey. *American Journal of Otolaryngology*, 1(2), 119–130.

Mishkin, M. (1978). Memory in monkeys severely impaired by combined but not by separate removal of amygdala and hippocampus. *Nature*, 273(5660), 297–298.

Mishkin, M., & Delacour, J. (1975). An analysis of short-term visual memory in the monkey. *Journal of Experimental Psychology: Animal Behavior Processes*, 1(4), 326–334.

Morris, J. S., Friston, K. J., & Dolan, R. J. (1998). Experience-dependent modulation of tonotopic neural responses in human auditory cortex. *Proceedings of the Royal Society B: Biological Sciences*, 265(1397), 649–657.

Murray, E. A., & Mishkin, M. (1998). Object recognition and location memory in monkeys with excitotoxic lesions of the amygdala and hippocampus. *Journal of Neuroscience*, 18(16), 6568–6582.

Neff, W., Diamond, I., & Cassedey, J. (1975). Behavioural studies of auditory discrimination: Central nervous system. In W. D. Keider & D. Neff (Eds.), *Handbook of sensory physiology*, Vol. V/2 (pp. 307–400). New York: Springer-Verlag.

Ng, C. W. (2011). *Behavioral and neural correlates of auditory encoding and memory functions in rhesus macaques*. Doctoral dissertation. Retrieved from ProQuest Dissertations and Theses. (879629871).

Niwa, M., Johnson, J. S., Marshall, E. A., O'Connor, K. N., & Sutter, M. L. (2009). Decision- and state-related activity in primary auditory cortex (A1) of rhesus macaques performing amplitude modulation (AM) discrimination [Abstract]. *Society for Neuroscience Abstracts*, 556, 2.

Orr, L. E., Russ, B. E., & Cohen, Y. E. (2008). Disruption of decision making capacities in the rhesus macaque by prefrontal cortex TMS [Abstract]. *Society for Neuroscience Abstracts*, 875, 24.

Overman, W. H., Jr., & Doty, R. W. (1980). Prolonged visual memory in macaques and man. *Neuroscience*, 5(11), 1825–1831.

Otazu, G. H., Tai, L. H., Yang, Y., & Zador, A. M. (2009). Engaging in an auditory task suppresses responses in auditory cortex. *Nature Neuroscience*, 12(5), 646–654.

Pasternak, T., & Greenlee, M. W. (2005). Working memory in primate sensory systems. *Nature Reviews Neuroscience*, 6(2), 97–107.

Pérez-González, D., Malmierca, M. S., & Covey, E. (2005). Novelty detector neurons in the mammalian auditory midbrain. *European Journal of Neuroscience*, 22(11), 2879–2885.

Plakke, B. (2010). *Auditory working memory: Contributions of lateral prefrontal cortex and acetylcholine in non-human primates*. Doctoral dissertation. Retrieved from ProQuest Dissertations and Theses (880271144).

Plakke, B., Ng, C. W., & Poremba, A. (2008). Scopolamine impairs auditory delayed matching-to-sample performance in monkeys. *Neuroscience Letters*, 438(1), 126–130.

Pollack, G. S. (1988). Selective attention in an insect auditory neuron. *Journal of Neuroscience*, 8(7), 2635–2639.

Polley, D. B., Steinberg, E. E., & Merzenich, M. M. (2006). Perceptual learning directs auditory cortical map reorganization through top-down influences. *Journal of Neuroscience*, 26(18), 4970–4982.

Poremba, A., & Mishkin, M. (2007). Exploring the extent and function of higher-order auditory cortex in rhesus monkeys. *Hearing Research*, 229(1–2), 14–23.

Poremba, A., Saunders, R. C., Crane, A. M., Cook, M., Sokoloff, L., & Mishkin, M. (2003). Functional mapping of the primate auditory system. *Science*, 299(5606), 568–572.

Poremba, A., Malloy, M., Saunders, R. C., Carson, R. E., Herscovitch, P., & Mishkin, M. (2004). Species specific calls evoke asymmetric activity in the monkey's temporal poles. *Nature*, 427(6973), 448–451.

Rainer, G., Rao, S. C., & Miller, E. K. (1999). Prospective coding for objects in primate prefrontal cortex. *Journal of Neuroscience*, 19(13), 5493–5505.

Rämä, P. (2008). Domain-dependent activation during spatial and nonspatial auditory working memory. *Cognitive Processing*, 9(1), 29–34.

Rämä, P., Poremba, A., Sala, J. B., Yee, L., Malloy, M., Mishkin, M., & Courtney, S. M. (2004). Dissociable functional cortical topographies for working memory maintenance of voice identity and location. *Cerebral Cortex*, 14(7), 768–780.

Rauschecker, J. P., & Tian, B. (2000). Mechanisms and streams for processing of "what" and "where" in auditory cortex. *Proceedings of the National Academy of Sciences of the USA*, 97(22), 11800–11806.

Recanzone, G. H., Schreiner, C. E., & Merzenich, M. M. (1993). Plasticity in the frequency representation of primary auditory cortex following discrimination training in adult owl monkeys. *Journal of Neuroscience*, 13(1), 87–103.

Romanski, L. M., & Averbeck, B. B. (2009). The primate cortical auditory system and neural representation of conspecific vocalizations. *Annual Reviews of Neuroscience*, 32, 315–346.

Romanski, L. M., & Goldman-Rakic, P. S. (2002). An auditory domain in primate prefrontal cortex. *Nature Neuroscience* 5(1), 15–16.

Romanski, L. M., Tian, B., Fritz, J., Mishkin, M., Goldman-Rakic, P. S., & Rauschecker, J. P. (1999). Dual streams of auditory afferents target multiple domains in the primate prefrontal cortex. *Nature Neuroscience*, 2(12), 1131–1136.

Russ, B. E., Orr, L. E., & Cohen, Y. E. (2008). Prefrontal neurons predict choices during an auditory same-different task. *Current Biology*, 18(19), 1483–1488.

Sakurai, Y. (1990). Cells in the rat auditory system have sensory-delay correlates during the performance of an auditory working memory task. *Behavioral Neuroscience*, 104(6), 856–868.

Sakurai, Y. (1994). Involvement of auditory cortical and hippocampal neurons in auditory working memory and reference memory in the rat. *Journal of Neuroscience*, 14(5 Pt 1), 2606–2623.

Scheich, H. (2011). Behavioral semantics of learning and crossmodal processing in auditory cortex: The semantic processor concepts. *Hearing Research*, 271(1–2), 3–15.

Shafi, M., Zhou, Y., Quintana, J., Chow, C., Fuster, J., & Bodner, M. (2007). Variability in neuronal activity in primate cortex during working memory tasks. *Neuroscience*, 146(3), 1082–1108.

Shamma, S. A., Elhilali, M., & Micheyl, C. (2010). Temporal coherence and attention in auditory scene analysis. *Trends in Neurosciences*, 34(3), 114–123.

Shea, S. D., & Margoliash, D. (2010). Behavioral state-dependent reconfiguration of song-related network activity and cholinergic systems. *Journal of Chemical Neuroanatomy*, 39(2), 132–140.

Suga, N., & Ma, X. (2003). Multiparametric corticofugal modulation and plasticity in the auditory system. *Nature Reviews Neuroscience*, 4(10), 783–794.

Sutter, M. L., & Shamma, S. A. (2010). The relationship of auditory cortical activity to perception and behavior. In J. A. Winer & C. E. Schreiner (Eds.), *The auditory cortex* (pp. 617–641). New York: Springer.

Ulanovsky, N., Las, L., & Nelken, I. (2003). Processing of low-probability sounds by cortical neurons. *Nature Neuroscience*, 6(4), 391–398.

Ungerleider, L. G., & Haxby, J. V. (1994). 'What' and 'where' in the human brain. *Current Opinion in Neurobiology*, 4(2), 157–165.

Ungerleider, L. G., & Mishkin, M. (1982). Two cortical visual systems. In D. J. Ingle, M. A. Goodale, & R. J. W. Masfield (Eds.), *Analysis of visual behavior* (pp. 49–586). Cambridge, MA: MIT Press.

Wang, H., Wang, X., Wetzel, W., & Scheich, H. (2006). Rapid-rate transcranial magnetic stimulation of animal auditory cortex impairs short-term but not long-term memory formation. *European Journal of Neuroscience*, 23(8), 2176–2184.

Watanabe, M. (1992). Frontal units of the monkey coding the associative significance of visual and auditory stimuli. *Experimental Brain Research*, 89(2), 233–247.

Weinberger, N. M. (2004a). Experience-dependent response plasticity in the auditory cortex: Issues, characteristics, mechanisms and functions. In T. N. Parks, E. W. Rubel, & R. R. Fay (Eds.), *Plasticity of the auditory system* (pp. 173–227). New York: Springer.

Weinberger, N. M. (2004b). Specific long-term memory traces in primary auditory cortex. *Nature Reviews Neuroscience*, 5(4), 279–290.

Weinberger, N. M., Javid, R., & Lepan, B. (1993). Long-term retention of learning-induced receptive-field plasticity in the auditory cortex. *Proceedings of the National Academy of Sciences of the USA*, 90(6), 2394–2398.

Werner-Reiss, U., Porter, K. K., Underhill, A. M., & Groh, J. M. (2006). Long lasting attenuation by prior sounds in auditory cortex of awake primates. *Experimental Brain Research*, 168(1–2), 272–276.

Winer, J. A., & Lee, C. C. (2007). The distributed auditory cortex. *Hearing Research*, 229(1–2), 3–13.
Winkowski, D. E., & Knudsen, E. I. (2006). Top-down gain control of the auditory space map by gaze control circuitry in the barn owl. *Nature*, 439(7074), 336–339.
Wollberg, Z., & Sela, J. (1980). Frontal cortex of the awake squirrel monkey: Responses of single cells to visual and auditory stimuli. *Brain Research*, 198(1), 216–220.
Woloszyn, L., & Sheinberg, D. L. (2009). Neural dynamics in inferior temporal cortex during a visual working memory task. *Journal of Neuroscience*, 29(17), 5494–5507.
Yin, P., Mishkin, M., Sutter, M., & Fritz, J. B. (2008). Early stages of melody processing: Stimulus-sequence and task-dependent neuronal activity in monkey auditory cortical fields A1 and R. *Journal of Neurophysiology*, 100(6), 3009–3029.
Yu, X. J., Xu, X. X., He, S., & He, J. (2009). Change detection by thalamic reticular neurons. *Nature Neuroscience*, 12(9), 1165–1170.
Zikopoulos, B., & Barbas, H. (2006). Prefrontal projections to the thalamic reticular nucleus form a unique circuit for attentional mechanisms. *Journal of Neuroscience*, 26(28), 7348–7361.

Chapter 9
Attention and Dynamic, Task-Related Receptive Field Plasticity in Adult Auditory Cortex

Jonathan B. Fritz, Stephen David, and Shihab Shamma

Keywords Adaptive coding • Bottom-up attention • Context • Contrast filter • Foreground-background segregation • Rapid plasticity • Spectrotemporal receptive field • Stimulus-specific adaptation • Task valence • Task-relevant behavior • Tone-in-noise • Top-down attention

Abbreviations

A1	primary auditory cortex
AC	auditory cortex
ACh	acetylcholine
BF	best frequency of cell receptive field
CA	conditioned avoidance
CS	conditioning stimulus
EEG	electroencephalographic
FC	frontal cortex
fMRI	functional magnetic resonance imaging
FRF	frequency response fields
GABA	γ-aminobutyric acid
IC	inferior colliculus
MEG	magentoencephalographic

J.B. Fritz (✉) • S. Shamma
Neural Systems Lab, Institute for Systems Research, Electrical and Computer Engineering, University of Maryland, College Park, MD 20742, USA
e-mail: ripple@isr.umd.edu; sas@isr.umd.edu

S. David
Oregon Hearing Research Center, Oregon Health and Science University, 3181 S.W. Sam Jackson Park Road, Portland, OR 97239-3098, USA
e-mail: davids@ohsu.edu

Y.E. Cohen et al. (eds.), *Neural Correlates of Auditory Cognition*, Springer Handbook of Auditory Research 45, DOI 10.1007/978-1-4614-2350-8_9,

© Springer Science+Business Media New York 2013

MGB	medial geniculate nucleus
NB	nucleus basalis
PFC	prefrontal cortex
PR	positive reinforcement
RF	receptive field
RFP	receptive field plasticity
SSA	stimulus-specific adaptation
STRF	spectrotemporal receptive field

9.1 Introduction

The brain is an extraordinarily adaptive and predictive machine, adapting to present demands and predicting the future, and reveals tremendous plasticity in multiple diverse forms, from birth through adulthood. Neuronal plasticity is a fundamental property of neurons and neuronal circuits because it facilitates adaptation to new environments, dynamically adjusts cortical sensory filters to improve processing of salient stimuli to optimize task performance, enables prediction of reward, and provides the basis for learning from experience. Depending upon the time scales and mechanisms involved in induction and persistence of receptive field (RF) changes, these changes may be described as ephemeral stimulus-driven adaptive plasticity, rapid attention-driven plasticity, or consolidated learning-induced plasticity. There are likely to be common molecular and synaptic mechanisms underlying all rapid RF transformations, but also some striking differences.

Rapid plasticity is generated to adapt to the current environmental context and accentuate responses to the most salient cues in the present moment to optimize processing of the most important stimuli. One of the major functions of rapid RF plasticity (RFP) is contrast enhancement of the attended auditory object against the acoustic background. There are many ways of achieving this—not only by simple response enhancement at the task-relevant or conditioning stimulus (CS) tone frequency—but also by more complex changes in RF shape, gain, and neuronal activity and connectivity, depending upon stimulus features, task, and behavioral context. Hence, it is important to emphasize that in the adult brain, there are likely to be multiple forms of plasticity, each with its own characteristic context of sensory experience, or behavioral challenge, and its unique set of cellular mechanisms and sites of plasticity. One of the major challenges for the field is to come up with a synthesis that explains the origins and mechanisms, roles, and interactions of the different kinds of plasticity.

The main focus of this chapter is on dynamic, RF transformations in auditory cortex (AC) that accompany top-down attentional modulation of auditory processing and reveal the ability of the brain to make nimble, adaptive changes from one moment to the next as acoustic context and task demands change. These transformations occur at the level of synapses, single-neuron RFs, and also at the level of brain networks. This chapter also discusses several other related topics including rapid, "automatic" RF adaptive plasticity that is not driven by attention, possible common mechanisms of RFP, contributions of the broader attentional network to task-driven RFP, and insights from human neuroimaging studies that may help to put the results of animal studies in perspective.

A key to understanding the extraordinary adaptive abilities of the adult brain has been insight into the role of top-down attention in directing some forms of learning and neural plasticity (Baluch & Itti, 2011). There are other engines of plasticity, and in the developing brain, long-term neural changes can arise over a period of days to weeks from passive exposure to environmental sound during sensitive periods of development (Hensch, 2005; Morishita et al., 2010). The topics of developmental plasticity in the auditory system and plasticity from sound exposure in the adult are issues discussed by Shepard, Kilgard, and Liu, (Chapter 10). However, in addition to the involuntary, adaptive changes arising from passive sound exposure, another form of plasticity arises in the adult brain during perceptual or associative learning (e.g., classical and operant conditioning, training on perceptual discriminations), which involves interactions of the animal with its environment in pursuit of goals. In mature animals, the behavioral task salience of acoustic stimuli can trigger top-down plasticity and thus sets task learning apart from the highly plastic and impressionable nature of the brain during early development, which largely reflects the passively experienced features of the environment. Thus, many such learning-associated changes in the adult are driven by top-down attention, and arise to sounds that are behaviorally relevant to the animals, are paired with positive or negative reinforcement and likely are shaped by neuromodulatory influences. Scheich and Brosch (Chapter 3) offer a thoughtful discussion of task-related activation of the AC, nonauditory components of the task representation in AC, and the semantic dimension in the transformation from sound to behavioral meaning in auditory perception.

Attention selects which aspects of sensory input are processed and brought to awareness. To promote survival and well-being, attention prioritizes stimuli both voluntarily, according to context-specific goals and involuntarily, through attentional capture driven by physical salience. Thus, within the larger set of all rapid, plastic, adaptive transformations, one may distinguish between automatic, preattentive adaptive bottom-up changes versus plasticity arising from top-down attention. Bottom-up effects arise from adaptation to the physical properties or sound statistics of an acoustic environment or from attention elicited by an intrinsically salient stimulus or an over-learned and meaningful sound (an ambulance siren, a loud explosion, thunder, the roar of a lion, the cry "fire," the call of one's name). In contrast, in top-down attention, response modulation arises from directed focus on, and expectations about the stimuli that are important for the organism at that moment, independent of physical salience. Such voluntary, top-down selective attention often occurs in the context of goal-directed behavior.

9.1.1 Rapid Plasticity Customizes Contrast Filters to Enable Separation of Foreground from Background

One way to think about these forms of rapid auditory plasticity is that the brain seeks to form an accurate, up-to-date, predictive model of the current acoustic environment, the salient stimuli within it, and the events associated with those stimuli—and seeks to optimize its ability to perform a given auditory task in this acoustic environment by dynamically reshaping cortical filters to extract relevant information. An overarching principle of such attention-driven RF transformation is the contrast filter, which enhances features of the relevant acoustic foreground and suppresses the background. Such spectral contrast filter transforms in the auditory system are key in behavioral reshaping of acoustic information processing. Some of the earliest examples of such contrast filter transforms came in the demonstration of rapid transformations of RFs in primary auditory cortex (A1) after classical conditioning, in which responses to the conditioning frequency were enhanced, and adjacent frequencies were suppressed (Edeline et al., 1993).

Such contrast filters have also been described in other sensory systems, such as the visual system (Treue and Martinez-Trujillo, 1999). One might argue that the contrast filter is the fundamental basis of selective attention and subsequent perceptual and cognitive processing, in which behaviorally relevant information is selected for further processing, while the rest is suppressed or ignored (Cherry, 1953). Such filters are observed in classical conditioning (Edeline et al., 1993), and they are also observed during plasticity arising from two-tone discrimination, in which there is a "push–pull" contrast, such that there is response enhancement at the target frequency (foreground) combined with suppression of activity at the frequency of the reference tone (background), presumably to enhance contrast between it and the attended target (see Fritz et al., 2005a,b, 2007a,b and Sections 9.3.4–9.3.6). This top-down contrast filter may also be complemented by a bottom-up process that also enhances responses to unexpected, brief stimuli as compared with repeated stimuli that are filtered out (Ulanovsky et al., 2003, 2004).

9.2 Rapid Bottom-Up Adaptation

Unlike top-down plasticity—arising from top-down attentional processes—most forms of bottom-up adaptive plasticity do not require attention, but largely arise from automatic, adaptive processes (Robinson & McAlpine, 2009; Rabinowitz et al., 2011). Hence, the responses and RFs of auditory neurons can be altered not only by behavioral relevance (top-down changes) but also by specific habituation (highly specific reduction in response to a repeated tone—to within 0.125 octave resolution) to repetitive stimulus presentation (Condon & Weinberger, 1991), or more generally by bottom-up adaptation to the pattern of stimulus statistics. Such changes can occur in "passive," nonattending hearing conditions, even in the anesthetized animal, and are caused by adaptive adjustments to the statistics of the stimulus input at multiple stages in the auditory pathway. These changes help maintain coding accuracy for prevailing sound conditions over a wide range of acoustic environments.

9.2.1 Adaptive Coding of Stimulus Statistics

In "bottom-up" adaptive coding, the sensitivity of auditory neurons can be adjusted so that the most frequently presented stimuli come to be represented more precisely.

This kind of matching to current stimulus statistics, and adaptive coding can be observed at low levels in sensory processing (such as the adaptation of retinal ganglion cells to the intensity statistics of visual stimuli to make best use of the cells' dynamic range). To use neurons' limited dynamic ranges more efficiently, adaptive adjustments can be made quickly within a split-second (as little as 100-150 ms!) of encountering a change in the composition of sensory input. A lovely example of this type of "adaptive coding" was described by Dean and colleagues in the anesthetized guinea pig (Dean et al., 2005, 2008), who showed that the relationship between the firing rate of IC neurons and sound level changed very quickly as stimulus statistics changed, presumably to improve the population coding of those intensity levels that occur with the highest probability. They also observed adaptation to stimulus variance as well as stimulus intensity, and also to the more complex statistics of bimodal intensity distributions. The time course of these adaptations was very rapid—on the order of 150–300 ms (Dean et al., 2008) and may be the auditory analog of rapid gain control in the visual system. Similarly rapid adaptation to the recent history of spectrotemporal contrast has also been shown in the ferret auditory cortex (Rabinowitz et al., 2011). Auditory spatial processing also adapts rapidly to stimulus statistics of interaural level differences in the anesthetized ferret IC (Dahmen et al., 2010) and shows the extraordinary capacity of the auditory brain to fine-tune its auditory filters to the local acoustic environment. Interestingly, Dahmen and colleagues also showed a close similarity between the behavioral effects of changing input statistics on human interaural level differences (ILD). Others have also shown that neurons at different processing levels can change with the composition of their input (Wen et al., 2009; Watkins & Barbour, 2011), such that the most frequently encountered stimuli are encoded most precisely, even to the level of adapting to complex statistics of the stimulus distribution, such as skewness or kurtosis. However, it is worth noting that many of these animal studies have been conducted in the anesthetized state, and that sensory adaptation is most prominent in the quiescent animal in states such as slow-wave sleep, anesthesia and awake immobility, but tends to decrease during states of high alertness, such as when an animal is learning a new behavioral task (as has been shown in the somatosensory system by Castro-Alamancos, 2004). Note that in the awake marmoset A1, not all neurons track stimulus statistics (Watkins & Barbour, 2008, 2011). Instead, some nonmonotonic neurons respond to loud sounds by adjusting their response rate-level functions away from high intensities and consequently remain sensitive to softer sounds.

Human psychophysical studies (Dahmen et al., 2010) are consistent with the animal studies—and suggest subcortical as well as cortical sites are involved in adaptation. Recent studies from Kraus and colleagues (Chandrasekaran et al., 2009; Skoe & Kraus, 2010) highlight the presence of rapid subcortical adaptive plasticity in the human brain stem. They have reported a form of plasticity that accompanies implicit pattern learning, and occurs rapidly (<2 min), without training, reinforcement, attention, or conscious awareness. They found that the degree of enhancement for a repeated speech syllable was correlated with (separate) performance on a speech-in-noise task. Hence, Kraus and colleagues speculate that these enhanced responses may act to "tag" relevant signal properties or acoustic objects.

9.2.2 Stimulus-Specific Adaptation: Novelty Responses

Another important form of adaptive plasticity is stimulus-specific adaptation (SSA) (Ulanovsky et al., 2003). The basic phenomenon in SSA is that neurons respond less strongly to a commonly occurring "standard" sound, than to the same sound when it occurs infrequently. Thus, normal habituation for repeated stimuli is observed, unless the stimulus is presented only occasionally, in which case the response is restored or even enhanced. For instance, neurons in AC respond more strongly to rarely presented stimuli ("oddballs") embedded in a regular sequence, a behavior that can be understood by positing that these neurons have adapted to the recent statistics of their inputs and that they preferentially represent sounds that are outliers with respect to these recent statistics. In this way, SSA facilitates novelty detection, detection of rare events, and sudden changes in the acoustic environment, by highlighting low-probability, unexpected events. It has been shown to occur for deviant tones or amplitudes in the belt areas of the inferior colliculus (IC), nonlemniscal medial geniculate body (MGB), and AC in gerbils, rats, and cats (Ulanovsky et al., 2003, 2004; Antunes et al., 2010; Bauerle et al., 2011). A broad range of SSA effects across a wide range of acoustic parameters (sound frequencies, binaural localization cues, sound amplitudes) has also been demonstrated in the tectum and arcopallium gaze fields of the barn owl (Reches and Gutfreund, 2008).

The activity of many neurons in A1 is strongly influenced by stimulus history, as described above for fast adaptive plasticity to stimulus statistics. A similar time course (approximately 100 ms–1 s) is observed in forward masking in tone pairs, for the effects of the first tone on the response to the second tone (which can be excitatory or inhibitory). The response to the second stimulus in a pair of pure tones is shaped by the interstimulus interval (ISI) and the frequency of the first tone (Brosch and Schreiner, 1997). As seen in the aforementioned examples, purely stimulus and context-dependent adaptive plastic events can occur in the mature auditory system over multiple timescales, ranging from a second or less to more than 10 minutes for SSA (Ulanovsky et al., 2004), such that the time course of adaptation scales with the time structure of the stimulus. Such context-dependent responses, modulated by changes in the acoustic landscape, occur at the level of single AC neurons (Ulanovsky et al., 2003) and also in human auditory perception (Rahne & Sussman, 2009).

9.2.3 Role of Bottom-Up Adaptation

These phenomena, SSA, habituation, and adaptive plasticity to stimulus-statistics, may operate in complementary ways in signal processing in the auditory pathways. Habituation decreases responses selectively to a repeated stimulus, adaptive coding shifts RF tuning to the population average distribution of the acoustic landscape, whereas SSA basically enhances the infrequent or oddball stimulus. These all seem like useful "first-order" automatic signal-processing mechanisms: basically, habituate to the repeated stimuli that are not likely to carry information (reduce responses selectively), adapt to the world around you (by shifting your tuning curves for most efficient coding at the prevailing stimulus conditions), but keep your inner ears wide open for unexpected events.

The time constants in adaptive coding cover a wide range. There appears to be a fast mechanism: the relatively short-term buildup and recovery from adaptive effects (100–300 ms). In contrast, SSA can occur over a longer time window, in rat A1 even for short tone pips separated by 2 s of silence. And in contrast to fast adaptive coding, there are relatively long-term "memory" persistent effects of SSA, in which neurons track stimulus history with time courses of up 10 minutes, suggesting that the system can adapt to any time scale present in the acoustic environment.

As mentioned, the listening brain seeks to extract the regularities or patterns of incoming sounds and uses the information to engage in predictive processing (Winkler et al., 2009; Berkes et al., 2011), which is important for acoustic scene analysis and for goal-directed behavior. Intriguing recent results suggest that the distinction between "bottom-up" and "top-down" may be too simple a dichotomy, which also oversimplifies the interaction between the two. For example, over-learned salient acoustic stimuli—such as the word "Fire!" or one's own name—may elicit strong, automatic preattentive responses. Recent studies also indicate that SSA and adaptive coding shifts are both affected by cortical cooling, suggesting "top-down" interactions with "bottom-up" phenomena. Such interactions have also been described in visual processing (McMains & Kastner, 2011).

9.3 Dynamic Plasticity of RFs During Attention-Driven Behavior

As described by Scheich and Brosch (Chapter 3), there have been many studies exploring changes in neural response to sounds, before and after they have been associated with behavioral meaning. Such task-related or reward-related changes in stimulus response strength are often paralleled by changes in RF gain or shape that can persist for hours, days, weeks, or even months afterwards (Galvan & Weinberger, 2002). One characteristic of the training procedure in most of these behavioral neurophysiology studies was that the animals were usually trained on only one task, with one limited set of stimuli. Without doubt, such previous studies were well designed to show dramatic long-term plastic effects in A1. But the learning-induced plasticity may have so successfully remodeled and specialized the system to focus on a single task involving a single discrimination between a very small set of stimuli that such highly "specialized" A1 neurons might not have as great a range for dynamic RF changes with attention as would generalist neurons from an animal trained to switch quickly between multiple-task conditions, and for which stimuli of all frequencies could be salient targets.

9.3.1 Time Resolution for Analysis of Dynamic RF Properties of Neurons in A1 During Behavior

What is the time course for the emergence of such dynamic RFP? One challenge in answering this question is developing the temporal resolution to follow on-line RF transformations because it takes time to map out RF tuning. A classic study in the field showed that RFP occurred rapidly following classical conditioning in A1 (Edeline et al., 1993). In that study, the frequency response fields (FRFs) were mapped before and after conditioning occurred. Edeline and colleagues (1993) found that RFs changed, with modulation of the relative responses to the CS+ frequency and CS– frequency, in as few as five trials (or pairings of CS frequency with shock). Because the protocol for conditioning for five pairings took about 10 minutes and it took about 20 minutes to complete RF analysis, the temporal resolution in this study was about 20–30 minutes. Typical changes in response of A1 neurons after classical conditioning show frequency-specific RF increases observed after 5 or 15 training trials. In some cases, the increased responses at the CS frequency, with concomitant decreased responses at the pretraining best frequency (BF), allowed the CS to become the new BF.

A more recent study with 10-fold higher temporal resolution, using a new technique for on-line RF measurements found rapid RFP within 1–2 minutes after the onset of attention to acoustic features of an auditory task (Fritz et al., 2003). Given the remarkable ability to shift attention rapidly from one acoustic object or stream to another, a reasonable conjecture is that RFs can change even more rapidly than current measurements allow us to detect—on the order of a few seconds as the focus of attention is shifted (Mesgarani & Chang, 2012). However, proof of this will require the development of further improvements in the temporal resolution of on-line RF measurements to capture such dynamic moment-to-moment RF changes.

9.3.2 Task-Related Plasticity in A1: Measuring Changes in Spectrotemporal RFs

To quantify dynamic, on-line adaptive RF changes in AC as animals switched attentional focus, it was necessary to develop procedures to study rapid task-related RFP in the awake, behaving animal model. We chose to use the ferret (*Mustela putorius*), an animal whose AC had been mapped and previously studied (Bizley et al., 2005). Using a modified version of the conditioned avoidance (CA) experimental paradigm (Heffner and Heffner, 1995), ferrets were trained to distinguish behaviorally between two variable and broad acoustic categories, "target" and "reference" sounds, which have different reward value (Fritz et al., 2003). The ferrets became auditory "generalists" because the stimuli and tasks varied from day to day. They learned to prune the incoming acoustic input and extract the salient acoustic cues for a given task, to differentiate target and reference sounds, and separate "target" or "foreground" sounds that are best to attend to, and ignored at risk of mild shock—from "reference" or "background" —sounds that could be safely ignored after initial detection. This combination of stimulus variability during training, and stimulus stability during testing, gives us the ideal opportunity to measure attention-driven neuronal plasticity in a cortical population that is poised for change, but has not already been biased by prior repetitive training with fixed target stimuli over previous weeks or months (e.g., Recanzone et al., 1993).

In addition to increasing the range of the animals' behavioral repertoire, so as to explore the effects of attention in different task and stimulus conditions, one other important departure from previous plasticity studies was purely methodological: the choice to use the spectrotemporal RF (STRF) to measure cortical response properties, rather than the more common isoamplitude RF or FRF (measured by responses to tone pips of different frequency and intensity). One advantage of this approach was that it was possible to build on knowledge of A1 RFs in the anesthetized and awake but nonbehaving ferret using these linear methods.

The essence of this approach was to record from single neurons in A1 while the animal performed a variety of different auditory tasks (Fig. 9.1), with the goal of quantitatively analyzing the nature and time-course of state-dependent, task-dependent adaptive plasticity in AC on a cellular and network level. Our approach to quantification of RF changes is illustrated (Fig. 9.2) with examples of RFP during tone detect and two-tone discrimination tasks.

Both negative and positive reinforcement (PR) behavioral paradigms have been used to explore plasticity under different reward valence and stimulus conditions. In all detection experiments, reference sounds were drawn from a class of 30 distinct noise stimuli called temporally orthogonal ripple combinations (TORCs) that are temporally and spectrally rich, broadband stimuli that also serve during physiological experiments to characterize the STRF of the cell under study. In each experiment, the STRFs were generated based upon the neural responses to rippled noise (for a detailed discussion see Klein et al., 2000; Depireux et al., 2001; Miller et al., 2002). In contrast, the target sound varied from one experiment to another over a broad range of spectral or temporal, or combined spectrotemporal features.

As shown in Fig. 9.1, the tasks were grouped by the salient features or the type of acoustic target that the animal needed to attend to in order to perform the detection task (e.g. tones, tones-in-noise, silent gaps in TORCS, tone duration, click rate, frequency-modulated [FM] sweep direction, acoustic streams, consonant-vowel (CV) phonemes, etc.).

9.3.3 *Dynamic Changes in Spectral Tuning in a Sea of STRF Stability*

In a nontask condition in a quiescent, listening ferret, STRFs are fairly stable (Elhilali et al., 2007). These results are consistent with an earlier studies by Kisley and Gerstein (2001) and by Galvan et al. (2001) that monitored cortical local field

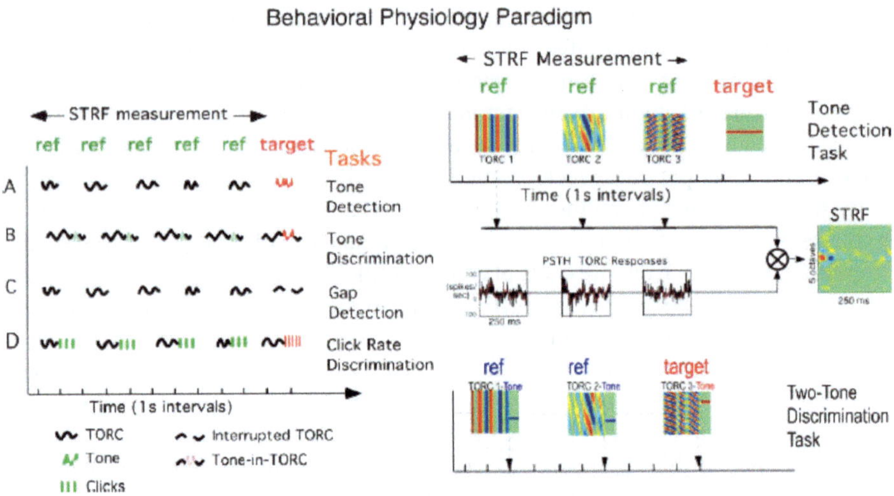

Fig. 9.1 Design of experimental stimulus presentations in CA tasks. (Upper right) On a given trial during a behavioral session, a random number of TORCs (1–6 noisy reference signals) were followed by a target tone. The panels illustrate spectrograms of three such TORCs and of the following tonal target. Responses to each TORC are collected in PST histograms that are cross-correlated with the TORC spectrograms to estimate the STRF. Although the animal behaves in anticipation of the target, all spike measurements to derive the STRF are made during the presentation of the reference TORCs. (Lower right) Similar design for a two-tone discrimination task in which the ferret is presented with a random number of TORCtone combinations (1–6 reference signals in which the reference tone is fixed in frequency) followed by a target TORC-tone combination (in which the tone component changes to a different frequency than that of the reference tone). (Left) Schematic of various possible experimental paradigms, including (**a**) tone detection, (**b**) two-tone discrimination, (**c**) gap detection, and (**d**) click rate discrimination. All follow the same basic design. The reference signals are (or include) TORCs used to measure the STRF. The target varies from one experiment to another. (Reprinted from Fritz et al., 2005b, with permission)

Fig. 9.2 (Top row) Comparison of a pre-behavior, quiescent STRF (left) and a behavioral STRF (middle). Each panel depicts an STRF with a color scale, representing increased (red) to suppressed (blue) firing about the (green) mean. The STRF in each panel is normalized. The dashed contours depict the regions with statistically significant fluctuations (color level is standard deviations from the mean). All excitatory (red) and inhibitory (blue) features of the STRF are statistically significant. The black arrow indicates the frequency of the target tone during the detection task. The difference between the normalized quiescent and detection STRF is shown in the right panel (STRFdiff). The asterisk marks the location of maximal change. (Second and third rows) Facilitative STRF plasticity in AI. STRFs from two single-units in A1 illustrate typical changes observed during performance of the tone detection task. (Second row) Pre-behavior STRF (left). Localized enhancement of an excitatory region in the STRF at the target frequency during behavior (middle). The post-behavior quiescent STRF (right) reverted immediately to a RF very close to its original pre-behavior shape. The maximum change was nearly coincident with the target frequency. (Third row) Pre-behavior STRF (left). Local decrease (near elimination) of lower inhibitory sideband at the target frequency in the detection STRF. (Fourth row) Three quiescent STRFs interleaved with two sequential two-tone discrimination tasks, all measured during recordings from the same neuron. The times at which STRFs were measured relative to the beginning of recording are shown on top of each panel. The arrows mark the reference (green) and target (red) frequencies used. Note the disappearance of the excitatory area near 250 Hz in the STRFs measured during the discrimination tasks. (Reprinted from Fritz et al., 2005b, with permission)

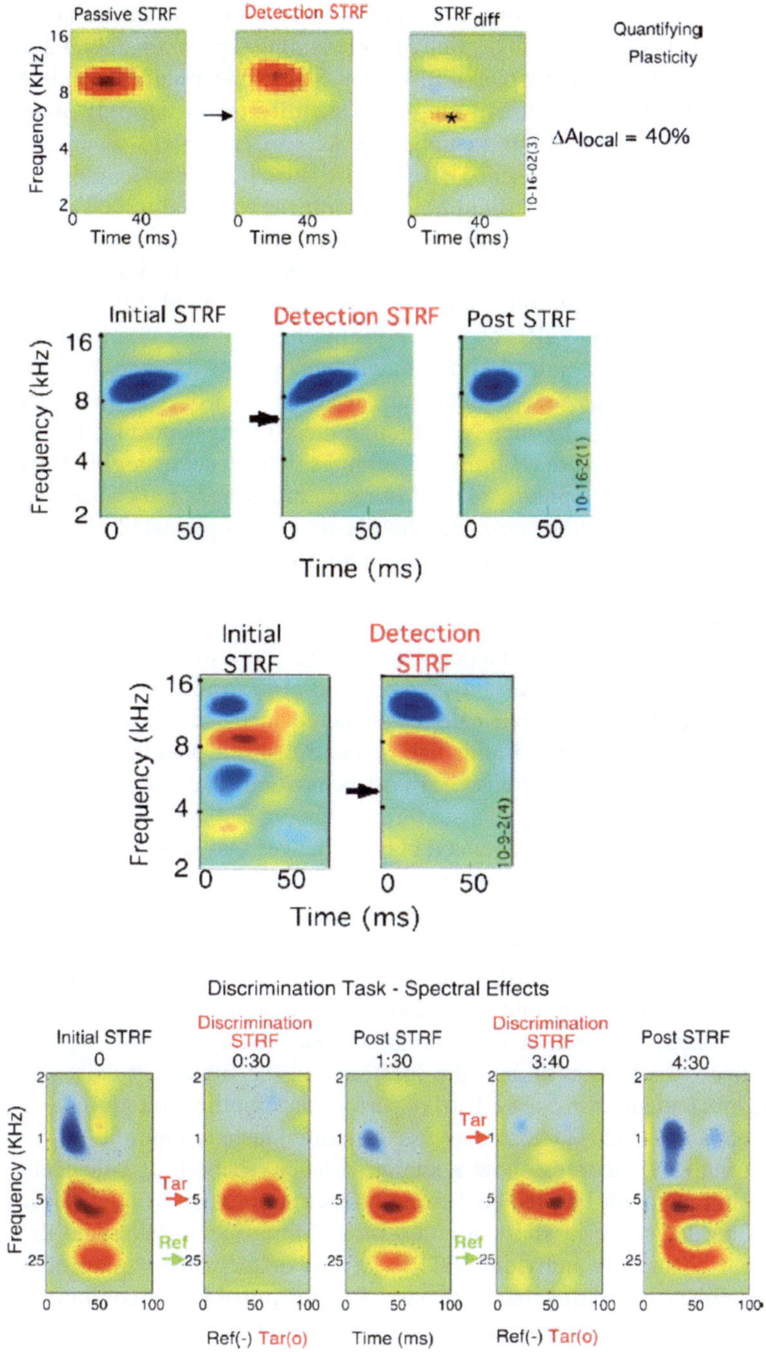

Fig. 9.2 (continued)

potentials (LFPs) over the course of several weeks in A1 of awake guinea pigs and reported only small daily changes in frequency tuning (approximately 0.2 octaves), suggesting that the tonotopic map is also fairly stable under natural conditions.

In contrast with this picture of long-term stability, an initial study of attention-driven plasticity revealed rapid changes in A1 cortical STRFs when the animal was asked to detect the presence of one particular pure tone stimulus, as compared to a broadband background noise stimulus (Fritz et al., 2003). The extent of intrinsic variation in STRFs during the quiescent state was insignificant compared with the magnitude of behaviorally induced STRF changes observed during performance of spectral auditory tasks (Elhilali et al., 2007). The most common change in an STRF observed during this task was "facilitation" at the target frequency as a result of either an enhancement of its excitatory field, or a weakening of the suppressive sidebands, leading to an overall enhancement of the RF at tone "target" frequency. Although these changes occurred rapidly, and could be reliably measured within 1–2 minutes in a subset of highly linear cells, the effects were often ephemeral, and in many cases were only observed during task performance. In other neurons, the plastic effects persisted after task completion, and only gradually faded away, with a time course of minutes to hours after task performance.

The overall effect of engaging attention in the tone detection task could be observed at a single cell and at a population level. There was a correlation between the magnitude of neuronal plasticity and behavioral performance (Fritz et al., 2003). Because more than half of all cells encountered in these experiments exhibited STRF changes that persisted after one or more tasks, a population analysis also confirmed the presence of long-term post-behavioral changes. The fact that the reshaped STRF continues in changed form following task completion suggests that attention triggers a set of task-related plastic changes that do not require the maintenance of active attention in order to persist.

Animals had no behavioral difficulty in transitioning from a single-tone detection task to a "chord" detection task, in which the target consisted of multiple simultaneously presented tones (Fritz et al., 2007b). Because it is not possible for ferrets or humans to extract individual tone frequency components from a random chord complex containing three or more tonal components, animals are not be able to focus selective attention on specific locations on the frequency axis, but instead used a global attention strategy. During performance of the multiple-tone detection (MTD) task, in which a chord composed of simultaneous multiple tones was the target stimuli, A1 neurons show a similar plasticity pattern of enhancement and suppression as might be predicted from a linear summation of the effects seen in single-tone detection, in short, a comb filter. Moreover, on average, there were significant changes in individual STRFs at two or more tone frequencies in the chord complex, suggesting that there was a faithful global imprint of much, or all of, the whole auditory chord object on the reshaped STRF. Thus, these effects were clearly seen at the population level (Fritz et al., 2007b).

9.3.4 Bidirectional Plasticity (Both Enhancement and Suppression)

When the animal has to discriminate one tone from another in a two-tone discrimination task, the RF changes to enhance selectively the target tone frequency and suppress the comparison tone frequency (Fritz et al., 2005a,b). This suggests that there may be at least two different and opposite types of modulatory influences that can shape cortical RFs.

When an animal is discriminating between the frequencies of two tones (as opposed to simply detecting the presence of one tone as earlier), a reasonable prediction is that the STRFs could change adaptively so as to improve performance by enhancing "foreground" over "background" by facilitating the STRF at the target (foreground) frequency while suppressing it at the reference (background) frequency. This hypothesis is consistent with earlier results (Edeline and Weinberger, 1993; Blake et al., 2002). In theory, another equally viable strategy to achieve discrimination might be to enhance "background" while suppressing "foreground" stimuli because the key point for the nervous system is to enhance the contrast between the stimuli.

Ferrets were trained on a two-tone discrimination task and soon learned to respond only when the tone frequency (background reference) changed, which occurred when the target tone was presented. An example of plasticity during the two-tone discrimination task is shown in Fig. 9.2.

In keeping with the results of the individual neuron shown in the example, the population data also show a consistent pattern of specific suppression of facilitatory fields of STRFs at the reference frequency during two-tone discrimination in contrast to the simultaneous enhancement observed at the target frequency, as in the single-tone detection task. Given these opposite effects of reference and target stimuli, it was possible to play one effect against another in successive experiments on the same neuron (Fig. 9.3), by recording from the same neuron in two different task conditions in which the same stimulus had two different meanings: as reference in the first task (two-tone discrimination) and as target in the second task (single tone detect).

As predicted, neurons show a chameleon-like adaptive ability to change their STRF based upon changing task conditions. In this case, in the same neuron, a single tone elicited net suppression in task 1 (in which the tone was a reference), and net enhancement in task 2 (in which the tone was a target). Similar effects were seen at a population level in multiple task sequences (see Fritz et al., 2005a,b).

9.3.5 Plasticity During Detection of Tone-in-Noise

One of the fundamental problems in auditory processing is how relevant acoustic signals are extracted from a noisy background. To explore neural plasticity arising from such a task, ferrets were trained to detect a tonal signal embedded in a

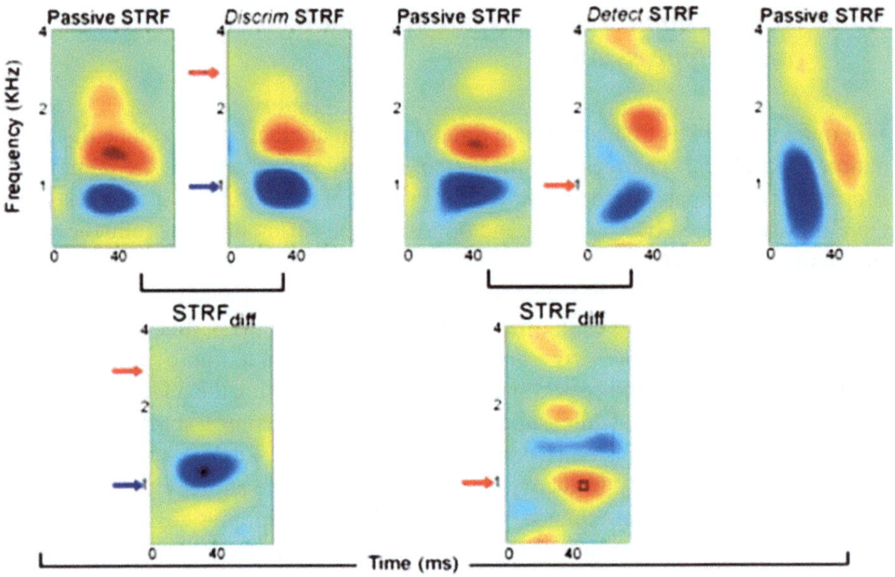

Fig. 9.3 STRF plasticity in a sequence of multiple tasks and behavioral contexts. Three passive STRFs (first, third, and fifth panels) interleaved with discrimination and detection tasks (second and fourth panels). In the two behavioral tasks, the 1-kHz tone first played the role of reference tone (in the discrimination task) and then in the second task, played the role of target tone (in the detection task). The effect on the STRF was different depending on the behavioral context and stimulus meaning (as seen in the STRFdiff panels below). As a reference tone, the 1-kHz tone (blue arrow) induced a strong depression in the STRF (second panel). As a target tone (red arrow), it facilitated the STRF at 1 kHz (weakening inhibition at 1 kHz). (Reprinted from Fritz et al., 2005a, with permission)

background noisy stimulus. Task difficulty in different behavioral sessions was varied by changing the signal-to-noise ratio (SNR) of the tonal target signal embedded in background noise. In order to perform this task successfully, the auditory system needs to both (1) filter out the background, reference noisy stimuli and (2) extract the tone in the target from the same background noise. The resulting RF transformation included enhancement at the tone frequency against a background of broadband suppression and also changes in gain (Atiani et al, 2009) using a gain fitting STRF model (David et al., 2008). The overall decrease in gain during attentive task performance in the tone-in-noise task is consistent with other findings of gain suppression during auditory behavior in A1 (Otazu et al., 2009). The combined effect of these adaptations is to increase the strength of the target tone representation relative to the noise. This is achieved by enhancing responses of near (to target frequency) cells to the tone during high SNR tasks while suppressing responses of far (from target frequency) cells to the masking noise in low SNR tasks. These adaptive STRF changes were greatest in high-performance sessions, confirming a close correlation of the magnitude of rapid RFP with behavior.

A key point is that the attention-driven plasticity changes observed cannot be explained as simply changes in gain, but must also require changes in auditory RF structure. Although previous studies in visual cortex have emphasized the importance of gain during attention, recent work has also confirmed that attention can also operate by reshaping visual RFs (Anton-Erxleben et al., 2009; McManus et al., 2011).

9.3.6 Unique, Persistent Plasticity 'Contrast Filter' Signatures for Each Combination of Task and Stimuli

Is there a general theory that can predict which RF changes will be observed when single tones are targets, and also when more complex stimuli (such as tone combinations or chords) are used (Fritz et al., 2003, 2005a,b, 2007a; Atiani et al., 2009)? In the case of chord detection, the resulting RFP is a linear sum of the predicted plasticity for each of the individual tones in the chord. A plausible conjecture is that when attention is directed to a complex stimulus, it enhances RF responses to the spectrotemporal features of the stimulus—convolving the RF with the imprint of the salient stimulus. A mathematical model, incorporating these insights, was developed to estimate RFs that enhance discrimination between any two arbitrary classes of sounds (Mesgarani et al., 2010). In this computational study, cortical RFs were modeled as filters that change their spectrotemporal tuning properties so as to respond best to the discriminatory features between foreground and background stimuli. The estimated RFP from this model closely resemble the experimentally observed changes in spectral discrimination tasks. Overall, these experimental and computational results suggest that each spectral task has a distinctive plasticity signature transformation of cortical RFs, depending upon the task structure, and the acoustic properties of the foreground and background stimuli that the animal discriminates.

One way to summarize the basic transformation in all of these results is by invoking a contrast filter, in which acoustic features of the foreground are enhanced and the acoustic features of the background are suppressed (Fritz et al., 2007a,b; Mesgarani et al., 2010). If this is how the brain extracts acoustic signals of interest from the environmental soundscape then it should be possible to predict the resulting plasticity transformation for optimal discrimination between any two arbitrary classes of sounds by enhancing the spectrotemporal features of the foreground and inhibiting the spectrotemporal features of the background to maximally enhance the contrast.

One important caveat that is important to make is that although spectral RFP is highly specific at a population level, there is considerable variability from neuron to neuron, with some cells showing little or no change and others showing a range of modulation, but in the same overall direction. The greatest modulation takes place when the target tone is close to the BF of the cell, where presumably there are the greatest number of modifiable synaptic inputs on a given neuron. A second important caveat is that this linear STRF analysis depends upon the presence of a significant linear component in neuronal responses - which is not found in all A1 neurons-and is found in even fewer neurons in higher auditory areas.

9.3.6.1 Persistence of RFP Changes After Spectral Tasks

As indicated above (in Sections 9.3, 9.33, 9.36), the plastic effects observed following performance of the tone-detect task could persist for hours after the completion of the single-tone detect task (Fritz et al., 2003). A similar degree of persistence of RF plasticity occurred after performance of other tasks such as complex chord detection and two-tone discrimination (Fritz et al., 2005a,b, 2007a). The time course of this persistence varied considerably from neuron to neuron, but at a population level had a half-life of about 30 minutes. Such persistent effects could last for even longer periods of time after classical conditioning (Galvan and Weinberger, 2002).

9.3.7 Importance of Task Valence in Shaping RFP

Another factor that strongly influences RFP is the valence of the task reward (Ilango et al., 2010; David et al., 2012). Ferrets were trained on an appetitive, mirror-image version of the tone detection task (Fritz et al., 2003), in which the acoustic discrimination was the same but the rewards associated with the noisy and tonal stimuli were opposite.

Opposite to the stimulus-action associations in the CA version of the task, in the appetitive or positive reinforcement (PV) version the ferrets learned to refrain from licking to the noisy stimuli, and lick only during presentation of the tonal target. In these conditions, the sign of the plasticity effect for the attended target tone frequency was inverted (Fig. 9.4), now consisting of overall suppression of the RF at that frequency (David et al., 2012).

As in earlier studies of task-related plasticity, these changes occurred rapidly, and could be reliably measured within 1–2 minutes. Incorporating the parameter of reward valence leads to a broader summary of the changes in RFs in A1 during task performance on a variety of tasks. The general principle that emerges from these studies is that there are signature changes in RF tuning for a bank of filters in A1 during different behavioral tasks with different reward valence and audiomotor links (Fig. 9.5).

Fig. 9.4 Behavior paradigms with opposite valence structure. (**a**) Schematic of PR and CA tasks. In both tasks, subjects were required to detect the occurrence of a pure tone target (red) after a random number of rippled noise reference sounds (blue). During PR, they were rewarded with water if they licked a water spout 0.1–1.0 s after the onset of the target (green bar, middle) and punished with a timeout if they licked before this window (red bar). During CA, they were rewarded by licking a continuously flowing stream of water during the presentation of references (green bar, bottom) and were punished with a mild tail shock if they failed to cease licking for a 0.4-s period beginning 0.4 s after target offset (red bar). The duration of each reference and target stimulus was fixed within an experimental task but generally was shorter during PR (0.5–1.0 s) than during CA (1.0–1.5 s). (**b**) Average licking behavior during PR (left) and CA (right). Lick rate is plotted as a function of time after onset of reference (blue) or target stimuli (red) for each 50-ms time bin. Shading indicates one standard error on the mean across sessions. Vertical lines indicate stimulus onset and offset.

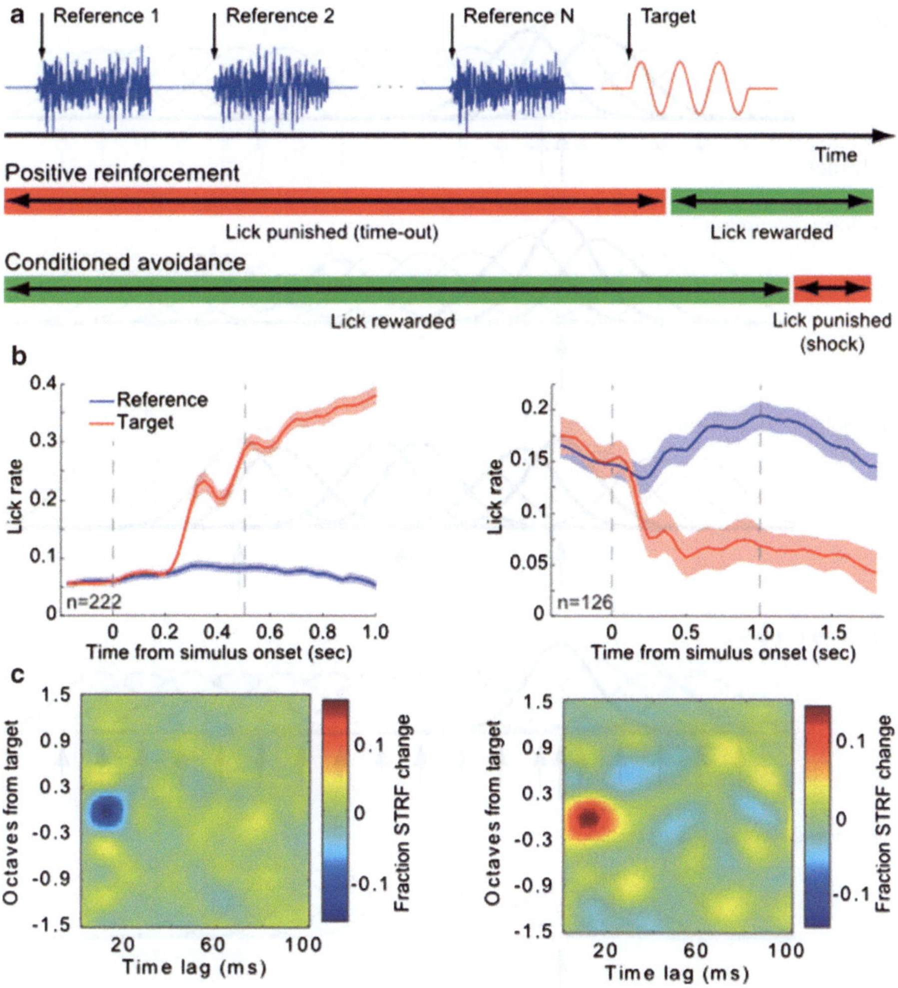

Fig. 9.4 (continued) During PR, animals showed minimal licking after the onset of reference stimuli, as trials when this occurred were punished as false alarms. Licking substantially increased after the onset of a target, and these trials were rewarded as hits (stereotyped licking at about 5 licks/s is reflected in the multiple peaks in the target curve). During CA animals maintained an elevated lick rate to retrieve water reward while reference stimuli were presented. Licking was attenuated within 0.2–0.4 s after target onset until after target offset (1.4–1.8 s), during which time licking resulted in punishment for a miss. (**c**) Average STRF difference between PR behavior and passive listening aligned at the target frequency and averaged over $n = 227$ neurons (left). Red regions indicate frequencies and time lags with increased responsiveness and blue regions indicate those with decreased responsiveness. The average difference shows a selective decrease in amplitude within one-quarter octave of the target frequency. The mean change across 75 significantly modulated neurons was –26% of peak passive STRF amplitude (median, –30%; both significant, $p < 0.0001$, randomized paired t-test). The average STRF difference for CA behavior is plotted in the right panel ($n = 235$ neurons). Here, the average difference shows an increase in amplitude near the target frequency. The mean change across the 83 significantly modulated neurons was a 20% increase at the target frequency (median, 23%; both significant $p < 0.0001$, randomized paired t-test)

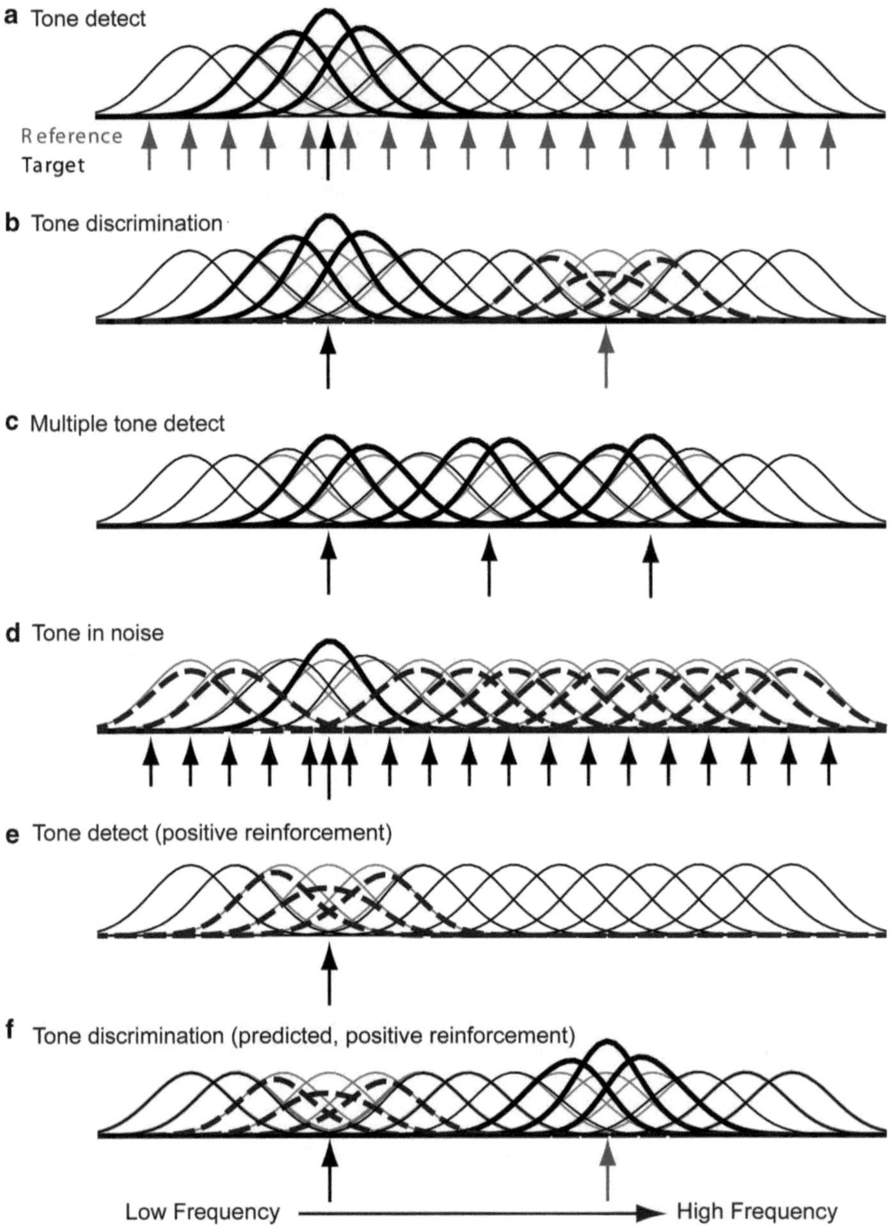

Fig. 9.5 Schematic of rapid STRF plasticity observed or predicted under different task conditions. At baseline, the population tuning of A1 neurons is modeled as a bank of Gaussian tuning curves, each with the same height and bandwidth, centered at a different best frequency (light gray). Neurons undergoing enhancement are plotted in solid black curves, and those undergoing suppression are in dashed gray (tuning curves of neurons plotted in a thin black line remain unchanged). Panels **a–d** summarize published effects for the CA paradigm. (**a**) During tone detection, reference stimuli consist of broadband noise that span the entire range of frequencies (gray arrows), and the target is

9.3.8 Temporal Tasks and Plasticity in the Temporal Dimension of the STRF

Sounds unfold in time. There are at least two ways in which the temporal dimension becomes important in shaping plastic neuronal responses: (1) temporal expectancies develop after learning the temporal structure and contingencies of the task, such as the ordering of a sound sequence, the temporal relationship of task stimuli, appropriate responses, and reward and (2) the representation of the temporal dimension of RFs of auditory neurons may also change or sharpen following development of greater temporal acuity in tasks that require discrimination of temporal features of the stimuli, such as repetition rate or duration of sounds. Several studies have explored the neuronal basis of temporal plasticity in both of the senses mentioned in the preceding text.

As the animal learns the temporal task structure, it also learns to anticipate and predict the timing of task events, which may help in developing proficiency in task performance. A recent paper (Jaramillo & Zador, 2011) addressed how AC mediates the perceptual effects of temporal expectation. Rats were trained to anticipate, in each trial (for a fixed block of trials) that the target sound (modulated tones) would appear either early or late in a sequence of multiple tones. The anticipatory effect in A1 was an increase in response to the tones that preceded the target, which was stronger the closer the frequency of the tones to the neuronal BF, and the closer the tones were in time to the anticipated occurrence of the target. There was also an increase in response to targets at the expected time (compared to the unexpected time) for cells that were tuned to the target frequency.

This finding of temporal expectancy in A1 is consistent with the results of earlier studies of reward time prediction in V1 (Shuler & Bear, 2006). They showed that when adult rats experience an association between visual stimuli and reward, the responses of nearly half of the neurons in V1 change from representing the physical attributes of the stimuli to accurately predicting the timing of reward. A computational model shows that local cortical networks in primary sensory cortex can learn

Fig. 9.5 (continued) a pure tone at a fixed frequency (black arrow). During behavior, responses at the target frequency are enhanced, while neurons with best frequency far from target show little change. (**b**) During tone discrimination, in which the reference sound contains a tone at a different fixed frequency (gray arrow), responses are suppressed at the reference frequency in addition to being enhanced at the target frequency. (**c**) For multitone detect tasks, the target is a chord composed of multiple pure tones, and during this behavior responses increase at each target frequency. (**d**) During tone-in-noise detection, the target tone is presented on top of noise identical to reference noise. In this case, responses are suppressed across all frequencies, although near the target frequency this is compensated for by enhancement similar to that observed in the tone detection task in (**a**). (**e**) For tone detection using the PR paradigm an opposite effect is observed, in comparison to that seen in CA tone detection (shown in (**a**)). Here neurons with tuning near the target frequency are suppressed. (**f**) Given the effects observed during the different tasks and behavioral paradigms, it is possible to predict STRF changes that would be observed during two tone discrimination in the PR paradigm. In this case there should be enhancement of responses at the reference frequency in addition to suppression at the target frequency

temporal representations through reward-dependent expression of synaptic plasticity (Gavornik et al., 2009).

In the other sense of RF temporal plasticity, previous research exploring RF plasticity in temporal tasks has shown that temporal rate coding plasticity can be induced in A1 by pairing electrical stimulation of cholinergic basal forebrain with tone pips at a high temporal rate (Kilgard & Merzenich, 1998a), or by auditory perceptual learning (Beitel et al., 2003; Bao et al., 2004), or even by structured noise exposure at a consistently low modulation rate (Zhou & Merzenich, 2008), which leads to an overall reduction in the temporal modulation transfer function and decreased the highest temporal following rate. However, the RF changes in these previous studies of temporal RFP took place over a long time course of weeks or months.

New studies have begun to explore the role of temporal expectation on cortical responses and RF plasticity (Zhou et al., 2010; Jaramillo & Zador, 2011). To study the rapid RFP in A1 that may result from engaging in temporal tasks, ferrets were trained on gap detection and click-rate discrimination tasks. Unlike the spectral tasks described previously in which the ferret attended to one or more spectral frequencies, in these temporal tasks the ferret needed to attend primarily to the salient temporal characteristics of the sounds, independent of spectral information, to perform well on the task. A reasonable conjecture is that that both temporal tasks should yield faster STRF dynamics, as evidenced, for example, by shortening peak latencies, shorter durations of facilitatory (or suppressive) fields, and/or by a concomitant sharpening of its facilitatory and suppressive fields along the temporal axis. All of these predicted effects have been observed (Fritz et al., unpublished data, 2012), and illustrative examples of STRF changes in both a gap task and a click-rate discrimination temporal task are given in Fig. 9.6. In both cases, there is a dynamic temporal shift in peak latency indicating faster temporal processing after the animal performed the temporal tasks.

9.3.9 Summary of Variables Contributing to RFP

As the preceding sections indicate, the umbrella phrase "cortical plasticity" actually covers a wide range of effects that can be modulated by multiple variables. The importance of motivation and task strategy is emphasized in recent studies (Rutkowski & Weinberger, 2005; Bieszczad & Weinberger, 2010a,b). Thus, even a partial list of variables that can affect RFP include: (1) stimuli used in a given task session, (2) long-term experience with task stimuli, (3) task design and strategy used by the animal to solve the task, (4) task familiarity, (5) task reward valence, (6) task reward magnitude, (7) motivation, (8) task performance. As this list makes clear, plasticity in AC may be critical in attention, reward, learning, and memory (Weinberger, 2011).

In addition to changes in neuronal RF tuning, other attention-driven changes have been described including gain changes and changes in background activity. One other key effect of attention is to increase the independence of cortical neurons,

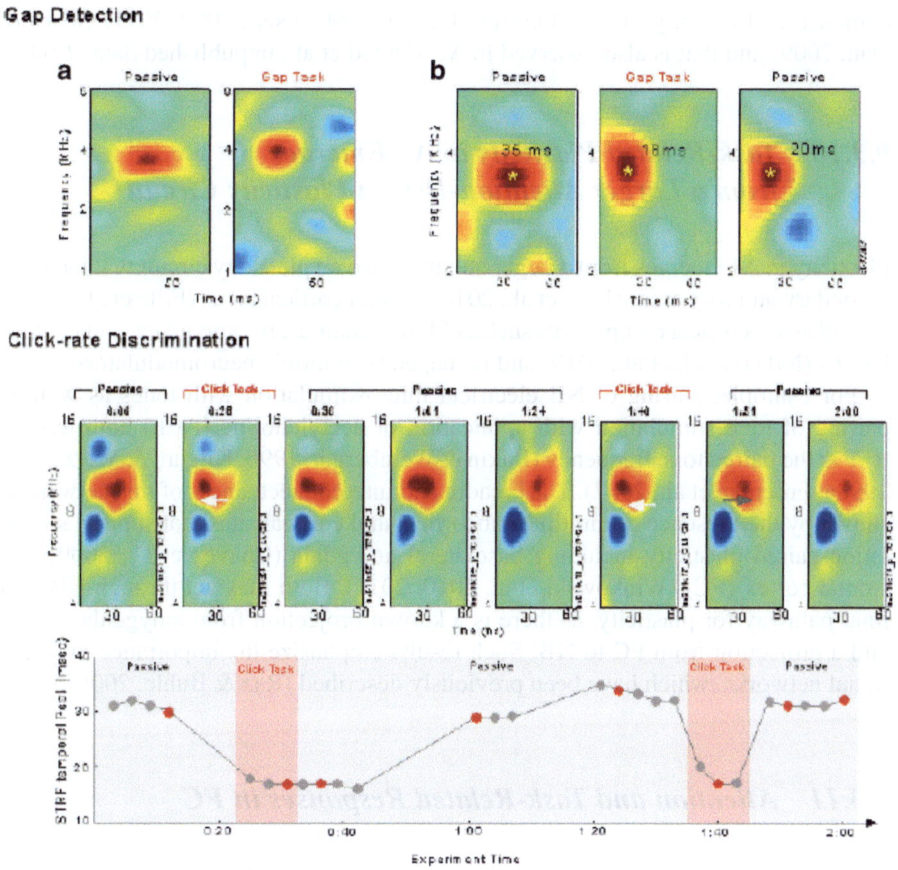

Fig. 9.6 STRF changes during temporal tasks. (Top) Two examples from a gap detection task. In this task the ferret had to detect a 100-ms gap in a TORC. (**a**) Example of STRF changes during gap detection task showing two effects: compacting (decreasing temporal duration of facilitatory field) and increasing contrast (sharper temporal delineation) in the facilitatory field, of the STRF during behavior. (**b**) Another example of STRF changes during a gap detection task showing a third type of effect, decrease in peak latency of the facilitatory field. In this case, the peak latency decreased from 35 ms (passive) to 18 ms (active task performance). This STRF temporal change persisted after task completion. (Bottom) Example of STRF changes during a sequence of two successive temporal click-rate discrimination tasks showing a shift to shorter peak latencies (approximately 15 ms) of the facilitatory field during behavior, followed by a rebound in the following passive STRFs to the long peak latency (approximately 30 ms) of the original passive STRF. The ferret had to distinguish between click rates of 36 Hz (reference) and 10 Hz (target) in the first task, and 32 and 8 Hz in the second task. Such dramatic changes in the temporal dimension of the STRF are characteristic of temporal tasks, such as click rate discrimination and gap detection. Note that in this case, during continuous recording from the same cell, the temporal changes gradually faded over a time course of hours, but then immediately re-occurred when the animal engaged in a second click rate discrimination task. (Reprinted from Fritz et al., 2007a, with permission)

demonstrated recently in visual cortex (Cohen & Maunsell, 2009, 2011; Goard & Dan, 2009) and that is also observed in AC (David et al., unpublished data, 2011).

9.3.10 Task-Related Plasticity in A1 Emerges from a Larger Attention-Driven Plasticity Circuit

Plasticity in A1 emerges from a larger attention-driven plasticity circuit that includes secondary auditory areas (Fritz et al., 2011), frontal cortical areas (Fritz et al., 2010), as well as subcortical components such as MGB thalamus, IC, amygdala, and nucleus basalis (NB) (Leach et al., 2010) and is shaped by multiple neuromodulators.

For example, pairing of NB electrical microstimulation with tones as well as pairing of tone stimulation with application of acetylcholine (ACh) both lead to RFP at the paired tone frequency (Bakin & Weinberger, 1996; Kilgard & Merzenich, 1998a; Froemke et al., 2007). The breadth and interconnectedness of this network is shown by the presence of multiple subcortical and cortical sites that support stimulation-paired plasticity including basolateral amygdala (Chavez et al., 2009) and frontal cortex (FC) (Winkowski et al. 2010, 2011). Both may act through NB as a final pathway for plasticity, as there is a known projection from amygdala to NB, and a projection from FC to NB. Such results emphasize the importance of attentional networks, which have been previously described (Raz & Buhle, 2006).

9.3.11 Attention and Task-Related Responses in FC

What are the sources of top-down influences on A1 that could mediate the effects of attention on auditory processing? Neuroimaging studies examining the common neural circuitry underlying the control of both visual and auditory attention have revealed that there is a common frontoparietal network with interactions between widespread neural regions that underlies attentionally driven, fluid, organized behavior.

One key component of any top-down signal in a selective attention task is its specificity of activation to the target category of stimuli. In a discrimination task in which the animal learns to discriminate between two categories of stimuli, one would predict that a top-down "source" should be able to monitor the task, and provide a biasing top-down signal when the appropriate target stimulus appears. Thus, one feature of a top-down "source" should be its ability to encode task-relevant categories, that is, perceptual categorization. Similar to the attention-activated areas, neural correlates of perceptual categories in the visual and auditory domains have been observed in the frontal, temporal and parietal cortices. Specifically, in the auditory domain, there is compelling evidence of neural correlates of category perception in auditory tasks in monkey ventrolateral PFC (Cohen et al., 2006; Russ et al., 2007; Lee et al., 2009) and monkey auditory associative cortex (Tsunada et al., 2011). The recent paper by Chang et al. (2010) on human higher order AC (pSTG) also demonstrates similar category (phonetic) boundaries in both neurometric and psychometric functions. Similar results have been found in recent neuroimaging studies, in which Desai et al. (2008) found overall increased activation of the left pSTG during categorical perception tasks and Raizada & Poldrack (2007) observed selective amplification of left supramarginal gyrus activity in response to the contrastive features of stimulus pairs spanning a phonetic (CV) category boundary, thus emphasizing categorical differences in speech sounds.

To explore possible top-down influence of FC on RFP in A1, simultaneous recordings were made from A1 and ferret FC (Duque & McCormick, 2010; Fritz et al., 2010) in the behaving animal. FC responses were behaviorally gated: FC neurons responded only in the context of a task, and showed very little, if any, response to the same acoustic stimuli in a non-task condition. Moreover, FC responses were highly selective in response to the class of "target" stimuli. Although responses in the FC were time locked to the onset and offset of target sounds, these neurons represented the task-related meaning of the sounds, rather than their acoustic properties (Fig. 9.7).

Neurons in FC were also remarkably nimble in adapting to changed meaning of acoustic stimuli. As shown in Fig. 9.8, when the "meaning" of a sound changed from "target" to "reference," there was a rapid and complete change in its representation in FC, reminiscent of the rapid changes in A1 plasticity in similar circumstances.

A further interesting parallel with our plasticity changes in A1 was the finding that some of the FC cells showed persistent changes in their target response after task completion. This persistence showed a time course similar to that of the persistent plasticity observed in AC (Fritz et al., 2010).

9.3.11.1 Specific Changes in Brain Oscillations May Influence RFP

Interareal coherence analysis of LFP activity recorded simultaneously in FC and A1 revealed a behaviorally driven reduction in synchronous activity between FC and A1 in the α- and β-frequency range of the local field potential during behavior. What was striking was that this reduction in coherence occurred only for pairs of electrodes where the A1 electrode was close to target frequency, suggesting a focal top-down connectivity from FC to the relevant area on the tonotopic map in A1 during attentive behavior (Fritz et al., 2010). This observed decrease in low-frequency LFP coherence is consistent with results in the visual cortex (Harris and Thiele, 2011) suggesting that when attention is directed within, rather than outside, the RF of recorded neurons near the recording electrode, there is a decrease in low-frequency power. This result suggests that attention can modulate the size of low-frequency LFPs at a local level in sensory cortex. However, high-frequency LFPs may also play a role in attention (Debener et al., 2003) and RFP in the AC. A recent study (Headley & Weinberger, 2011) found that the strength of tone-induced gamma activation in rat A1 predicted RFP that specifically enhanced representation of the signal tone as well as predicting RFP persistence.

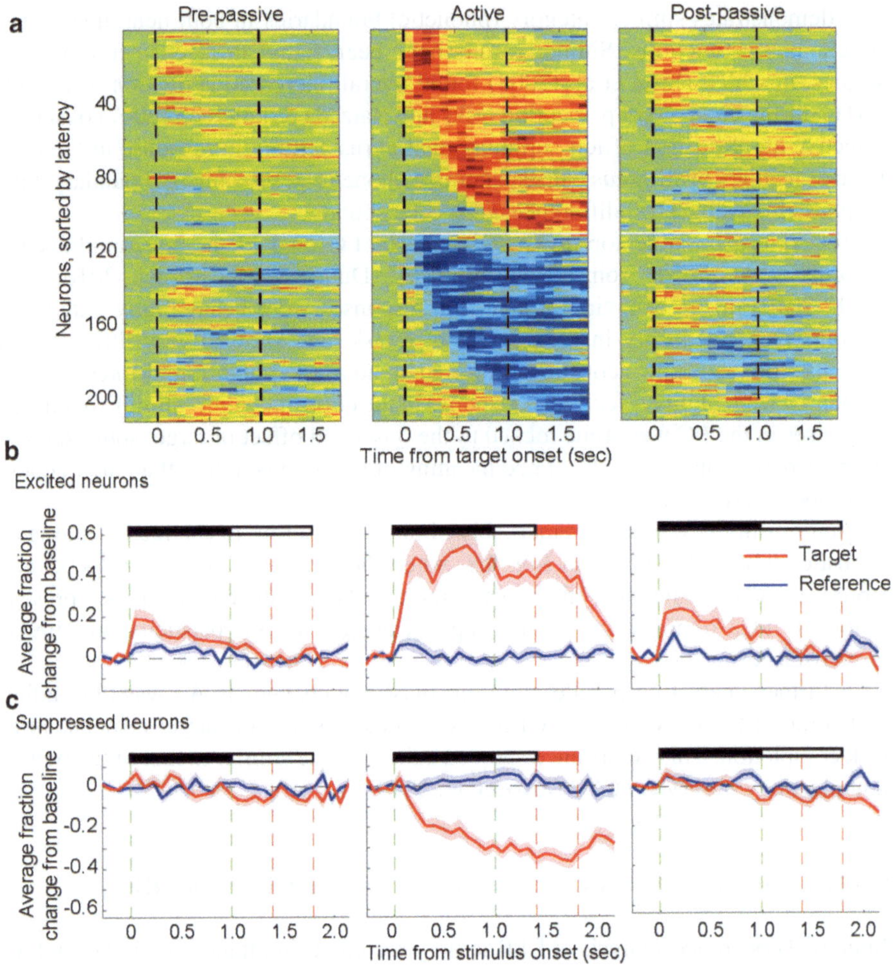

Fig. 9.7 FC population target responses during tone-detection behavior. (**a**) Three heat maps show population responses ($n = 208$) to target sounds during presentation of tone-detection stimuli in pre-behavior passive condition (left), during active behavior (middle) and in post-behavior passive condition (right). Each horizontal line illustrates modulation in a single neuron for 400 ms before target onset, 1 s during the stimulus and 800 ms after offset. Neurons were normalized to have the same peak modulation, grouped by sign (enhanced vs. suppressed) and ordered by onset latency of modulation during behavior. Red indicates activity enhanced from baseline and blue indicates suppressed activity. After behavior, there were some faint, persistent post-behavioral passive effects (right). In contrast, the little modulation present in the pre-behavioral passive heat map (left) is likely a result of the imprint of previous behavioral sessions on the same recording day. (**b**) Each panel shows the population average target and reference PSTH for neurons with enhanced target responses during behavior in a. Stimulus and shock windows are labeled as in Fig. 9.2. During behavior, population activity increased rapidly after the target onset (black line) and was sustained through the shock window before returning rapidly to baseline. Very little modulation to reference stimuli was observed (gray line). In addition to being weaker, persistent modulation after behavior returned to baseline soon after target offset. (**c**) Average target and reference modulation for target-suppressed cells shown in (**a**). During behavior, response latency was slightly slower than for enhanced neurons but otherwise showed similar persistence. (Reprinted from Fritz et al., 2010, with permission)

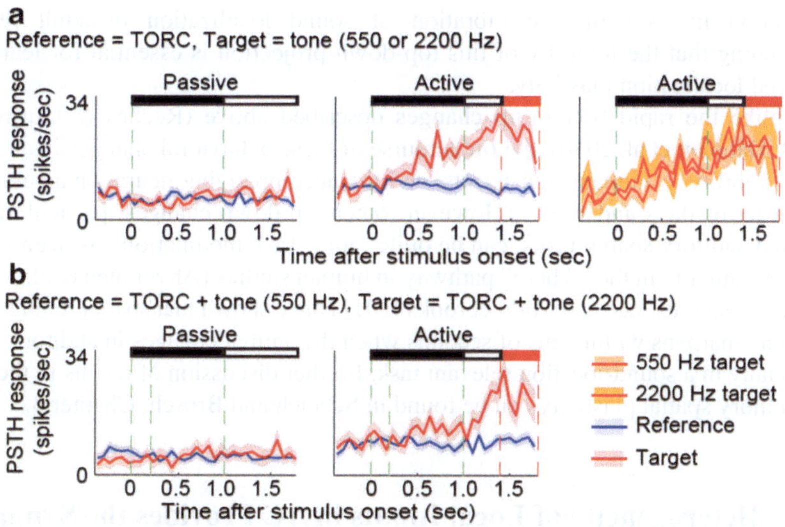

Fig. 9.8 FC class-specific responses to task-relevant sounds. (**a**) A single unit's response to two randomly alternating target tones in a tone-detection task. This cell responded equally well to either of the two target tones (550 or 2200 Hz) but not to any of the 30 different reference noise stimuli. TORC, temporally orthogonal ripple combination. (**b**) During two-tone discrimination, the 550-Hz (low) tone became a reference sound and the unit stopped responding to this tone but maintained its response to the 2200-Hz (high) tone, which remained a target. (Reprinted from Fritz et al., 2010, with permission)

9.4 Learning-Induced Plasticity in Sound Localization and Auditory Spatial Learning

As mentioned above (in Sections 9.2.1 and 9.2.2), rapid, bottom-up adaptive plasticity to stimulus statistics has been observed for auditory spatial information (Reches & Gutfreund, 2008; Dahmen et al. 2010). An example of learning-induced plasticity on a much longer time course, has recently been observed in behavioral studies in adult humans (Van Wanrooijand & Van Opstal, 2005; Kumpik et al., 2010) in relation to auditory spatial plasticity for relearning sound localization with a new ear. Similar findings have been reported in ferrets (Kacelnik et al., 2006), which have demonstrated that the mature auditory system is capable of adapting to markedly altered spatial cue values. In their study, adult animals with unilateral earplugs remained severely impaired in their ability to localize sound for at least several weeks if they receive no behavioral training during this time. By contrast, if the animals were trained on a sound localization task, adaptive changes took place within a few days, and the extent and rate at which their localization judgments improved was determined by the frequency of training (Kacelnik et al., 2006; King et al., 2011). A recent study explored the neural basis for this improvement in behavior (Bajo et al., 2010) and has shown the critical importance of the corticocollicular projection in mediating recalibration of sound localization in adult ferrets, concluding that the integrity of this top-down projection is essential for learning-induced localization plasticity.

Unlike the rapid bottom-up changes described above (Reches & Gutfreund, 2008; Dahmen et al. 2010), the time-course of these behavioral changes in auditory spatial ability in humans and ferrets, and the accompanying neural changes, is on the order of days and weeks. However, other top-down changes, particularly for learned auditory spatial tasks, can be quite rapid. Task modulation has been shown to occur quickly in the "where" pathway in human studies (Ahveninen et al., 2006). In animal studies, Lee and Middlebrooks (2011) have shown that spatial tuning of A1 neurons sharpens within tens of seconds when the animal engages in auditory tasks, especially in a sound-location relevant task. Further discussion of results in the field of auditory spatial plasticity can be found in Scheich and Brosch, Chapter 3.

9.5 Heterogeneity of Local Inputs in AC Provides the Synaptic Substrate for Dynamic Modulation During RFP

Cortical pyramidal cells typically receive tens of thousands of synaptic inputs across an extensive dendritic tree. Because the task-related plasticity changes observed in A1 occur within 1–2 minutes (and likely much more quickly, on the order of 1–10 s) and can be highly frequency specific, a plausible conjecture is that the gain of individual synaptic inputs is changed during attention. This suggests that synapse modulation is the elemental basis of attention-driven RFP.

The RF changes observed during task performance have a small overall gain component (Atiani et al., 2009) but also a highly selective set of effects that can swiftly change RF shape, particularly at the target frequency (Fritz et al., 2003, 2005a). These changes can most parsimoniously be interpreted as arising from selective changes in synaptic input weightings for the synaptic input matrix of cortical neurons. This could take the form of enhanced excitatory and/or diminished inhibitory inputs at the salient frequencies (Froemke et al., 2007).

Because task-related plasticity changes observed in A1 occur quite rapidly after task onset (Fritz et al., 2003), it seemed unlikely that the underlying mechanism would involve dendrite growth or formation of new synapses or plastic remodeling of connections, all of which are important components of longer-term learning. Thus, a reasonable conjecture is that during selective attention, the gain of individual synaptic spine inputs on cortical sensory neurons is dynamically changed in real time under exquisite top-down control. An assumption underlying the hypothesis of rapidly and selectively adjustable synaptic weights on individual neurons in A1, was that these modulations must build upon a heterogeneous substrate, in other words, that there must be a synaptic palette of frequencies that could be used to paint the new RF. A recent study (Chen et al., 2011) functionally mapped the frequency tuning of single spines on the basal and apical dendrites of layer II–III A1 cortical neurons in mice and discovered compelling evidence for substantial heterogeneity in the synaptic input matrix by demonstrating: (1) the presence of both narrowly and widely tuned spines side-by-side on dendritic trees of individual neurons and (2) the presence of neighboring spines tuned to widely different frequencies.

These results from mapping the dendritic distribution of auditory inputs are in keeping with the idea that thalamic inputs (from MGBv to A1) are largely stable, frequency-specific, and tonotopically organized, but that cortical–cortical inputs from elsewhere in A1 and AC may show considerable plasticity, display great range in frequency tuning, and are weakly tonotopically organized. Thus, we conjecture that the great variability in frequency tuning of spines in the synaptic input matrix of A1 pyramidal cells may provide the substrate for cortical RFP. The RFs of individual cortical neurons may be computed by integrating spatially distributed and heterogeneous synaptic inputs across the dendritic field, coding for multiple acoustic features, such as bandwidth, frequency, intensity, and spatial and temporal characteristics of sound. Rapid components of plasticity may be mediated by multiple functional and structural mechanisms, such as structural rearrangements of synapses and spines (Hensch, 2005). A key component of plasticity is stimulus-timing-dependent plasticity (STDP), which is an up- or down-regulation of synaptic efficacy that is dependent on the order and timing of presynaptic and postsynaptic activity (Caporale & Dan, 2008). A recent study showed that STDP can be elicited in A1 by repetitive pairing of tones of different frequencies leading to a shift in neuronal frequency selectivity (Dahmen et al., 2008). STDP has been proposed as a mechanism underlying auditory perceptual learning (Mossbridge et al., 2006).

This view is somewhat different than the hypothesis that has been advanced emphasizing the primary contribution of thalamocortical projections in cortical plasticity (Metherate et al., 2005; Jafari et al., 2007; Xiong et al., 2009) as compared to the contribution of diverse corticocortical connections. For example, Yan and colleagues write (p.5 in Liu et al., 2011) "… The underlying synaptic mechanism for the plasticity of acquired sound cortical neurons is likely straight-forward: it may be directly interpreted by NMDA dependent LTP of thalamocortical inputs …. The high frequency-specificity and precise information relayed by the thalamocortical system supports this system's decisive role in learning-induced and experience-dependent auditory cortex plasticity." Yan and colleagues go on to discuss more complex circuitry that may underlie cortical plasticity, but emphasize the central importance of modulation of the thalamocortical inputs. In experimental support for this perspective, it is true that the greatest attention-driven plasticity effects tend to occur near BF for a given cell (Atiani et al., 2009), suggesting the importance of plasticity of the thalamocortical input or the greatest number of modifiable synapses near BF. However, there is also evidence for RF modulation at distances far from the BF of a given cell (>1 octave from BF), much broader than the thalamocortical projection (Fritz et al., 2003), suggesting that modification of thalamocortical inputs alone is not sufficient to explain these results. In fact, the neuroanatomical and neurophysiological evidence suggests that thalamocortical connections may be limited to a range of about 1/3 octave near the cortical BF (Miller et al., 2002) and that thalamocortical inputs comprise less than 3–5% of the total synaptic inputs in the cortex (and are only about 15% of the synaptic inputs to layer IV!). This point is also made by Scheich and Brosch (Chapter 3) who observe that AC receives only a small fraction of its inputs from the ascending auditory system and that most of the inputs are from extremely diverse other subcortical and especially other cortical sources (Budinger & Scheich, 2009).

Studies of cortical learning-induced plasticity also provide support for the idea that changes in cortical maps are due to modulation in intracortical lateral connectivity—synaptic modulation that is critical for plasticity and learning (Rioult-Pedotti et al., 1998, 2000). More recent research (Xu et al., 2009; Wang et al., 2011) has shown that within an hour of learning new motor tasks there are profound changes in cortical connectivity, suggesting strong associations between learned tasks and specific, nonoverlapping groups of novel synapses whose strength might remain relatively constant once formed, and optimized specifically for that task. However, there may also be a complementary system that modulates the strength of individual synapses over a continuum that is critical for attentional modulation of learned circuits in which there must be dynamic regulation of a basic connectivity with changes in task conditions and context (Scheich et al., 2007, 2011).

It is worth noting that several studies have shown the importance of γ-aminobutyric acid-ergic (GABAergic) inhibitory circuitry in cortical plasticity (Froemke et al., 2007; Feldman, 2009). Fast-spiking inhibitory interneurons receive strong glutaminergic drive from the thalamus and send powerful feedforward inhibition to all cortical cell types, and specifically inhibit synaptic inputs to supragranular pyramidal cells (Cruikshank et al., 2010). However, cholinergic input from NB drives another class of inhibitory interneurons that inhibit the fast-spiking interneurons, thus removing an inhibitory brake during attention (Letzkus et al., 2011; Arroyo et al., 2012).

Thus, cortical plasticity is likely to arise as the vector sum of changes in synaptic weighting and dynamics at multiple cortical layers and levels involving thalamocortical, corticocortical, and neuromodulatory excitatory and inhibitory synaptic and presynaptic modulatory connections. Moreover, such plasticity is likely to exhibit different forms in the different cell types in AC (Reyes, 2011). Finding the underlying molecular, synaptic, and cellular mechanisms underlying rapid RFP at a neuronal and circuit level provides an enormous challenge for the future.

9.6 Relation of Task-Related RFP to Results of Earlier Classical and Instrumental Conditioning Studies

The chapters by Scheich and Brosch (Chapter 3) and Shepard et al. (Chapter 10) provide an excellent overview of the short-term and long-term RF changes arising from classical conditioning and instrumental conditioning experiments. These plastic changes include local changes in RF shape and gain, global changes in cortical maps, and global changes in ensemble dynamics and are fully compatible with the studies of rapid RFP summarized in Section 9.3 above.

It is important to emphasize that the rapid attentional modulation of cortical RFs may also lead to the longer-term changes observed in learning-induced plasticity. However, it is highly unlikely that global cortical map plasticity is the only substrate of perceptual and skill learning in the adult brain (Weinberger, 2004; Kover & Bao, 2010). There are many forms of neural plasticity that are likely to play vital roles. However, developmentally, there **is** good evidence that the size of the A1 representation of a given stimulus covaries with the long-term frequency of the stimulus. In the adult, such over-representation has also been shown to correlate with the behavioral salience of that stimulus (Rutkowski & Weinberger, 2005). Such results have lead to the view that the cortical representation of perceptual bias may take a long-term form (increased cortical real estate and elevated baseline activity), and a different dynamic, short-term encoding mechanism that is based on top-down modulation of firing patterns on a trial-by-trial basis.

However, other studies have failed to replicate the findings of map expansion and global plasticity accompanying learning in the adult brain. In fact, such map expansion may not necessary for learning, as it is possible to achieve improved behavioral performance in frequency discrimination without an increase in map representational area (Brown et al., 2004; Bieszczad & Weinberger, 2010a,b). As described in more detail by Shepard et al. (Chapter 10), recent results from the auditory system (Reed et al., 2011) provide evidence that although cortical map plasticity can enhance perceptual learning, such A1 map plasticity fades over time even though tone discrimination performance remains stable. This observation is consistent with other recent reports that cortical map expansions associated with perceptual and motor learning are followed by a period of map renormalization without a decrement in performance (Molina-Luna et al., 2008; Yotsumoto et al., 2008). These results indicate that cortical map plasticity is not necessary to maintain improved discriminative ability.

One way to look at these results is that under some circumstances, there may be a short-term increase in representational area in sensory or motor cortex for perceptual or motor learning. This temporary map expansion may be a brief phase, followed by sparsening of the representation and consolidation, a more efficient long-term representation of the new learning (Molina-Luna et al., 2008; Yotsumoto et al., 2008; Reed et al., 2011).

This makes sense at a logical level. If the only thing you do in life is a perceptual discrimination task of pure tones near 2 kHz, then it would be useful to devote more cortical real estate in A1 to the 2-kHz region. However, most animals are cognitive generalists rather than specialists, and are engaged daily in complex acoustic tasks that span a wide range of frequencies and call upon auditory processing over the animal's full hearing range. Thus, although the demonstration of global tonotopic map change is highly significant in dramatically illustrating the plastic capabilities of the system, it is quite likely that in the "real world," there would be no net change in the tonotopic map needed because behaviorally salient stimuli come in all flavors and frequencies. Thus, in this perspective, global map change is not likely to play a very great role in learning in the adult, whereas dynamic changes arising from attention and learning are likely to play a very important role in modulating synaptic responses and functional connectivity against the framework of a relatively stable tonotopic map.

An interesting recent result emphasizes the dynamic nature of the cortical map (Zhou et al., 2010). In trained rats that had learned tone pair (A, B) associations, they observed a selective amplification of the response to the second tonal element (B) in a trained pair at its expected time of occurrence if and only if it was preceded by the first pair element (A). If stimulus A occurred, neurons were more strongly excited by stimulus B, over a larger neuronal cell assembly, with a more coordinated distributed response and with greater specificity. This leads to the view of an evolving, dynamically changing tonotopic representation that may come closer to what actually happens in the real world when learned sound associations shape expectations and cortical responses.

9.7 Role of ACh and Other Neuromodulators in Rapid Plasticity in the Adult

ACh and other neuromodulators including dopamine, serotonin, and noradrenaline have all been shown to be important in evoking plasticity (Edeline, 1999; 2003; Edeline et al., 2011). Maternal hormones have also been shown to play a role in modulation of auditory cortical processing and plasticity (Miranda & Liu, 2009). The topic of the role of neuromodulators as engines of plasticity in the auditory system is thoughtfully discussed by Shepard et al. (Chapter 10). However, one aspect worth expanding is the role of neuromodulators, particularly ACh, in attention and rapid RFP (Soto et al., 2006).

As mentioned, the cholinergic system has been the most extensively studied of all of the neuromodulator systems (Metherate, 2011). Coupling Ach release with sound stimulation is sufficient to evoke massive reorganization of the A1 tonotopic map (Bakin & Weinberger, 1996; Kilgard & Merzenich, 1998a). Moreover, relatively short pairing of the cholinergic NB electrical microstimulation with tones (<10 s) as well as pairing of tone stimulation with application of ACh both lead to RFP at the paired tone frequency (Froemke et al., 2007). Although there is considerable insight into the role of cholinergic projections in cortical plasticity, there are still many puzzling questions. For example, the interactions of ACh with other co-released neurotransmitters from the NB, and the collective effect of multiple NB transmitters on cortical activity, remain a mystery (Lin & Nicolelis, 2008). The relative contributions of muscarinic and nicotinic receptors in plasticity are also still an open question. Moreover, an intriguing recent study has even shown that glia as well as neurons, contribute to cholinergic-induced cortical plasticity (Takata et al., 2011).

Early classic views held that ACh was a slowly acting neuromodulator and acted as a state variable that gradually changed arousal level. However, evidence from experiments using new techniques for measuring rapid changes in cholinergic neurotransmission give a novel perspective on the cholinergic regulation of cognitive processes, indicating that rapid phasic changes in cholinergic modulation occur on a timescale of seconds (particularly in projection to PFC), and is triggered by sensory input cues and serves to facilitate cue detection and attentional performance (Sarter et al., 2009; Hasselmo & Sarter, 2011). In fact, recent results suggest that enhancing cholinergic pairing with salient stimuli (through NB paired stimulation) improves auditory perception and behavioral performance for detection and discrimination of the paired stimuli (Yuan et al., 2011).

It is known that the PFC receives inputs from sensory neocortical regions and projects heavily to GABAergic neurons in the NB and other basal forebrain (BF) cholinergic structures (Gyengesi et al., 2008) and has a tonic facilitatory effect on BF cholinergic neurons (Rasmusson et al., 2007). Zaborzsky and colleagues (Golmayo et al., 2003) adduced evidence for their hypothesis that pathways from sensory cortical regions via the PFC-BF projection and from the BF back to specific sensory cortical areas (a cortico–prefrontal–basalo–cortical circuitry) could modulate sensory responses and play a decisive role in cortical plasticity and selective attention. Sensory or electrical stimulation of sensory pathways evokes an increase in cholinergic release in the corresponding cortical areas. This is blocked by cooling of PFC (Rasmusson et al., 2007). PFC also modulates release of ACh in posterior parietal cortex (Nelson et al., 2005), suggesting a nonsensory, cognitive dimension for the role of ACh. Although the cholinergic projection from BF has traditionally been thought of as diffuse, there is some evidence for functional specificity (Fournier et al., 2004) so that there are specific topographic projections to different sensory cortical areas from different parts of the NB and BF. However, it is somewhat unlikely that the projections from NB have sufficient spatial specificity to target any small cortical area that represents an attended stimulus, so the effects may be more general or permissive. It has been suggested (Herrero et al., 2008; Deco & Thiele, 2011; Harris & Thiele, 2011) that the cholinergic drive is important for setting the network to be maximally responsive to focal glutaminergic top-down feedback from higher cortical areas during attention. This "reverse hierarchy" top-down feedback is likely to play a key role in attention and learning, and in shaping expectations and tuning of auditory neurons at lower cortical levels features, so that irrelevant features are ignored, while permitting the learning of relevant ones (Jaaskelainen et al., 2007, 2011; Roelfsema et al., 2010). It may also be important in encoding complex sound patterns and sequences (Kudoh et al., 2004).

In keeping with the Zaborzsky hypothesis, there is now evidence from PFC cooling experiments that PFC is necessary for sensory pathway evoked ACh release (Rasmusson et al., 2007). Thus PFC may orchestrate attention and sensory cortical plasticity through a combination of direct projections and a loop projection through BF. This projection is shown as part of a larger attention-driven plasticity network in Fig. 9.9.

A prediction of this hypothesis, that paired tone-PFC stimulation would elicit A1 plasticity at the paired frequency, has recently been validated (Winkowski et al., 2011). Of course, the network is likely to be much more complex, involving many other cortical and subcortical structures (such as amygdala, sensory thalamus, where different cholinergic receptors are distributed over lemniscal and nonlemniscal pathways (Mooney et al., 2004), reticular nucleus of the thalamus (Zikopoulos & Barbas, 2006), and midbrain (Mysore & Knudsen, 2011).

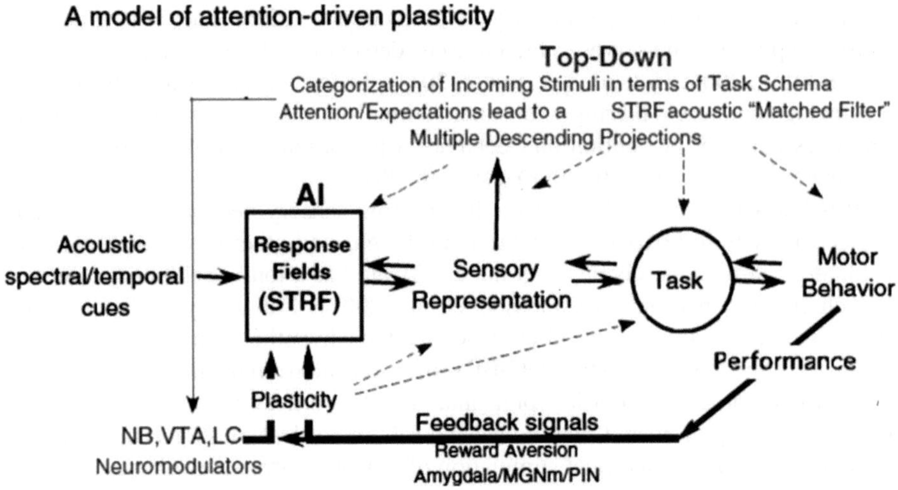

Fig. 9.9 Network model for rapid STRF plasticity in A1. A key point of the model is the presence of a set of descending projections from auditory association and/or FC. Once these higher areas decode the incoming sound stream, and categorize stimuli as either "reference" or "target" (or neither), one hypothesis is that an appropriate top-down signal is sent to subcortical nuclei such as NB, locus coeruleus (LC), and the ventral tegmental area (VTA). These three nuclei send neuromodulatory projections (acetylcholine, noradrenaline, and dopamine respectively) throughout neocortex that have been shown to enhance cortical plasticity. Thus, during task performance, neuromodulator activity increases selectively during reference and target stimuli to suppress (reference) or enhance (target) responses to the incoming signals. In this way, attention could gate receptive field plasticity in A1. (Reprinted from Fritz et al., 2007a, with permission)

9.8 Studies of Plasticity in Human AC

Most fMRI (functional magnetic resonance imaging), electroencephalographic (EEG), and magnetoencephalographic (MEG) studies find overall enhancement of AC activity by selective attention to sound. Attention-driven plasticity has been shown to occur in the human AC for tasks calling upon auditory spectral, temporal, complex spectrotemporal, spatial processing and auditory scene analysis (Spierer et al., 2011). Although there is some controversy over the relative amounts of regional modulation, there is general agreement that attentional enhancement occurs throughout primary and secondary AC for auditory feature and spatial attention (Ahveninen et al., 2006; Paltoglou et al., 2009). In parallel with animal studies, ACh appears to play a powerful role in plasticity in human AC (Thiel et al., 2002; Thiel, 2007).

Because of their superior time resolution, MEG and EEG are excellent techniques to study rapid plasticity in AC. Studies of the neural basis of rapid perceptual learning (Alain et al., 2007, 2010) found selective, rapid changes in sensory-evoked ERP amplitude within the first hour of training, consistent with top-down modulation. Other studies have shown that auditory attention in a two-tone discrimination task rapidly changed the tonotopic map in A1, expanding the distance between the two discriminant tone pair frequencies (Spierer et al., 2007). Another MEG study explored the representation of foreground and background in an auditory scene (Elhilali et al., 2009), showing a selective increase in the neural representation of the foreground stimulus. Further, the target's perceptual detectability improved over time, correlating with the target representation's neural buildup. A recent study of learning-induced plasticity (Kluge et al., 2011) demonstrated multiple auditory evoked cortical responses, arising from different hierarchical levels of cortical processing, all of which rapidly acquired the distinction between CS+ and CS– within less than 20 minutes. Only the response associated with secondary AC could flexibly reverse (within a time course of 1–2 minutes!) with a change in contingency, consistent with top-down mediation. The role of secondary AC in auditory attention (mediating the extraction of signal from noise) has been emphasized in a recent, combined MEG/EEG/fMRI study (Ahveninen et al., 2011) that also demonstrated rapid changes in auditory tuning that occurred during the first few seconds after attention switching and correlated with behavioral discrimination performance. A direct neurophysiological approach to studying human auditory attention in a more complex acoustic scene segregation task, through recording responses with multielectrode arrays has been used recently in an elegant study to show that population neural activity can encode attended speech to individual voices in the presence of simultaneous speakers (Mesgarani & Chang, 2012), also with a rapid time course of seconds as subjects switched attention between speakers.

9.9 Summary

The literature reviewed in this chapter shows the dynamic landscape in AC that is shaped by attention-driven and learning-induced plasticity, as well as RF adaptation to changing stimulus statistics. RFs of A1 neurons in the adult are in a state of rapid flux that is modulated by continuous "top-down" biasing as a function of changing salience and task-relevance of auditory stimuli, and ongoing predictions. Auditory attention involves a distributed network of auditory cortical and subcortical structures that are activated selectively in a task-specific manner during auditory processing, which also integrate with a generalized multisensory attentional network that includes parietal, frontal, and anterior cingulate cortical regions. It is likely that RFP can be generated by this attentional network, may underlie selective attention, and optimizes auditory cortical filters for task performance, by enhancing their selectivity for task-relevant information (Fritz et al., 2007a,c; Jääskeläinen et al., 2007, 2011).

9.9.1 Open Questions and Future Directions in the Field of Cortical Plasticity

A common set of cellular and synaptic mechanisms is likely in participate in plasticity across all sensory neocortex (Feldman, 2009). For example, in the same way that classical conditioning with tone stimuli can increase A1 responses to trained frequencies, so can classical conditioning using whisker stimuli expand the representation of trained whiskers in S1 (Siucinska & Kossut, 1996, 2004) and orientation discrimination or visual spatial training can change V1 tuning for the trained feature (Dan & Poo, 2006; Gilbert & Sigman, 2007; McManus et al., 2011). Open questions for the field of cortical plasticity include:

1. How are behavioral goals represented and communicated to auditory cortical neurons that are tuned to specific features via top-down projections?
2. What are the molecular, synaptic, and cellular mechanisms involved?
3. As noted by Feldman (2009), multiple forms of cortical plasticity have been discovered, each with different synaptic loci, different cellular mechanisms, and different time scales. What are the broad principles that underlie this exuberant variety of experience-dependent plasticities?
4. How are these different plasticities at multiple levels orchestrated?
5. How is perceptual stability achieved given changing RF properties in cortical neurons during behavior?
6. How does RFP contribute to auditory attention, learning and memory?
7. Are there forms of task-related plasticity that enhance feature space at a higher level, such as the plasticity for a complex contrast filter driven by selective attention to the features of an individual human voice in the context of a noisy bar? How is this effortless attentional switch in focus achieved within seconds?

References

Ahveninen, J., Jääskeläinen, I. P., Raij, T., Bonmassar, G., Devore, S., Hämäläinen, M., et al. (2006). Task-modulated "what" and "where" pathways in human auditory cortex. *Proceedings of the National Academy of Sciences of the USA*, 103, 14608–14613.

Ahveninen, J., Hämäläinen, M., Jääskeläinen, I. P., Ahlfors, S. P., Huang, S., Lin, F. H., et al. (2011). Attention-driven auditory cortex short-term plasticity helps segregate relevant sounds from noise. *Proceedings of the National Academy of Sciences of the USA*, 108, 4182–4187.

Alain, C., Snyder, J. S., He, Y., & Reinke, K. S. (2007). Changes in auditory cortex parallel rapid perceptual learning. *Cerebral Cortex*, 17, 1074–1084.

Alain, C., Campeanu, S., & Tremblay, K. (2010). Changes in sensory evoked responses coincide with rapid improvement in speech identification performance. *Journal of Cognitive Neuroscience*, 22, 392–403.

Anton-Erxleben, K., Stephan, V. M., & Treue, S. (2009). Attention reshapes center-surround receptive field structure in macaque cortical area MT. *Cerebral Cortex*, 19, 2466–2478.

Antunes, F. M., Nelken, I., Covey, E., & Malmierca, M. S. (2010). Stimulus-specific adaptation in the auditory thalamus of the anesthetized rat. *PLoS ONE*, 5(11), e14071.

Arroyo, S., Bennett, C., Aziz, D., Brown, S.P., & Hestrin, S. (2012) Prolonged disynaptic inhibition in the cortex mediated by slow, non-a7 nicotinic excitation of a specific subset of cortical interneurons. *Journal of Neuroscience*, 32, 3859-3864.

Atiani, S., Elhilali, M., David, S. V., Fritz, J. B., & Shamma, S. A. (2009). Task difficulty and performance induce diverse adaptive patterns in gain and shape of primary auditory cortical receptive fields. *Neuron*, 61, 467–480.

Bajo, V. M., Nodal, F. R., Moore, D. R., & King, A. J. (2010). The descending corticocollicular pathway mediates learning-induced auditory plasticity. *Nature Neuroscience*, 13, 253–260.

Bakin, J. S., & Weinberger, N. M. (1996). Induction of a physiological memory in the cerebral cortex by stimulation of the nucleus basalis. *Proceedings of the National Academy of Sciences of the USA*, 93, 11219–11224.

Baluch, F., & Itti, L. (2011) Mechanisms of top-down attention. *Trends in Neurosciences*, 34, 210–224.

Bao, S., Chang, E. F., Woods, J., & Merzenich, M. M. (2004). Temporal plasticity in the primary auditory cortex induced by operant perceptual learning. *Nature Neuroscience*, 7, 974–981.

Bauerle, P., von der Behrens, W., Kossl, M., & Gaese, B. H. (2011). Stimulus-specific adaptation in the gerbil primary auditory thalamus is the result of a fast-frequency-specific habituation and is regulated by the corticofugal system. *Journal of Neuroscience*, 31, 9708–9722.

Beitel, R. E., Schreiner, C. E., Cheung, S. W., Wang, X., & Merzenich, M. M. (2003). Reward-dependent plasticity in the primary auditory cortex of adult monkeys trained to discriminate temporally modulated signals. *Proceedings of the National Academy of Sciences of the USA*, 100, 11070–11075.

Berkes, P., Orban, G., Lengyel, M., & Fiser, J. (2011). Spontaneous cortical activity reveals hallmarks of an optimal internal model of the environment. *Science*, 331, 83–87.

Bieszczad, K. M., & Weinberger, N. M. (2010a). Learning strategy trumps motivational level in determining learning-induced cortical plasticity. *Neurobiology of Learning and Memory*, 93, 229–239.

Bieszczad, K. M., & Weinberger, N. M. (2010b). Remodeling the cortex in memory: Increased use of a learning strategy increases the representational area of relevant acoustic cues. *Neurobiology of Learning and Memory*, 94, 127–144.

Bizley, J. K., Nodal, F. R., Nelken, I., & King, A. J. (2005). Functional organization of ferret auditory cortex. *Cerebral Cortex*, 15, 1637–1653.

Blake, D. T., Strata, F., Churchl, A. K., & Merzenich, M. M. (2002). Neural correlates of instrumental learning in primary auditory cortex. *Proceedings of the National Academy of Sciences of the USA*, 99, 10114–10119.

Brosch, M., & Schreiner, C. E. (1997). Time course of forward masking tuning curves in cat primary auditory cortex. *Journal of Neurophysiology*, 77, 923–943.

Brown, M., Irvine, D. R., & Park, V. N. (2004). Perceptual learning on an auditory frequency discrimination task by cats: Association with changes in primary auditory cortex. *Cerebral Cortex*, 14, 952–965.

Budinger, E., & Scheich, H. (2009). Anatomical connections suitable for the direct processing of neuronal information of different modalities via the rodent primary auditory cortex. *Hearing Research*, 258, 16–27.

Caporale, N., & Dan, Y. (2008). Spike timing-dependent plasticity: A Hebbian learning rule. *Annual Review of Neuroscience*, 31, 25–46.

Castro-Alamancos, M. A. (2004). Absence of sensory adaptation in neocortex during information processing states. *Neuron*, 41, 455–464.

Chandrasekaran, B., Homickel, J., Skoe, Nicol, T., & Kraus, N. (2009). Context-dependent encoding in the human auditory brainstem relates to hearing speech in noise: implications for developmental dyslexia. *Neuron*, 64, 311–319.

Chang, E. F., Rieger, J. W., Johnson, K., Berger, M. S., Barbaro, N. M., & Knight, R. T. (2010). *Nature Neuroscience*, 13, 1428–1432.

Chavez, C. M., McGaugh, J. L., & Weinberger, N. M. (2009). The basolateral amygdala modulates specific sensory memory representations in the cerebral cortex. *Neurobiology of Learning and Memory*, 91, 382–392.

Chen, K., Leischner, U., Rochefort, N., Nelken, I., & Konnerth, A. (2011). Functional mapping of single spines in cortical neurons in vivo. *Nature*, 475, 501–505.

Cherry, E. C. (1953). Some experiments on the recognition of speech, with one and two ears. *Journal of the Acoustical Society of America*, 25, 975–979.

Cohen, M. R., & Maunsell, J. H. (2009). Attention improves performance primarily by reducing interneuronal correlations. *Nature Neuroscience*, 12, 1594–1600.

Cohen, M. R., & Maunsell, J. H. (2011). Using neuronal populations to study the mechanisms underlying spatial and feature attention. *Neuron*, 70, 1192–1204.

Cohen, Y. E., Hauser, M. D., & Russ, B. E. (2006). Spontaneous processing of abstract categorical information in the ventrolateral prefrontal cortex. *Biological Letters*, 2, 261–265.

Condon, C. D., & Weinberger, N. M. (1991). Habituation produces frequency-specific plasticity of receptive fields in the auditory cortex. *Behavioral Neuroscience*, 105, 416–430.

Cruikshank, S. J., Urabe, H., Nurmikko, A. V., & Connors, B. W. (2010). Pathway-specific feedforward circuits between thalamus and neocortex revealed by selective optical stimulation of axons. *Neuron*, 65, 230–245.

Dahmen, J. C., Hartley, D. E., & King, A. J. (2008). Stimulus-timing-dependent plasticity of cortical frequency representation. *Journal of Neuroscience*, 28, 13629–13639.

Dahmen, J. C., Keating, P., Nodal, F. R., Schulz, A. L., & King, A. J. (2010). Adaptation to stimulus statistics in the perception and neural representation of auditory space. *Neuron*, 66, 937–948.

Dan, Y., & Poo, M. M. (2006). Spike timing-dependent plasticity: From synapse to perception. *Physiological Reviews*, 86, 1033–1048.

David, S. V., Hayden, B. Y., Mazer, J. A., & Gallant, J. L. (2008). Attention to stimulus features shifts spectral tuning of V4 neurons during natural vision. *Neuron*, 59, 509–521.

David, S. V., Fritz, J. B., & Shamma, S. A. (2012). Task reward structure shapes rapid receptive field plasticity in auditory cortex. *Proceedings of the National Academy of Sciences of the USA*, 109, 2144-2149.

Dean, I., Robinson, B. L., Harper, N. S., & McAlpine, D. (2005). Neural population coding of sound level adapts to stimulus statistics. *Nature Neuroscience*, 8, 1684–1689.

Dean, I., Robinson, B. L., Harper, N. S., & McAlpine, D. (2008). Rapid neural adaptation to sound level statistics. *Journal of Neuroscience*, 28, 6430–6438.

Debener, S., Herrmann, C. S., Kranczioch, C., Gembris, D., & Engel, A. K. (2003). Top-down attentional processing enhances auditory evoked gamma band activity. *NeuroReport*, 14, 683–686.

Deco, G., & Thiele, A. (2011). Cholinergic control of cortical network interactions enables feedback-mediated attentional modulation. *European Journal of Neuroscience*, 34, 146–157.

Depireux, D. A., Simon, J. Z., Klein, D. J., & Shamma, S. A. (2001). Spectro-temporal response field characterization with dynamic ripples in ferret primary auditory cortex. *Journal of Neurophysiology*, 85, 1220–1334.

Desai, R., Liebenthal, E., Waldron, E., & Binder, J. R. (2008). Left posterior temporal regions are sensitive to auditory categorization. *Journal of Cognitive Neuroscience*, 20, 1174–1188.

Duque, A., & McCormick, D. A. (2010). Circuit based localization of ferret prefrontal cortex. *Cerebral Cortex*, 20, 1020–1036.

Edeline, J.M., Pham, P., & Weinberger, N.M. (1993) Rapid development of learning induced receptive field plasticity in the auditory cortex. *Behavioral Neuroscience*, 107, 539–551.

Edeline, J. M. (1999). Learning-induced physiological plasticity in thalamo-cortical sensory systems: A critical evaluation of receptive field plasticity, map changes and their potential mechanisms. *Progress in Neurobiology*, 57, 165–224.

Edeline, J. M. (2003). Thalamo-cortical auditory receptive fields: Regulation by the states of vigilance, learning and the neuromodulatory systems. *Experimental Brain Research*, 153, 554–572.

Edeline, J. M., Pham, P., & Weinberger, N. M. (1993). Rapid development of learning induced receptive field plasticity in the auditory cortex. *Behavioral Neuroscience*, 107, 539–551.

Edeline, J. M., Manunta, Y., & Hennevin, E. (2011). Induction of selective plasticity in the frequency tuning of auditory cortex and auditory thalamus neurons by locus coeruleus stimulation. *Hearing Research*, 274, 75–84.

Elhilali, M., Fritz, J. B., Chi, T.-S., & Shamma, S. A. (2007). Auditory cortical receptive fields: Stable entities with plastic abilities. *Journal of Neuroscience*, 27, 10372–10382.

Elhilali, M., Xiang, J., Shamma, S. A., & Simon, J. Z. (2009). Interaction between attention and bottom-up saliency mediates the representation of foreground and background in an auditory scene. *PLoS Biology*, 7(6), e1000129.

Feldman, D. (2009). Synaptic mechanisms for plasticity in neocortex. *Annual Review of Neuroscience*, 32, 33–55.

Fournier, G. N., Semba, K., & Rasmusson, D. D. (2004). Modality- and region-specific acetylcholine release in the rat neocortex. *Neuroscience*, 126, 257–262.

Fritz, J. B., Shamma, S. A., Elhilali, M., & Klein, D. J. (2003). Rapid task-related plasticity of spectrotemporal receptive fields in primary auditory cortex. *Nature Neuroscience*, 6, 1216–1223.

Fritz, J. B., Elhilali, M., & Shamma, S. (2005a). Differential dynamic plasticity of A1 receptive fields during multiple spectral tasks. *Journal of Neuroscience*, 25, 7623–7635.

Fritz, J. B., Elhilali, M., & Shamma, S. A. (2005b). Active listening: Task-dependent plasticity of receptive fields in primary auditory cortex. *Hearing Research*, 206, 159–176.

Fritz J. B., Elhilali, M., David, S. V., & Shamma, S. A. (2007a). Does attention play a role in dynamic receptive field adaptation to changing acoustic salience in A1? *Hearing Research*, 229, 186–203.

Fritz, J. B., Elhilali, M., & Shamma, S. A. (2007b). Adaptive receptive field changes during detection of complex spectral targets. *Journal of Neurophysiology*, 98, 2337–2346.

Fritz, J. B., Elhilali, M., David, S. V., & Shamma, S. A. (2007c). Auditory attention: Focusing the searchlight on sound. *Current Opinion in Neurobiology*, 17, 437–455.

Fritz, J. B., David, S. V., Radtke-Schuller, S., Yin, P., & Shamma, S. A. (2010). Adaptive, behaviorally-gated, persistent encoding of task-relevant auditory information in ferret frontal cortex. *Nature Neuroscience*, 13, 1011–1019.

Fritz, J. B., Atiani, S., David, S. V., Locastro, M., Elgueda, D., Radtke-Schuller, S., & Shamma, S. A. (2011). Comparison of task-related plasticity in ferret primary and secondary auditory cortex. Society for Neuroscience Meeting (November, 2011 in Washington, DC), Abstract 173.06.

Froemke, R. C., Merzenich, M. M., & Schreiner, C. E. (2007). A synaptic memory trace for cortical receptive field plasticity. *Nature*, 450, 425–429.

Galvan, V. V., & Weinberger, N. M. (2002). Long-term consolidation and retention of learning-induced tuning plasticity in the auditory cortex of the guinea pig. *Neurobiology of Learning and Memory*, 77, 78–108.

Galvan, V. V., Chen, J., & Weinberger, N. M. (2001). Long-term frequency tuning of local field potentials in the auditory cortex of the waking guinea pig. *Journal of Association for Research in Otolaryngology*, 2, 199–215.

Galvan, V.V., & Weinberger, N.M. (2002) Long-term consolidation and retention of learning-induced tuning plasticity in the auditory cortex of the guinea pig. *Neurobiology of Learning and Memory*, 77, 78–108.

Gavornik, J. P., Shuler, M. G., Loewenstein, Y., Bear, M. F., & Shouval, H. Z. (2009). Learning reward timing in cortex through reward dependent expression of synaptic plasticity. *Proceedings of the National Academy of Sciences of the USA*, 106, 6826–6831.

Gilbert, C. D., & Sigman, M. (2007). Brain states: top-down influences in sensory processing. *Neuron*, 54, 677–696.

Goard, M., & Dan, Y. (2009). Basal forebrain activation enhances cortical coding of natural scenes. *Nature Neuroscience*, 12, 1444–1449.

Golmayo, L., Nunez, A., & Zaborszky, L. (2003). Electrophysiological evidence for the existence of a posterior cortical-prefrontal-basal forebrain circuitry in modulating sensory responses in visual and somatosensory rat cortical areas. *Neuroscience*, 119, 597–609.

Gyengesi, E., Zaborszky, L., & Detari, L. (2008). The effect of prefrontal stimulation on the firing of basal forebrain neurons in the urethane anesthetized rat. *Brain Research Bulletin*, 75, 570–580.

Harris, K. D., & Thiele, A. (2011). Cortical state and attention. *Nature Reviews Neuroscience*, 12, 509–523.

Hasselmo, M. E., & Sarter, M. (2011). Modes and models of forebrain cholinergic neuromodulation of cognition. *Neuropsychopharmacology*, 36, 52–73.

Headley, D. B., & Weinberger, N. M. (2011). Gamma-band activation predicts both associative memory and cortical plasticity. *Journal of Neuroscience*, 31, 12748–12758.

Heffner, H. E., & Heffner, R. S. (1995). Conditioned avoidance. In G. Klump, R. Dooling, R. R. Fay, & W. C. Stebbins (Eds.), *Methods in comparative psychoacoustics* (pp. 79–94). Basel: Birkhauser Verlag.

Hensch, T. K. (2005). Critical period plasticity in local cortical circuits. *Nature Reviews Neuroscience*, 6, 877–888.

Herrero, J. L., Roberts, M. J., Delicato, L. S., Gieselmann, M. A., Dayan, P., & Thiele, A. (2008). Acetylcholine contributes through muscarinic receptors to attentional modulation in V1. *Nature*, 454, 1110–1114.

Ilango, A., Wetzel, W., Scheich, H., & Ohl, F. W. (2010). The combination of appetitive and aversive reinforcers and the nature of their interaction during auditory learning. *Neuroscience*, 166, 752–762.

Jääskeläinen, I. P., Ahveninen, J., Belliveau, J. W., Raij, T., & Sams, M. (2007). Short-term plasticity in auditory cognition. *Trends in Neuroscience*, 30, 653–661.

Jääskeläinen, I. P., Ahveninen, J., Andermann, M. L., Belliveau, J. W., Raij, T., & Sams, M. (2011). Short-term plasticity as a neural mechanism supporting memory and attentional functions. *Brain Research*, 1422, 66–81.

Jafari, M. R., Zhang, Y. F., & Yan, J. (2007). Multiparametric changes in the receptive fields of cortical auditory neurons induced by thalamic activation in the mouse. *Cerebral Cortex*, 17, 71–80.

Jaramillo, S., & Zador, A. M. (2011). The auditory cortex mediates the perceptual effects of acoustic temporal expectation. *Nature Neuroscience*, 14, 246–251.

Kacelnik, O., Nodal, F. R., Parson, C. H., & King, A. J. (2006). Training-induced plasticity of auditory localization in adult mammals. *PLoS Biology*, 4, 627–638.

Kilgard, M., & Merzenich, M. M. (1998a). Cortical map reorganization enabled by nucleus basalis activity. *Science*, 279, 1714–1718.

Kilgard, M., & Merzenich, M. M. (1998b). Plasticity of temporal information processing in the primary auditory cortex. *Nature Neuroscience*, 1, 727–731.

King, A. J., Dahmen, J. C., Keating, P., Leach, N. D., Nodal, F. R., & Bajo, V. M. (2011). Neural circuits underlying adaptation and learning in the perception of auditory space. *Neuroscience and Biobehavioral Reviews*, 35, 2129–2139.

Kisley, M. A., & Gerstein, G. L. (2001). Daily variation and appetitive conditioning-induced plasticity of auditory cortex receptive fields. *European Journal of Neuroscience*, 13, 1993–2003.

Klein, D. J., Depireux, D. A., Simon, J. Z., & Shamma, S. A. (2000). Robust spectrotemporal reverse correlation for the auditory system: Optimizing stimulus design. *Journal of Computational Neuroscience*, 9, 85–111.

Kluge, C., Bauer, M., Leff, A. P., Heinze, H. J., Dolan, R. J., & Driver, J. (2011). Plasticity of human auditory-evoked fields induced by shock conditioning and contingency reversal. *Proceedings of the National Academy of Sciences of the USA*, 108, 12545–12550.

Kover, H., & Bao, S. (2010). Cortical plasticity as a mechanism for storing Bayesian priors in sensory perception. *PLoS One, 5*, e10497.

Kudoh, M., Seki, K., & Shibuki, K. (2004). Sound sequence discrimination learning is dependent on cholinergic inputs to the rat auditory cortex. *Neuroscience Research*, 50, 113–123.

Kumpik, D. P., Kacelnik, O., & King, A. J. (2010). Adaptive reweighting of auditory localization cues in response to chronic unilateral earplugging in humans. *Journal of Neuroscience*, 30, 4883–4894.

Kvale, M. N., & Schreiner, C. E. (2004). Short-term adaptation of auditory receptive fields to dynamic stimuli. *Journal of Neurophysiology*, 91, 604–612.

Leach, N. D., Bajo, V. M., King, A. J., David, S. V., Shamma, S. A., & Fritz, J. B. (2010). Selective responses to salient acoustic stimuli in nucleus basalis of the behaving ferret. Society for Neuroscience Meeting (November, 2010 in San Diego, California, USA). Abstract 18.7

Lee, C. C., & Middlebrooks, J. C. (2011). Auditory cortex spatial sensitivity sharpens during task performance. *Nature Neuroscience*, 14, 108–114.

Lee, J. H., Russ, B. E., Orr, L. E., & Cohen, Y. E. (2009). Prefrontal activity predicts monkeys' decisions during an auditory category task. *Frontiers in Integrative Neuroscience*, 3, 16.

Letzkus, J.J., Wolff, S.B., Meyer, E.M., Tovote, P., Courtin, J., Herry, C., Luthi, A. (2011) A disinhibitory microcircuit for associate fear learning in the auditory cortex. *Nature*, 480, 331-335.

Lin, S. C., & Nicolelis, M. A. (2008). Neuronal ensemble bursting in the basal forebrain encodes salience irrespective of valence. *Neuron*, 59, 138–149.

Liu, X. Basavaraj, S., Krishnan, R., & Yan, J. (2011). Contributions of the thalamocortical system towards sound. *Neuroscience and Biobehavioral Reviews*, 35, 2155–2161.

McMains, S., & Kastner, S. (2011). Interaction of top-down and bottom-up mechanisms in human visual cortex. *Journal of Neuroscience*, 31, 587–597.

McManus, J. N., Li, W., & Gilbert, C. D. (2011). Adaptive shape processing in primary visual cortex. *Proceedings of the National Academy of Sciences of the USA*, 108, 9739–9746.

Mesgarani, N., Fritz, J. B., & Shamma, S. A. (2010). A computational model of rapid task-related plasticity of auditory cortical receptive fields. *Journal of Computational Neuroscience*, 28, 19–27.

Mesgarani, N., & Chang, E. F. (2012). Robust cortical representation of attended speaker in multi-talker speech perception. *Nature* (published online 18 April 2012) doi:10.1038/nature11020

Metherate, R., Kaur, S., Kawai, H., Lazar, R., Liang, K., Rose, H.J. (2005) Spectral integration in auditory cortex: mechanisms and modulation. *Hearing Research*, 206, 146-158.

Metherate, R. (2011). Functional connectivity and cholinergic modulation in auditory cortex. *Neuroscience and Biobehavioral Reviews*, 35, 2058–2063.

Miller, L. M., Escabi, M. A., Read, H. L., & Schreiner, C. E. (2002). Spectrotemporal receptive fields in the lemniscal auditory thalamus and cortex. *Journal of Neurophysiology*, 87, 516–527.

Miranda, J. A., & Liu, R. C. (2009). Dissecting natural sensory plasticity: Hormones and experience in a maternal context. *Hearing Research*, 252, 21–28.

Molina-Luna, K., Hertler, B., Bultrago, M. M., & Luft, A. R. (2008). Motor learning transiently changes cortical somatotopy. *NeuroImage*, 40, 1748–1754.

Mooney, D. M., Zhang, L., Basile, C., Senatorov, V.V., Ngsee, J., Omar, A., & Hu, B. (2004). Distinct forms of cholinergic modulation in parallel thalamic sensory pathways. *Proceedings of the National Academy of Sciences of the USA*, 101, 320–324.

Morishita, H., Miwa, J.M., Heintz, N., Hensch, T.K. (2010) Lynx1, a cholinergic brake, limits plasticity in adult visual cortex. *Science*, 330, 1238-1240.

Mossbridge, J. A., Fitzgerald, M. B., O'Connor, E. S., & Wright, B. A. (2006). Perceptual-learning evidence for separate processing of asynchrony and order tasks. *Journal of Neuroscience*, 26, 12708–12716.

Mysore, S. P., & Knudsen, E. I. (2011). The role of a midbrain network in competitive stimulus selection. *Current Opinion in Neurobiology*, 21, 653–660.

Nelson, C. L., Sarter, M., & Bruno, J. P. (2005). Prefrontal cortical modulation of acetylcholine release in posterior parietal cortex. *Neuroscience*, 132, 347–359.

Otazu, G. H., Tai, L. H., Yang, Y., & Zador, T. Z. (2009). Engaging in an auditory task suppresses responses in auditory cortex. *Nature Neuroscience*, 12, 646–654.

Paltoglou, A. E., Sumner, C. J., & Hall, D. A. (2009). Examining the role of frequency specificity in the enhancement and suppression of human cortical activity by auditory selective attention. *Hearing Research*, 257, 106–118.

Rabinowitz, N. C., Willmore, B. D. B., Schnupp, J. W. H., & King, A. J. (2011). Contrast gain control in auditory cortex. *Neuron*, 70, 1178–1191.

Rahne, T., & Sussman, E. (2009). Neural representations of auditory input accommodate to the context in a dynamically changing acoustic environment. *European Journal of Neuroscience*, 29, 205–211.

Raizada, R. D., & Poldrack, R. A. (2007). Selective amplification of stimulus differences during categorical processing of speech. *Neuron*, 56, 726–740.

Rasmusson, D. D., Smith, S. A., & Semba, K. (2007). Inactivation of prefrontal cortex abolishes cortical acetylcholine release evoked by sensory or sensory pathway stimulation in the rat. *Neuroscience*, 149, 232–241.

Raz, A., & Buhle, J. (2006). Typologies of attentional networks. *Nature Reviews Neuroscience*, 7, 367–379.

Recanzone, G., Schreiner, C., & Merzenich, M. M. (1993). Plasticity in the frequency representation of primary auditory cortex, following discrimination training in adult owl monkeys. *Journal of Neuroscience*, 13, 87–103.
Reches, A., & Gutfreund, Y. (2008). Stimulus-specific adaptations in the gaze control system of the barn owl. *Journal of Neuroscience*, 28, 1523–1533.
Reed, A., Riley, J., Carraway, R., Carrasco, A., Perez, C., Jakkamsetti, V., & Kilgard, M. P. (2011). Cortical map plasticity improves learning but is not necessary for improved performance. *Neuron*, 70, 121–131.
Reyes, A. D. (2011). Synaptic short-term plasticity in auditory cortical circuits. *Hearing Research*, 279, 60–66.
Rioult-Pedotti, M. S., Friedman, D., Hess, G., & Donoghue, J. P. (1998). Strengthening of horizontal cortical connections following skill learning. *Nature Neuroscience*, 3, 230–234.
Rioult-Pedotti, M. S., Friedman, D., & Donoghue, J. P. (2000). Learning-induced LTP in the neocortex. *Science*, 290, 533–536.
Robinson, B. L., & McAlpine, D. (2009). Gain control mechanisms in the auditory pathway. *Current Opinion in Neurobiology*, 19, 402–407.
Roelfsema, P. R., Van Ooyen, A., & Watanabe, T. (2010). Perceptual learning rules based on reinforces and attention. *Trends in Cognitive Neuroscience*, 14, 64–71.
Russ, B. E., Lee, Y. S., & Cohen, Y. E. (2007). Neural and behavioral correlates of auditory categorization. *Hearing Research*, 229, 204–212.
Rutkowski, R. G., & Weinberger, N. M. (2005). Encoding of learned importance of sound by magnitude of representational area in primary auditory cortex. *Proceedings of the National Academy of Sciences of the USA*, 102, 13664–13669.
Sarter, M., Parikh, V., & Howe, W. M. (2009). Phasic acetylcholine release and the volume transmission hypothesis: time to move on. *Nature Reviews Neuroscience*, 10, 383–390.
Scheich, H., Brechmann, A., Brosch, M., Budinger, E., & Ohl, F. W. (2007). The cognitive auditory cortex: Task-specificity of stimulus representations. *Hearing Research*, 229, 213–224.
Scheich, H., Brechmann, A., Brosch, M., Budinger, E., Ohl, F. W., Selezneva, E., et al. (2011). Behavioral semantics of learning and crossmodal processing in auditory cortex: The semantic processor concept. *Hearing Research*, 271, 3–15.
Shuler, M. G., & Bear, M. F. (2006). Reward timing in the primary visual cortex. *Science*, 311, 1606–1609.
Siucinska, F., & Kossut, M. (1996). Short-lasting classical conditioning induces reversible changes of representational maps of vibrissae in mouse SI cortex—a 2 DG study. *Cerebral Cortex*, 6, 506–513.
Siucinska, F., & Kossut, M. (2004). Experience-dependent changes in cortical whisker representation in the adult mouse: A 2-deoxyglucose study. *Neuroscience*, 127, 961–971.
Skoe, E., & Kraus, N. (2010). Hearing it again and again: On-line sub-cortical plasticity in humans. *PLoS One, 5*, e13645.
Soto, G., Kopell, N., & Sen, K. (2006). Network architecture, receptive fields, and neuromodulation: Computational and functional implications of cholinergic modulation in primary auditory cortex. *Journal of Neurophysiology*, 96, 2972–2983.
Spierer, L., Tardif, E., Sperdin, H., Murray, M. M., & Clarke, S. (2007). Learning-induced plasticity in auditory spatial representations revealed by electrical neuroimaging. *Journal of Neuroscience*, 27, 5474–5483.
Spierer, L., De Lucia, M., Bernasconi, F., Grivel, J., Bourquin, N. M., Clarke, S., & Murray, M. M. (2011). Learning-induced plasticity in human audition: Objects, time and space. *Hearing Research*, 271, 88–102.
Takata, N., Mishima, T., Hisatsune, C., Nagai, T., Ebisui, E., Mikoshiba, K., & Hirase, H. (2011). Astrocyte calcium signaling transforms cholinergic modulation to cortical plasticity *in vivo*. *Journal of Neuroscience*, 31, 18155–18165.
Thiel, C. M. (2007). Pharmacological modulation of learning-induced plasticity in human auditory cortex. *Restorative Neurology and Neuroscience*, 25, 435–443.

Thiel, C. M., Friston, K. J., & Dolan, R. J. (2002). Cholinergic modulation of experience-dependent plasticity in human auditory cortex. *Neuron*, 35, 567–574.
Treue, S., & Martinez-Trujillo, J. C. (1999). Feature-based attention influences motion processing gain in macaque visual cortex. *Nature*, 399, 575–579.
Tsunada, J., Lee, J. H., & Cohen, Y. E. (2011). Representation of speech categories in the primate auditory cortex. *Journal of Neurophysiology*, 105, 2634–2646.
Ulanovsky, N., Las, L., & Nelken, I. (2003). Processing of low-probability sounds by cortical neurons. *Nature Neuroscience*, 6, 391–398.
Ulanovsky, N., Las, L., Farkas, D., & Nelken, I. (2004). Multiple time scales of adaptation in auditory cortex neurons. *Journal of Neuroscience*, 24, 10440–10453.
Van Wanrooij, M. M., & Van Opstal, A. J. (2005). Relearning sound localization with a new ear. *Journal of Neuroscience*, 25, 5413–5424.
Wang, L., Conner, J. M., Rickert, J., & Tuszynski, M. H. (2011). Structural plasticity within highly specific neuronal populations identifies a unique parcellation of motor learning in the adult brain. *Proceedings of the National Academy of Sciences of the USA*, 108, 2545–2550.
Watkins, P. V., & Barbour, D. L. (2008). Specialized neuronal adaptation for preserving input sensitivity. *Nature Neuroscience*, 11, 1259–1261.
Watkins, P. V., & Barbour, D. L. (2011). Level-tuned neurons in primary auditory cortex adapt differently to loud versus soft sounds. *Cerebral Cortex*, 21, 178–190.
Weinberger, N. M. (2004). Specific long-term memory traces in primary auditory cortex. *Nature Reviews Neuroscience*, 5, 279–290.
Weinberger, N. M. (2011). Reconceptualizing the primary auditory cortex: Learning, memory and specific plasticity. In J. Winer & C. Schreiner (Eds.), *The auditory cortex* (pp. 465–491). New York: Springer.
Wen, B., Wang, G. I., Dean, I., & Delgutte, B. (2009). Dynamic range adaptation to sound level statistics in the auditory nerve. *Journal of Neuroscience*, 29, 13797–13808.
Winkler, I., Denham, S. L., & Nelken, I. (2009). Modelling the auditory scene: predictive regularity representations and perceptual objects. *Trends in Cognitive Sciences*, 13, 532–540.
Winkowski, D. E., Bandyopadhyay, S., Shamma, S. A., & Kanold, P. O. (2012). Rapid plasticity of auditory cortical circuits induced by prefrontal cortex activation (submitted)
Xiong, Y., Zhang, Y., & Yan, J. (2009). The neurobiology of sound-specific auditory plasticity: A core neural circuit. *Neuroscience and Biobehavioral Reviews*, 33, 1178–1184.
Xu, T., Yu, X., Perlik, A. J., Tobin, W. F., Zweig, J. A., Tennant, K., et al. (2009). Rapid formation and selective stabilization of synapses for enduring motor memories. *Nature*, 462, 915–919.
Yotsumoto, Y., Watanabe, T., & Sasaki, Y. (2008). Different dynamics of performance and brain activation in the time course of perceptual learning. *Neuron*, 57, 827–833.
Yuan, K., Barker, A. J., Seybold, B., Carcea, I., Martins, A. R. O., Wachs, M., et al. (2011). Long-term modification of cortical synapses improves sensory perception. Society for Neuroscience Meeting, November, 2011, Washington, DC Abstract 869.10.
Zhou, X., & Merzenich, M. (2008). Enduring effects of early structured noise exposure on temporal modulation in the primary auditory cortex. *Proceedings of the National Academy of Sciences of the USA*, 105, 4423–4428.
Zhou, X., de Villers-Sidani, E., Panizzutti, R., & Merzenich, M. M. (2010). Successive-signal biasing for a learned sound sequence. *Proceedings of the National Academy of Sciences of the USA*, 107, 14839–14844.
Zikopoulos, B., & Barbas, H. (2006). Prefrontal projections to the thalamic reticular nucleus form a unique circuit for attentional mechanisms. *Journal of Neuroscience*, 26, 7348–7361.

Chapter 10
Experience-Dependent Plasticity and Auditory Cortex

Kathryn N. Shepard, Michael P. Kilgard, and Robert C. Liu

Keywords Acetylcholine • Adult plasticity • Associative learning • Catecholamine • Cortical plasticity • Estrogen • Map plasticity • Neuromodulator • Operant conditioning • Receptive field • Sensitive period • Sensory plasticity • Serotonin • Sound exposure • Tonotopic map

10.1 Introduction

Throughout their lifetimes, individuals are constantly engaged by acoustic environments that leave lasting impressions, such as the sound of a familiar voice or the rhythm of a catchy tune. A fundamental question in auditory neuroscience concerns how and where such acoustic knowledge is acquired and stored in the brain. Such declarative forms of memory, including episodic memories from personal experience, are usually believed to rely on hippocampal and amygdalar processing, but a growing

---

K.N. Shepard
Neuroscience Graduate Program and Department of Biology, Emory University,
1510 Clifton Rd. NE; and Center for Behavioral Neuroscience, Atlanta, GA 30322, USA
e-mail: knshepa@emory.edu

M.P. Kilgard
School of Behavioral and Brain Sciences,
University of Texas at Dallas, GR 41, Richardson, TX 75083, USA
e-mail: kilgard@utdallas.edu

R.C. Liu (✉)
Department of Biology, Emory University, 1510 Clifton Rd. NE, Atlanta, GA 30322, USA

Center for Behavioral Neuroscience, Georgia State University,
PO Box 3999, Atlanta, GA 30302, USA

Center for Translational Social Neuroscience, Yerkes National Primate
Research Center, Emory University, 954 Gatewood Rd., Atlanta, GA 30322, USA
e-mail: robert.liu@emory.edu

Y.E. Cohen et al. (eds.), *Neural Correlates of Auditory Cognition*, Springer Handbook of Auditory Research 45, DOI 10.1007/978-1-4614-2350-8_10,
© Springer Science+Business Media New York 2013

literature has also implicated plasticity within auditory cortex as one of the key contributors to establishing long-term auditory memories. Indeed, the fact that electrophysiological activity within auditory cortex can change as a result of auditory experience has been known since Galambos et al. (1956) first demonstrated pronounced effects on the sound-evoked electroencephalogram after sound–shock pairing. Research since then has attempted to uncover the precise rules and roles for such plasticity in both auditory learning and memory.

This chapter guides readers through some of the literature that has helped inform the current debate about the nature of this long-term plasticity and the mechanisms that produce it. Work on auditory cortical plasticity has helped dispel the idea that cortical areas responsible for higher level cognitive functions such as learning and memory are cleanly separated from sensory processing areas. In addition, this research has helped counter dogma about the relative inflexibility of the adult brain, compared to the highly malleable brain of the developing animal. This progress has left some unanswered questions though, in part because of gaps arising from the manner by which long-term, experience-dependent auditory cortical plasticity has been traditionally investigated. For example, the auditory cortical plasticity literature is divided into studies on "active" paradigms in which acoustic associations are acquired (e.g., sound leads to shock or reward) or "passive" paradigms in which exposure to a sound occurs continuously in the background. Yet until recently, these paradigms have been pursued almost entirely in adults or developing individuals, respectively, leaving "siloed" the potential explanations about the function and underlying mechanisms for plasticity. Further, characteristic frequency (CF) map plasticity, explained in more detail in Section 10.2, has often been the index of choice for documenting cortical changes. Yet map plasticity may not be the final or only endpoint for storing auditory memories (even if it is involved in learning), and changes in a neuron's response properties aside from CF should be considered further.

This chapter covers these topics from the perspective that long-term changes in auditory cortical neuronal single unit or population activity arising from any acoustic experience likely result from the differential engagement of similar neural mechanisms. The parallel thrusts of research on "classic" paradigms involving active sound associations or passive sound exposure are first described, and then a discussion about neuromodulatory mechanisms is used to suggest how these manifestations of plasticity may be related, irrespective of their occurrence in development or adulthood. For reasons of space and cohesion, this chapter likely misses some important contributions, but other chapters in this volume do more to address some of those topics, such as short-term, experience-dependent plasticity (Fritz, David, and Shamma, Chapter 9) and task-related auditory plasticity (Scheich and Brosch, Chapter 3).

10.2 Cortical Plasticity "Actively" Engaged by Sound Associations

The coarse pattern of evoked activity in primary auditory cortex is dictated to a large extent by the tonotopic distribution of characteristic frequencies. Nevertheless, exactly how cortical neurons respond to specific sounds depends on more than just

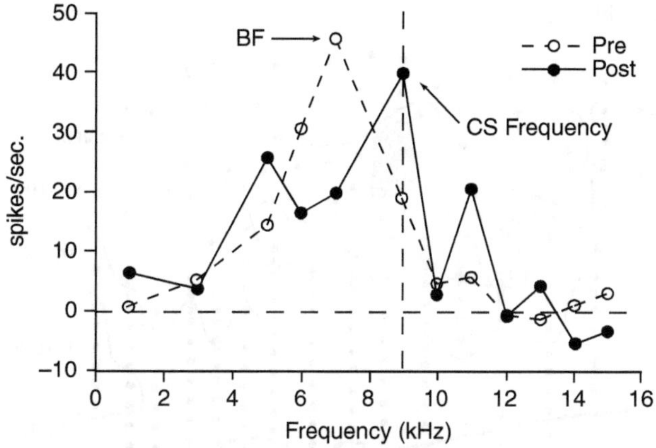

Fig. 10.1 An example tuning curve shift in a multiunit cluster of an adult guinea pig. The awake animal was restrained in a headpost and harness, and frequency tuning curves were measured before (pre) and after (post) 30 paired presentations of a 9-kHz tone and footshock. After conditioning, this cluster's response to its initial best frequency (BF) was attenuated whereas the response to the paired tone was enhanced to the extent that it became the new BF. (Reprinted from Edeline et al. [1993] with permission from the American Psychological Association)

sound frequency, and those responses can in fact be remodeled by behaviorally relevant acoustic experiences on an ongoing basis. This remodeling has traditionally been studied at either the neuronal single-unit or population level. For instance, evoked firing rates in single units are typically potentiated in an enduring, frequency-specific manner (Fig. 10.1) after that frequency is "actively" paired with a high-valence stimulus (e.g., footshock, as in a fear conditioning paradigm, or reward, as in an operant training paradigm) (Bakin & Weinberger, 1990; Ohl & Scheich, 1996; Gao & Suga, 2000). In some cases, the response to the conditioned frequency comes to exceed the response to the original CF of the single unit, such that the conditioned frequency actually becomes the new CF. When this occurs, a single unit is said to have been "retuned." At the population level, pairing a tone with punishment or reward can result in a long-lasting increase in the proportion of units that are tuned to the paired frequency (Recanzone et al., 1993). When analyzed spatially across the surface of an auditory cortical field, this type of plasticity is referred to as tonotopic map expansion (Fig. 10.2). One can intuitively see how these manifestations are related to one another: the combined retuning of a *population* of single units produces map expansion. However, it is important to note that individual single units could undergo tuning curve changes without generating large-scale map plasticity. For example, individual single units in a local population may not all retune to the same final CF; in this case, map plasticity would be absent. Even for just one single unit, the response to a salient frequency could undergo plasticity without "retuning" the neuron's CF, much less changing the CF map. Thus "single-unit plasticity" is not synonymous with "map plasticity."

Because retuning occurs in an associative context, in which the events a sound signifies become associated with the features of the sound itself, it has been postulated that retuning may provide a physiological trace for an auditory memory

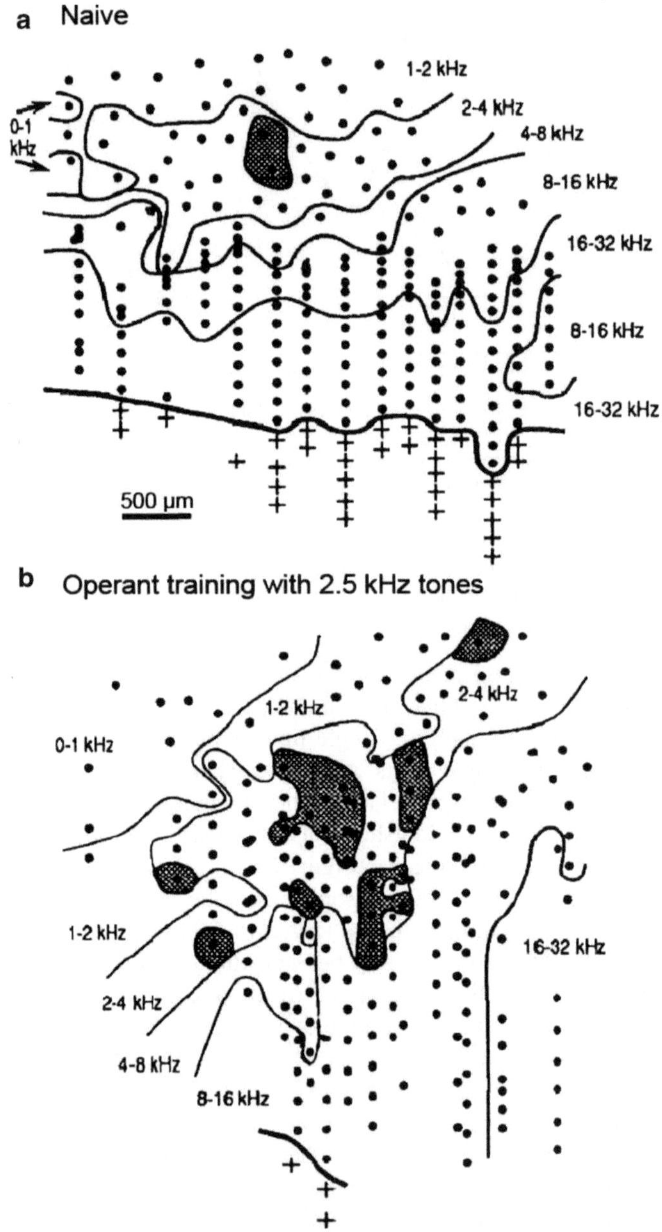

Fig. 10.2 Map plasticity in adult owl monkeys (*Aotus*). The animal in (**B**) was trained to discriminate among different frequencies in a range around 2.5 kHz in exchange for a food reward. After 97 sessions of 400–750 discrimination trials, the monkey's auditory cortex was mapped. Shaded areas of the cortical map show a larger area tuned to 2.5 kHz compared to naive control monkey (**A**). Black points represent locations of electrode penetrations within primary auditory cortex; + symbols indicate recording sites outside of primary auditory cortex. (Reprinted from Recanzone et al. [1993] with permission from the Society for Neuroscience)

(Weinberger, 2004). Research has indeed shown that the long-term auditory cortical plasticity that follows an arousing acoustic experience shares many qualities with long-term memory. First, the plasticity that arises is frequency specific. Like an animal that can distinguish in a behavioral test between frequencies that have or have not been paired with foot shock, single units that undergo plasticity exhibit a selective potentiation of their responses to frequencies at or near the paired stimulus (Bakin & Weinberger, 1990; Edeline & Weinberger, 1993). Second, single-unit plasticity can be induced rapidly. A hallmark of long-term memories is that they can be produced after a single salient experience (McGaugh, 1966), and indeed, the receptive fields of single units may shift following five or fewer tone-shock pairings (Edeline et al., 1993). In addition, auditory cortical plasticity can, under some conditions, last indefinitely, as long-term memories do. Though experimental restraints limit the time period over which plasticity can be studied, one study found that frequency tuning shifts observed within a cortical column could endure for up to 8 weeks (Weinberger et al., 1993). Finally, some work has suggested that receptive field shifts in auditory cortex during behavior result from Hebbian synaptic plasticity (Ahissar et al., 1998), the phenomenon that is believed to underlie long-term memory in a number of systems.

If auditory cortical plasticity is involved in the formation of auditory memories, it should, like memory formation, be elicited by a number of behavioral contexts and be capable of representing other acoustic features in addition to frequency. Indeed, a wealth of studies has by now shown considerable flexibility in the circumstances under which plasticity is induced and the forms it takes.

For example, initial studies relied on associative fear conditioning to evoke plasticity (without actually testing for learning), but subsequent research showed that auditory cortical plasticity can be elicited in operant paradigms in which learning the significance of the acoustic stimulus would confer a clear behavioral benefit. For instance, frequency-specific response potentiation has been demonstrated in both aversive (Bakin et al., 1996) and appetitive (Recanzone et al., 1993; Kisley & Gerstein, 2001) instrumental auditory learning tasks, wherein recognition of a specific sound (usually a tone of a particular frequency) prompts the animal to perform a behavior that forestalls a punishment or draws a reward, respectively. Because an associative memory is at the root of instrumental learning, and because the neural changes observed in associative conditioning and instrumental learning contexts are comparable, it may be that these different paradigms are two different means of achieving similar physiological effects, rather than distinct forms of plasticity.

Further, auditory cortical plasticity is flexible in the sense that it can alter the representation of various facets of a sound, depending in part on which acoustic features best predict high-valence stimuli. Though much research focuses on changes in the frequency tuning of single or multi-units after training with tones, additional work has shown neural plasticity in entrainment rate (Kilgard & Merzenich, 1998a; Percaccio et al., 2005; Liu et al., 2006), sensitivity to sound intensity (Polley et al., 2006), and responsiveness to frequency modulation (Moucha et al., 2005; van Wassenhove & Nagarajan, 2007). For instance, in an operant conditioning task in which reward is predicted by the frequency or intensity of a tone in separate cohorts of rats, neural responses to tones are potentiated in a frequency-specific or intensity-specific fashion, depending on which property is reinforced (Polley et al., 2006). Likewise, when the reinforced stimulus is a train of tones with constant repetition rate but variable frequency (Kilgard & Merzenich, 1998a), neurons in primary auditory cortex become entrained to the repetition rate without undergoing a change in frequency tuning. However, when both repetition rate and frequency are held constant in the same paradigm, neurons exhibit plasticity in the frequency response, but not the following rate, suggesting that some acoustic properties may naturally be more salient than others.

Importantly, although the aforementioned work (like the majority of work on auditory cortical plasticity) focuses on primary auditory cortex, several studies have shown that secondary auditory areas show a capacity for plasticity that perhaps exceeds that of primary areas. Researchers have shown that, as in primary cortex, secondary auditory fields can undergo both frequency map plasticity (Polley et al., 2006) and single-unit retuning (Diamond & Weinberger, 1986) in such a way that potentiates responses to reinforced stimuli after associative conditioning. In an early demonstration of this phenomenon that compared plasticity in secondary and primary cortices, a greater proportion of neurons in secondary cortex exhibited tuning shifts, suggesting that secondary areas may be more sensitive to the behavioral associations a sound has than primary cortical areas (Diamond & Weinberger, 1984; Weinberger et al., 1984). Subsequent work has lent support to this notion, showing in one case that the magnitude of plasticity is greater in secondary cortex than in primary (Puckett et al., 2007). Still, the issue of plasticity in secondary areas remains relatively underexplored, and future work will be critical in explaining what secondary cortical plasticity achieves and how it differs from primary cortical plasticity.

In summary, the auditory cortex undergoes "retuning" after an encounter with a behaviorally relevant sound, and features of this plasticity—the time course over which it appears and is maintained, the contexts that elicit it, and its flexibility in representing different stimulus features—suggest that the changes themselves are involved in the formation or maintenance of long-term auditory memories. However, as discussed in Section 10.2.1, recent work has revealed considerable complexity in the situational "rules" that determine how auditory cortical plasticity proceeds, and in particular, whether map plasticity itself always occurs. It is now apparent that although auditory cortical plasticity may be involved in the formation of long-term auditory memories, it may not represent the final, single site of memory storage.

10.2.1 Contextual Modulation of Cortical Map Plasticity

Subtle details about the environment in which an auditory association is learned can have a great impact on the manifestation of auditory cortical plasticity. These details generally fall into two categories to be discussed here: those concerning the acoustic stimulus itself and those relating to the behavioral environment in which learning takes place.

First, developmental plasticity studies (discussed further in Section 10.3.1) have been instrumental in demonstrating that temporal aspects of the acoustic environment are key modulators of auditory plasticity. Intriguingly, these studies suggest that sounds with ethologically plausible temporal features may trigger plasticity more readily than those that do not. Evidence from Kim and Bao (2009) indicates that if tones of a particular frequency are presented at a rate at which rats emit vocalizations, but not at higher or lower rates, rat pups will dedicate more space in primary auditory cortex to processing sounds at that frequency. Further, playback of a continuous pure tone fails to induce the map expansion one observes after exposing developing animals to pulsed tones (Zhou et al., 2008). This effect may be evidence of a mechanism for animals to ignore continuous background noises (e.g., the continuous trickle of a nearby stream) and tune in to behaviorally relevant stimuli, which are often more transient in nature (e.g., same-species vocalizations). However, while background noises themselves may not be sufficiently meaningful to induce plasticity in the adult cortex, the presence of unreinforced sounds during associative learning may impact how reinforced sounds come to be represented. In one study that used nucleus basalis stimulation for reinforcement (see Section 10.4.1.1), animals were presented with upward and downward frequency sweeps, though only downward sweeps were reinforced. These animals showed shorter response latencies and wider tuning bandwidths than a group of animals who underwent sweep/nucleus basalis stimulation pairing without also hearing unpaired background sounds (Moucha et al., 2005). Taken together, these studies argue that the features of the acoustic environment other than the qualities of the reinforced stimulus itself can have a significant impact on how plasticity is manifested.

Second, the behavioral conditions when an association is learned are themselves a strong modulator of auditory cortical map plasticity. Although the term "behavioral condition" refers to internal states that guide an animal's behavior, it is important to note that these states are often influenced by the animal's external environment and may be controlled in the laboratory. Motivation level, for instance, is an internal state that can be manipulated by an experimenter. An animal that is made to be highly motivated to receive a reward or avoid a punishment will learn the behavioral task associated with reinforcement more quickly than one that is less motivated. This was evident in a study where water-deprived animals were taught to press a bar for water upon hearing a tone of a particular frequency. More highly motivated animals learned the task more quickly and also came to devote a larger amount of their auditory cortex to processing the reinforced frequency, as assessed immediately after training (Rutkowski & Weinberger, 2005). Whether map expansion develops alongside improved behavioral responses or contributes to it directly is unclear, but a more recent study suggests that map expansion may support faster learning of an auditory association (Reed et al., 2011).

Another example of how behavioral conditions affect plasticity is the finding that *how* an animal learns an association, and not simply whether it learns, appears to dictate the presence of map plasticity. One paradigm that is often used in instrumental training involves rewarding behavioral responses that fall within some time period relative to the onset of the reinforced sound. In order to be rewarded consistently, an animal may learn to respond only upon hearing the onset of the reinforced sound, or may respond continuously starting at sound onset and lasting until presentation of the error signal that indicates the end of the reward period. A number of studies have now shown that while each strategy supports a comparable level of performance, auditory cortical map plasticity occurs only in animals that use the latter strategy (Berlau & Weinberger, 2008; Bieszczad & Weinberger, 2010a). Further, the extent to which any animal uses the tone-onset-to-error strategy correlates with the magnitude of map plasticity they exhibit (Bieszczad & Weinberger, 2010b). Though the behavioral implications of this work are still being established, they have made it clear that plasticity is shaped not only by external factors such as the stimulus an animal receives, but also by internal influences.

10.2.2 Cortical Map Plasticity in Learning versus Memory

The finding that learning strategy, and not simply the occurrence of learning, can dictate whether auditory cortical map plasticity is observed suggests that map plasticity may not always be necessary for the long-term storage of an auditory memory. Indeed, learning may induce subtle changes in single unit tuning that fall short of the retuning necessary for map plasticity to be evident, or map expansion may constitute a stage of the memory formation process, though not its endpoint. This notion that the auditory cortical map is not the final and/or only way of storing an auditory memory is supported by other studies where performance on auditory learning tasks improves in the absence of any enduring map plasticity in primary auditory cortex (Brown et al., 2004; Reed et al., 2011).

Even if long-term auditory cortical map plasticity is not necessary for the *expression* of auditory memories, evidence suggests that plasticity within auditory cortex is nevertheless memory related. Blocking long-term plasticity in primary auditory cortex, either through injection of a protein synthesis inhibitor (Kraus et al., 2002) or antagonism of a key plasticity-related gene (Carpenter-Hyland et al., 2010a), prevents animals from attaining typical levels of performance on an auditory learning task. Importantly, a decrement in performance is observed only when plasticity-blocking agents are administered *before* animals reach an asymptotic level of performance. If they are given after the animal's accuracy plateaus, no effect on behavior is observed. Therefore, although auditory cortical plasticity may not be required for memory storage, it may be centrally involved in memory formation.

Using alternative methods, Reed et al. (2011) have lent support to this hypothesis. In this study, map expansion at a particular frequency was artificially induced (by pairing sounds with nucleus basalis stimulation; see Section 10.4.1.1) and performance on a reinforced auditory discrimination task was evaluated. Animals that were made to devote larger proportions of their primary auditory cortex to the reinforced frequency learned the task more quickly than animals that had not, but long

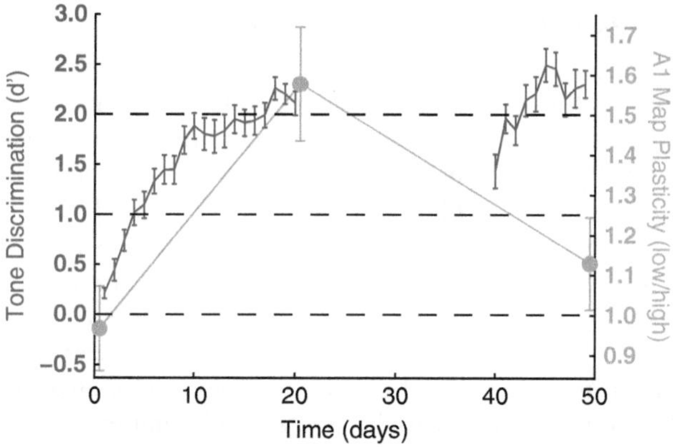

Fig. 10.3 Map plasticity develops with learning of an auditory discrimination task, but reverses once behavior has been learned (light gray line). High performance is maintained even in the absence of plasticity (dark gray line). (Based on data from Reed et al. [2011])

after they initially mastered the task, the animals' accuracy remained high despite an observed re-normalization of cortical topography. That is, though plasticity helped the animals achieve a high level of performance during the memory-formation portion of the task, it was not necessary for maintaining their perceptual accuracy over a long period of time (Fig. 10.3). This research reinforces the idea that auditory cortical map plasticity may not be involved in memory storage so much as the initial formation of a memory, and therefore may recede once the memory has been stably encoded. This might explain why some studies have failed to show map plasticity after training in an auditory learning task: perhaps the interval between training and examination of tuning properties in auditory cortex was too wide to observe the transient period when map plasticity would be evident (Brown et al., 2004).

Finally, it may also be that other neural changes due to acoustic experiences endure as memory traces even if map changes fade. Sound-specific response strength (Polley et al., 2004), response latency (Kilgard et al., 2001; Liu & Schreiner, 2007), and neuronal synchronization (Kilgard et al., 2007) can also be plastic features of auditory cortical activity—features that CF mapping alone would not reveal. This emphasizes the need in plasticity studies to measure features of activity aside from just the local neural population's CF. In this regard, a unique early study published by Ahissar and colleagues (1998) measured pairwise correlations between single units recorded from awake macaque monkeys undergoing "cellular classical conditioning" (Baranyi & Feher, 1981; Fregnac et al., 1988). The strength of neuronal correlations was changed to varying degrees after conditioning, depending on whether or not the animal was behaviorally engaged in listening to a sound. These changes were observed on a short timescale, so although they may just be part of the dynamic nature of cortical processing during behavior, it is also possible they represent a necessary first phase in the formation of long-term memories.

10.2.3 Ethologically Relevant Associations

Based on studies such as those cited in Sections and 10.2.1 and 10.2.2, there is no longer any doubt that long-lasting, behaviorally contingent neural changes do occur in auditory cortex. This leads though to the new question of how the thousands of complex sounds mammals can plausibly remember (Pilley & Reid, 2011) are stored simultaneously. If the memory traces were simply an expansion in the number of units tuned to a single frequency of interest, as expected from map plasticity, the cortex would presumably become overtaxed. Moreover, how complex sounds consisting of multiple frequencies (e.g., many behaviorally relevant vocalizations) would be retained in such a model of auditory memory is unclear. The research described in Sections 10.2.1 and 10.2.2 has yielded innumerable insights into the mechanisms of sensory memory, with a key point being that plasticity is shaped by seemingly small details about the experimental environment. This observation thus highlights the need to move toward ethological paradigms that attempt to replicate the more typical conditions under which auditory memories are naturally acquired.

Ethological approaches have indeed proven to be valuable in characterizing plasticity-related phenomena that would not have been apparent using the conventional methodologies. For example, one group used maternal and pup-naïve female mice to demonstrate differential activation of secondary auditory cortical areas upon presentation of tone bursts in the same frequency range as pup calls, which are highly relevant only to mothers (Fichtel & Ehret, 1999; Geissler & Ehret, 2004). This work suggests that a main locus of long-term plasticity for communication sounds may not lie in primary auditory cortex, where most of the recent work on auditory cortical plasticity has focused, but rather in secondary areas. Even within primary auditory cortex, however, plasticity in excitatory tuning may not be the most enduring change. Work using the same maternal mouse model revealed considerable long-term plasticity in call-evoked inhibition in maternal females compared to pup-naïve virgins. Specifically, mothers showed a stronger duration of pup ultrasound call–evoked inhibition in neurons tuned to frequencies outside the band occupied by single-frequency pup whistles, thereby enhancing the population level signal-to-noise ratio upon pup call presentation in mothers but not naïve females (Galindo-Leon et al., 2009). This work highlights an unconventional way of representing the salience of a sound through auditory cortical plasticity—by suppressing background activity rather than enhancing the auditory foreground. Like plasticity in secondary cortical areas, inhibitory plasticity warrants further investigation in its own right, but it also hints that continued exploration of ethologically evoked plasticity will yield new insights and propel the field forward.

10.3 Cortical Plasticity "Passively" Engaged by Sound Exposure

Until this point, this chapter has focused on experience-dependent plasticity as evoked by paradigms that, for the most part, call for the animal to make an auditory association. However, plasticity can also occur in more passive environments in which an animal is simply exposed to repeated acoustic stimulation, and an overt response is not engaged. This has been studied at the greatest depth in developing animals, during the auditory cortical sensitive period.

10.3.1 Sensitive Period for Plasticity from Sound Exposure

During early life, an animal's brain comes to represent the common features of its sensory environment via exposure-induced plasticity. Although thalamic afferents and intracortical projections innervate the mammalian sensory cortex very early in life (Metherate, 2004), it is believed that during a "sensitive" period (or alternatively, "critical" period), enhanced impressionability in response to sensory stimulation allows the animal to customize its wiring according to the type of environment in which it finds itself (Knudsen, 2004). The seminal example of sensitive period plasticity was characterized in visual cortex by Hubel and Wiesel in the 1960s: They found that the visual cortices of kittens reared with one eye occluded became insensitive to stimulation of that eye upon its uncovering, as long as the occlusion occurred during a circumscribed sensitive period (Hubel & Wiesel, 1970). Importantly, cortical blindness to stimulation of the occluded eye lasted into adulthood, and was not apparent when monocular occlusion occurred outside the sensitive period. In other words, during a period of heightened sensitivity to the environment, cortical processing space was diverted from inputs that carried no information and reallocated to richer sources of visual input. This mechanism provides a way for an individual to adapt to the environment it is born into on a much faster timescale than evolutionary mechanisms would allow. That such a mechanism is advantageous is evident in the fact that it has been adopted in other systems, to aid in language learning (Kuhl, 2004), auditory development (de Villers-Sidani et al., 2007), and acquisition of normal social behaviors (Freedman et al., 1961; Harlow et al., 1965) in addition to visual development. Sensitive periods presumably entrain these systems to receive the highest possible quality information about the environment to make advantageous behavioral responses.

Early in the 2000s, research on sensitive period plasticity in the auditory cortex greatly expanded. Analogous to the classic studies in visual cortex that exploited its ocular dominance organization to uncover plasticity from monocular deprivation, many studies in auditory cortex have taken advantage of the most robust feature of auditory cortical topography—tonotopic organization—and used long-term sound exposure during development to drive map expansion in the rat. By rearing rat pups and their mothers in the presence of pulsed tones of a particular frequency at a moderate intensity (60–70 dB sound pressure level [SPL]) for 10–16 hours per day, Zhang et al. (2001) demonstrated that the experience was reflected in the cortices of adult animals who had received stimulation as pups. Specifically, the amount of primary auditory cortex dedicated to processing the frequency of exposure was approximately double what it was in naïve animals. This was true only of the animals that had been exposed as pups; the mothers of exposed pups, which had received the stimulation in adulthood, did not show evidence of map expansion.

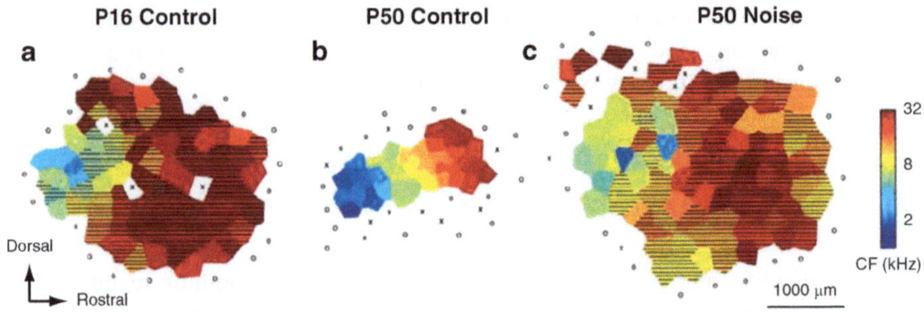

Fig. 10.4 Exposure to continuous white noise prolongs the auditory critical period. (**a**) Cortical map from a 16-day-old (P16) naive rat. Hatched area indicates sites where the tuning bandwidth is greater than 1.5 octaves at 20 dB above threshold. By P50 (**b**), the critical period has ended and primary auditory cortex has taken on mature characteristics (e.g., smaller area, narrower bandwidths). In contrast, the rat in (**c**) was reared in the presence of continuous white noise. Though it is the same age as the rat in (**b**), cortical organization is more similar to that of an immature animal (**a**). Open circles indicate recording sites in nonprimary auditory cortex; x symbols indicate recording sites outside auditory cortex. (Reprinted from Chang & Merzenich [2003] with permission from AAAS)

Later work replicated these findings and delineated an auditory sensitive period for tone exposure–induced cortical map plasticity in the rat as a several-day period beginning at the onset of hearing (de Villers-Sidani et al., 2007; Han et al., 2007). Interestingly, plasticity in the auditory system for other response features (e.g., binaurality) can arise from subcortical changes during developmental sound experience, or may have different critical time windows for plasticity (Popescu & Polley, 2010). Finally, even though sound exposure with pure tones may not be that ethological, other studies have confirmed that exposure to more complex sounds, such as frequency-modulated sweeps, also generate plasticity within a sensitive period (Razak et al., 2008; Insanally et al., 2009).

However, although the sensitive period may be defined in terms of an animal's age, it may not be age but rather experience that determines how maturation of auditory cortical responses occurs. In particular, it appears that patterned inputs are required for acquisition of normal auditory cortical responses during the sensitive period. Removing patterned input by exposing animals to continuous, moderate-intensity white noise rather than pulsed tones delays closure of the auditory sensitive period (Fig. 10.4), essentially stranding the animal's cortex in an immature state (Chang & Merzenich, 2003). Specifically, primary auditory cortex remains large in total area, tuning bandwidths are wide, and distribution of CFs is nonuniform, with more sites tuned to frequencies at the higher end of the audible spectrum. These immature-like qualities persist until exposure is discontinued and the animal is returned to a normal acoustic environment, at which time cortical organization renormalizes. That the sensitive period closes only after continuous noise exposure ends is evident when the cortical organization of a tone-exposed pup is compared to that of an animal reared in continuous noise into early adulthood, then moved to a tone exposure environment. The frequency of the exposed tone is overrepresented in the primary auditory cortices of both groups, suggesting that though the noise-exposed animals were mature at the time of tone exposure, their auditory cortices were as plastic as those of pups. Further emphasizing the importance of temporally patterned inputs to the maturation of auditory cortical responses, a similar profile of extended immaturity was observed when animals were exposed to continuous, rather than pulsed, tones from the sensitive period through early adulthood (Zhou et al., 2008).

Studies that instead introduce temporal regularity during the sensitive period show that experience with patterned inputs drives sensitive period closure. Introducing regular temporal patterning to a rat pup's acoustic environment by exposing it to repeated white noise pulses distorts tonotopy, receptive field breadth, and response timing in primary auditory cortex neurons, but unlike the abnormalities of animals reared in the presence of continuous noise, these changes *cannot* be reversed by reintroduction to a normal acoustic environment (Zhang et al., 2002; Zhou & Merzenich, 2007, 2008). This work implies that exposure to patterned inputs may trigger the end of the sensitive period, and a subsequent study by Dorrn et al. (2010) offers further support for this hypothesis. Using in vivo intracellular electrophysiology, they first established that a key difference between the adult and developing auditory cortices concerns the balance of excitatory and inhibitory tuning. Specifically, although the amplitude and specificity of excitatory and inhibitory frequency tuning are closely matched in adulthood, inhibitory tuning is markedly less frequency specific in developing animals. Hypothesizing that patterned input drives the inhibitory–excitatory matching that characterizes the mature auditory cortex, the authors then measured inhibitory and excitatory tuning curves before and after presenting a bout of pulsed tones. Indeed, patterned stimulation strengthened the correlation between inhibitory and excitatory conductances, inching the profile of cortical activity closer to that of the mature animal (Fig. 10.5). This work strongly suggests that progress through the sensitive period is not just a result of aging, but rather the incremental fulfillment of a "quota" of exposure to temporally patterned input.

It was previously thought that during the sensitive period, the cortex was akin to hardening cement: malleable until it is set, but then only modifiable via a "jackhammer" approach (e.g., behaviorally engaging, associative plasticity described in Section 10.2). Recent work, however, has challenged this notion. Noting that exposure to continuous white noise could prolong the sensitive period into maturity for young rats, Zhou et al. (2011) went a step further and asked whether the same stimulus could reinstate the sensitive period-like conditions in mature rats. In fact, subjecting colony-reared rats to uninterrupted, moderate-intensity white noise for 6 weeks *was* sufficient to return the cortex to a highly plastic state, even in animals that were relatively senescent (>1 year old). This contribution challenges the conventional wisdom about sensitive period dynamics, indicating that the transition from the highly plastic conditions during the sensitive period to a relatively inflexible state is not unidirectional as it was once assumed to be, but rather bidirectional. In addition, it suggests that patterned inputs are not only important for sensitive period closure, but perhaps also the maintenance of existing response properties.

Taken together, this research demonstrates that, as in the visual system, the auditory cortex passes through a developmental sensitive period wherein it briefly

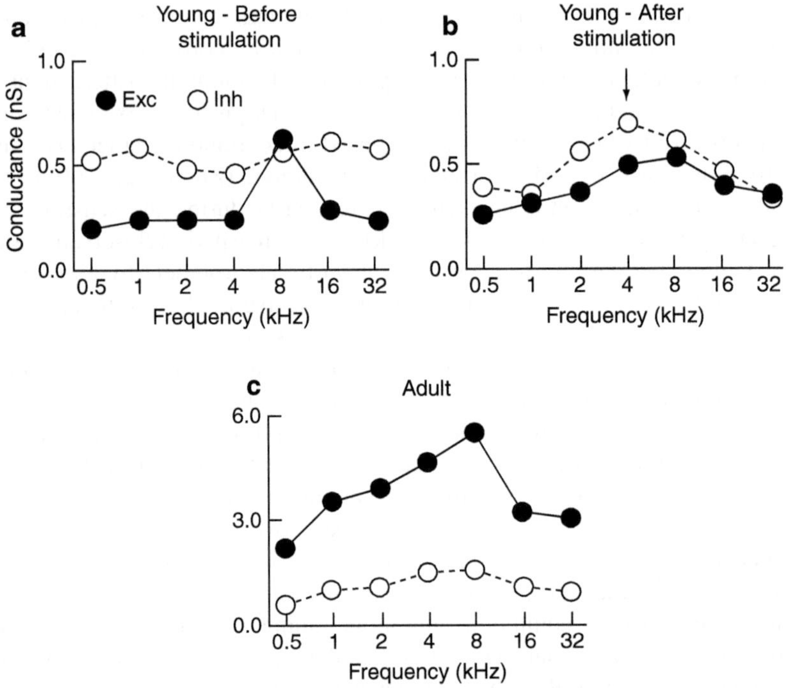

Fig. 10.5 Patterned stimulation drives response properties of developing auditory cortical neurons closer to those of mature animals. (**a**) Excitatory and inhibitory tuning curves from an example cell in the auditory cortex of a developing (P18) rat. The correlation between excitatory and inhibitory conductances is weak ($r = 0.27$). (**b**) Excitatory and inhibitory tuning curves from the same cell as in (**a**) after repeated stimulation with 4-kHz tones. The conductances have become more strongly correlated ($r = 0.82$), and therefore more similar to those seen in the mature auditory cortex. (**c**) In this example cell from an adult rat, there is a strong correlation between excitatory and inhibitory conductance ($r = 0.87$). (Reprinted from Dorrn et al. [2010] with permission from Macmillan Publishers Ltd)

becomes profoundly impressionable. The animal's cortex thus comes to preferentially represent those sounds that are most likely to be encountered in its environment. Closure of this period may not simply be dictated by age but rather experience. Each increment of temporally patterned input impels cortical organization closer to its mature, relatively inflexible state. Through this mechanism, developmental experience influences how subsequent acoustic environments are perceived, though under certain circumstances, the adult cortex can still be modified.

10.3.2 Adult Plasticity from Sound Exposure

Traditionally, research on auditory cortical plasticity has focused on one of two scenarios in which plasticity is commonly elicited: during the sensitive period when exposure alone can retune the cortex, or after a behaviorally relevant auditory experience in adulthood. Because previous investigations of sensitive period plasticity used yoked adults (often the pups' mothers) as controls and turned up no evidence that they undergo exposure-induced plasticity, it was believed that plasticity in adulthood can arise only when a sound is imbued with meaning via associative processes. However, as the work of Zhou et al. (2011) indicates, it is increasingly apparent that this is not the case. Though passive sound exposure may not lead to the same profile of plastic changes in adults and developing animals, the adult cortex is sufficiently plastic under baseline conditions to react in a lasting manner to passive exposure, given that duration of exposure is sufficiently long.

Much of the work that has been instrumental in demonstrating exposure-induced plasticity in adults was conducted in the interest of exploring how environmental enrichment impacts auditory cortical responses. Enrichment of an animal's environment, through exposure to and interaction with stimuli it would not experience under normal laboratory housing conditions, has been associated with improved learning and memory, and enhanced synaptogenesis and gliogenesis in the cortex (van Praag et al., 2000). These findings served as an impetus for workers in the field of audition to ask whether enrichment stimulates plasticity in auditory cortex. Indeed, Percaccio et al. (2007) found that merely exposing adult rats to an enriched acoustic environment consisting of moderate-intensity modulated and unmodulated tones, noise bursts, communication sounds, music, and nature sounds for approximately 1 month was effective in potentiating evoked responses to tones and increasing paired pulse depression in their primary auditory cortices.

However, making use of an alternative enrichment procedure, Noreña and colleagues (2006) showed a strikingly different pattern of exposure-induced plasticity in mature cats. Pulsed, pseudorandom presentation of tone pips between 5 and 20 kHz at 80 dB SPL, 24 hours per day for approximately 140 days led to a dramatic suppression of evoked cortical responses to frequencies within the exposed range, and a potentiation of responses to frequencies lateral to the exposed band (Fig. 10.6). Follow-up studies replicated this effect even after reducing the duration of exposure and lowering sound intensity (Pienkowski & Eggermont, 2009, 2010b). This pattern of plasticity is consistent with the notion that the auditory cortical neurons of exposed animals are becoming habituated to the exposed frequencies. Just as repeated exposure to an irrelevant stimulus reduces behavioral responsiveness to that stimulus, repeated stimulation with a sound that carries no particular meaning suppresses auditory cortical responses to that sound (Condon & Weinberger, 1991). However, when cats are exposed to tone pips drawn from two separate frequency bands, cortical responses to frequencies both within and between the exposed bands are suppressed (Pienkowski & Eggermont, 2010a). Because the habituation model would predict that neuronal responses would only be suppressed in response to frequencies *within* the exposed band, a mechanism more complex than simple habituation is likely to be at work here.

Regardless of the specific effects mediated by "enriched" exposure environments, all of the work discussed earlier in this section supports the notion that auditory cortical plasticity can be evoked by passive exposure in adult animals. Although

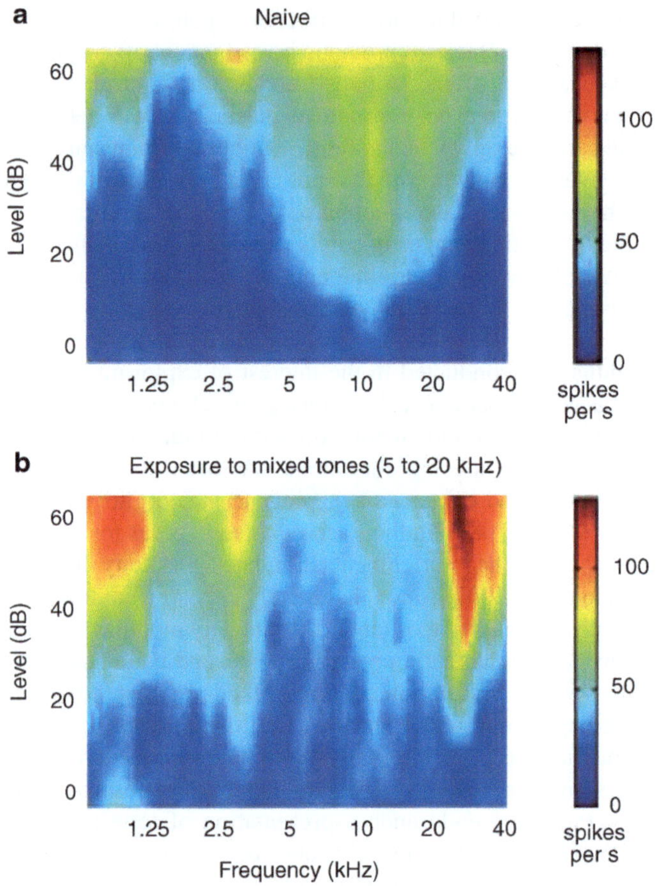

Fig. 10.6 Long-term exposure to tone pips alters auditory cortical response properties in mature animals. Plots show average firing rates of primary auditory cortex multiunits in response to tones of varying frequencies and intensities. (**a**) A group of naive control cats (*n* = 435 multiunits). (**b**) A group of cats exposed to 5- to 20-kHz tone pips presented at 80 dB SPL, 24 hours per day for 19–22 weeks (*n* = 207 multiunits). Responses to frequencies between 5 and 20 kHz were suppressed in animals that had been subject to the exposure environment, while responses to frequencies flanking that range were potentiated. (Reprinted from Noreña et al. [2006] with permission from Macmillan Publishers Ltd)

this work is somewhat at odds with previous studies that argued against the possibility of exposure-induced plasticity in adults, the methodological differences in exposure protocols between laboratories are a likely source of variability here. For instance, Zhang et al. (2001) exposed rats to pulsed pure tones for 10–16 hours per day for 19 days and did not observe plasticity in adult animals. Percaccio et al. (2007), on the other hand, did observe exposure-induced plasticity after exposing adult rats to a highly variable acoustic environment continuously for more than 30 days. With so many variables between studies, it is difficult to determine the locus of differences in plasticity, just as in the cases cited earlier of experimental variations producing differences in "active" paradigms of adult plasticity. Although further investigation will pin down the precise variables that induce one profile of plasticity over another, what is clear is that adult animals can undergo plasticity in response to passive acoustic exposure.

It is important to note that although the auditory cortices of adults and developing animals alike are sensitive to background stimulation in their acoustic environments, the physiological changes that they undergo upon exposure are not equivalent. First, the threshold to plasticity induction may be lower in developing animals. If adults underwent plasticity as readily as developing animals, Zhang and colleagues (2001) would have observed plastic changes in both adults and pups, as groups of animals containing adults and pups were exposed to identical acoustic environments before electrophysiological mapping of the cortex. Second, though an exposure routine that induces plasticity in adults has yet to be tested on a cohort of young animals, it does not appear that the type of plasticity induced by exposure is comparable between age groups. Even in response to moderate-intensity stimulation, adults react with a suppression of evoked activity within and, under certain conditions, lateral to the exposed frequency range (Pienkowski & Eggermont, 2010a). This can be contrasted against the frequency-specific potentiation of cortical responses observed in developing animals after moderate-intensity stimulation (Zhang et al., 2001; Han et al., 2007). Adults may be capable of undergoing exposure-induced plasticity, but that does not mean exposure-induced plasticity proceeds the same way in adults and developing animals.

Further exploration of these differences will be critical in understanding the mechanisms that enable plasticity in both infants and adults, and may also hint at the behavioral significance of such changes. To date, very few studies have investigated the behavioral advantages conferred by exposure-induced plasticity in developing animals (cf. Han et al., 2007), and none have done so in adults. Doing so will be critical in justifying the importance of auditory cortical plasticity in learning and cognitive processes.

10.4 Mechanisms of Experience-Dependent Cortical Plasticity

Until this point, this chapter has focused on two types of plasticity that have so far been treated as distinct phenomena: plasticity that accompanies active associative learning and plasticity that follows passive sound exposure. This reflects the tendency on display in the field at large, with plasticity classified either as under behavioral control or not. Although this implicitly suggests the involvement of two different plasticity-control systems, this section argues that at the cellular level, the same mechanisms regulate plasticity in passive exposure and active learning scenarios. What differs between those scenarios and dictates how plasticity proceeds is relative activation of those mechanisms.

Auditory cortical plasticity is the likely result of synaptic plasticity which, although directly caused by excitatory signaling events, is under the regulatory control of neuromodulator systems. Neuromodulators can influence receptor trafficking, postsynaptic conductances, and neurotransmitter release in a way that impacts the likelihood that plasticity will develop. For instance, on binding with β-adrenergic receptors, norepinephrine (NE) encourages delivery of GluR1-containing α-amino-3-hydroxy-5-methyl-4-isoxazolepropionic acid (AMPA) receptors to the postsynaptic cell membrane (Hu et al., 2007). This has the effect of enhancing the evoked excitatory current and decreasing the threshold for long-term potentiation (LTP), a key form of synaptic plasticity and the putative basis of long-term memory. Another neuromodulator, acetylcholine (ACh), increases likelihood of neurotransmitter release in γ-aminobutyric acid-ergic (GABAergic) neurons when it binds with muscarinic receptors (González et al., 2011). Enhancing GABAergic transmission would decrease the likelihood of synaptic plasticity in a postsynaptic cell by preventing the postsynaptic excitation necessary for LTP to develop. Importantly, both of these effects are highly receptor dependent. One cannot conclude, then, that NE increases the likelihood of plasticity and ACh decreases it; the signaling events that occur upon neuromodulator exposure depends on which neurotransmitter receptors are expressed in the tissue of interest.

Like the rest of the cortex, auditory cortex benefits from access to diverse neuromodulatory systems that likely enable its highly plastic nature. The cortex is innervated diffusely by noradrenergic, dopaminergic, serotonergic (Jacobs & Azmitia, 1992), cholinergic, and histaminergic (Watanabe et al., 1984) afferents. Further, lower auditory structures and other subcortical structures that feed into auditory cortex are themselves subject to neuromodulation by these systems, and may modify cortical response patterns via plasticity in feedforward projection neurons. Though the understanding of neuromodulation of auditory cortical plasticity is perhaps not as fully developed as that in visual cortex, a growing body of work is elucidating the roles played by the various neuromodulator systems in auditory cortex and beginning to clarify how they each contribute to plasticity.

10.4.1 Neuromodulator Systems Involved in Auditory Cortical Plasticity

10.4.1.1 ACh

By far, the most extensively studied neuromodulator in the field of auditory cortical plasticity is ACh. Release of ACh into auditory cortex appears to be the proximate mechanism enabling learning-induced plasticity, as cortical blockade of cholinergic muscarinic receptors prevents retuning under circumstances that otherwise permit it (Bakin & Weinberger, 1996; Ji et al., 2001; Ji & Suga, 2003). In addition to these local effects, cholinergic signaling at the level of the inferior colliculus is also required for the expression of plasticity in auditory cortex (Ji et al., 2001), indicating that neuromodulatory activity outside of cortex is as important as local activity in regulating plasticity. In addition to being necessary for expression of auditory cortical plasticity, release of ACh into the cortex is sufficient for induction of plasticity when paired with tone presentation. In primary and secondary cortices, retuning can be induced by pairing tone presentation with stimulation of the nucleus basalis (NB), which provides the forebrain with ACh (Bakin & Weinberger, 1996; Kilgard & Merzenich, 1998b). As in the case of learning-induced plasticity, paired NB + tone stimulation results in potentiation of the firing rate response to the paired frequency, but produces no change or a decrease in firing rate at other frequencies. In other words, NB stimulation can stand in as a proxy for the aversive stimulus in establishing the salience of an otherwise neutral sound. Consistent with the notion that cholinergic induction of plasticity represents the same processes that occur during learning, ACh-mediated plasticity is accompanied by behavioral responses similar to those expressed during acquisition of an auditory association (e.g., changes in heart or breathing rate in response to the paired frequency) (Miasnikov et al., 2006; Weinberger et al., 2006).

The idea that cholinergic signaling facilitates the formation of auditory associations is consistent with broad hypotheses about the role of ACh in learning and memory. Nicotine, an agonist of the nicotinic ACh receptor, has long been known to enhance learning in adults (Garg, 1969; Mangan, 1983), and its absence impairs performance on a conditioned avoidance task, mastery of which is dependent on successful encoding of an associative memory (Picciotto et al., 1995). Antagonism of muscarinic ACh receptors also disrupts learning and memory processes (Hagan et al., 1987; Taffe et al., 2002). It has been proposed that these behavioral effects are mediated by the action of ACh on local cortical networks (Hasselmo & Bower, 1993). Specifically, data from in vitro and in silico preparations indicate that ACh can suppress activity in intracortically projecting neurons without altering input from ascending afferents (Hasselmo et al., 1992; Hasselmo & Bower, 1992). The ultimate effect of this is enhanced receptivity to afferent stimulation without interference from ongoing intracortical firing patterns.

Given the well-researched link between ACh and active learning, it may not be surprising that plasticity induced by passive exposure or peripheral injury in adults may not be dependent on cholinergic signaling. Percaccio et al. (2007) found that lesioning the NB had no effect on the plasticity that arose from environmental enrichment in adult rats. Likewise, the redistribution of cortical territory following focal lesion to the cochlea was not impeded by destruction of the NB in adult cats (Kamke et al., 2005). It may be that attentional processes recruit cholinergic activity during associative learning, but because these processes are presumably attenuated during passive exposure or recovery from insult, cholinergic signaling takes a "back seat" to other plasticity-generating mechanisms.

Regardless, the aforementioned studies have made such a strong case for the involvement of ACh in auditory cortical plasticity that it is often referred to as the final common pathway for induction of plasticity. Its actions are so well characterized that it has been exploited to enable study of plasticity-related phenomena (Kilgard et al., 2001; Miasnikov et al., 2011) and probe the behavioral effects of auditory cortical plasticity (Reed et al., 2011). However, natural acquisition of auditory associations is likely to involve multiple neuromodulatory systems acting

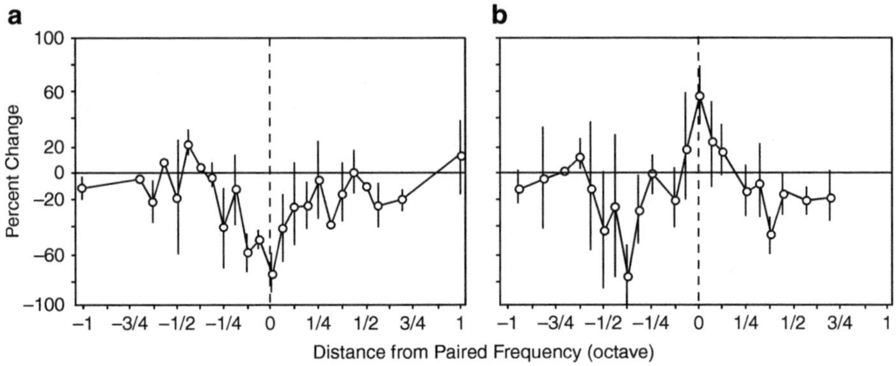

Fig. 10.7 Tuning curves of auditory cortical neurons were measured before and after 100 pairings of tone presentation and NE microinfusion. Differences between pre- and post-stimulation tuning curves are plotted. (**a**) One subset of single units (n = 15/67) exhibited a selective decrease in the response to the paired frequency. (**b**) Another group (n = 4/67) showed a frequency-specific potentiation in response to the paired tone. Remaining neurons developed general increases or decreases in firing rate responses or underwent no change. (Reprinted from Manunta & Edeline [2004] with permission from the American Physiological Society)

in concert, as it is believed they do in the visual system (Kirkwood et al., 1999). A small but growing group of researchers is now working in auditory cortex to uncover the contributions of other neurotransmitter systems to plasticity processes.

10.4.1.2 Norepinephrine

Work in visual cortex initially suggested a strong role for NE in regulating experience-dependent plasticity. Work by Kasamatsu and colleagues indicated that NE was required for the development of ocular dominance columns, which are formed depending on patterns of sensory input during the sensitive period in visual cortex (Kasamatsu & Pettigrew, 1976; Pettigrew & Kasamatsu, 1978). However, later studies refuted a critical role of NE (Bear & Daniels, 1983; Bear & Singer, 1986) and seemed to indicate that although NE was involved in sensitive period plasticity in visual cortex, its role was somehow redundant. This issue remains unresolved today.

Although dependence of sensitive period plasticity on NE in auditory cortex was never explored, others asked whether pairing tone presentation with activation of noradrenergic systems would induce plasticity in adult animals (Bollinger, 2006), as NB + tone stimulation did. Indeed, pairing tones with pulses of NE microinfusion leads to enduring changes in frequency tuning in auditory cortical neurons (Manunta & Edeline, 2004). Unlike ACh-induced plasticity, though, NE-induced plasticity is primarily suppressive, producing frequency-specific depression of evoked responses to the paired frequency (Fig. 10.7). Follow-up work replaced local NE infusion with stimulation of the locus ceruleus, a brain stem nucleus that serves as the only source of NE to the forebrain. This study showed more heterogeneity of plasticity responses, with frequency-specific enhancement occurring about as often as frequency-specific response suppression (Edeline et al., 2011). Cortical tuning shifts were enduring, lasting at least through the end of the recording (>15 minutes after pairing), whereas tuning curve shifts recorded in the thalamus dissipated quickly.

Although the data gleaned from studies of NE and auditory cortical plasticity may be difficult to interpret, they nevertheless indicate that noradrenergic signaling has the capacity to alter auditory cortical response properties. This observation is consistent with the widely posited hypothesis that noradrenergic signaling via the locus ceruleus serves as an indicator of stimulus salience or relevance (Foote et al., 1980; Aston-Jones et al., 1991). When activated by an arousing stimulus, the locus ceruleus releases NE diffusely throughout the forebrain. Its targets include auditory cortex (Harper & Wallace, 1995), as well as the NB (Haring & Wang, 1986; España & Berridge, 2006) and basolateral nucleus of the amygdala (Radwanska et al., 2010), each of which may participate in auditory plasticity, should a meaningful acoustic stimulus accompany arousal (Bakin & Weinberger, 1996; Kilgard & Merzenich, 1998b; Chavez et al., 2009).

10.4.1.3 Dopamine

Because of its close involvement in reward signaling, a number of studies have explored dopaminergic influences on auditory cortical plasticity. Stimulation of the dopaminergic ventral tegmental area (VTA) is known to produce a pleasurable feeling in animals, to the extent that it can serve as a reward in operant learning paradigms. Auditory system researchers have taken advantage of this, yoking a particular behavioral contingency to VTA stimulation in some auditory learning tasks (Hui et al., 2009). Indeed, in animals that learned the association between the conditioned stimulus and the VTA stimulation reward, an increase in cortical area devoted to the reinforced frequency occurred. However, such plasticity may not be dependent on a behavioral contingency. Others have shown that simply pairing VTA stimulation with tone presentation results in retuning in primary and secondary auditory cortices (Bao et al., 2001; Kisley & Gerstein, 2001). Additional work is needed to clarify how dopaminergic signaling contributes to such plasticity, and whether it is the proximate event leading to sensory retuning.

10.4.1.4 Serotonin

The link between 5-hydroxytryptamine (5-HT, serotonin) and long-term plasticity is well established in simple invertebrate preparations. For instance, it is known to facilitate long-term potentiation via closure of K^+ leak channels in a monosynaptic reflex pathway in *Aplysia* (Brunelli et al., 1976; Siegelbaum et al., 1982). Further, at the systems level, 5-HT is released under circumstances where plasticity is likely to take place: upon exposure to a novel environment (Bickerdike et al., 1993), following an aversive event (Boutelle et al., 1990), or during a social encounter

(Korzan et al., 2000). Despite this, a role for 5-HT in plasticity has been hard to define in the auditory cortex. There, 5-HT decreases neural responsiveness in a dose-dependent manner, but enables behaviorally induced tuning shifts that would otherwise be subthreshold at low doses (Ji & Suga, 2007). High doses of 5-HT, on the other hand, impede the development of plasticity. These somewhat contradictory effects are presumed to be mediated by differential effects on the various 5-HT receptors. Indeed, ritanserin, a 5-HT_{2A} receptor antagonist, always facilitated tuning curve shifts when applied to auditory cortex in the same study, suggesting activation of those receptors ordinarily works against plasticity induction.

Of the monoaminergic neuromodulators, 5-HT is the most poorly studied by workers in auditory cortical plasticity. However, serotonergic effects on plasticity have been studied at lower points in the auditory processing stream, particularly in the inferior colliculus (Hurley & Hall, 2011). There, as in auditory cortex, 5-HT exerts variable functions on neural response properties depending on which receptor populations it activates (Hurley & Pollak, 2001; Ramsey et al., 2010). Though evidence for serotonergic regulation of auditory cortical plasticity may be scarce, the work described in the preceding sentences has established it as a key modulator of auditory function in subcortical nuclei. Through feedforward connections, 5-HT–modulated plasticity in lower areas may impact cortical processing, or at least impact the likelihood of plasticity in cortical areas via enhancement of feedforward responses.

10.4.1.5 Estrogen

The sex hormone estrogen has far-reaching effects on brain and body systems, so its involvement in learning processes is unsurprising (Packard, 1998). In fact, estrogen has been linked to the maintenance of cognitive function in aging women, particularly for verbal memory (Sherwin, 2003), potentially protecting against neurodegenerative diseases (Gibbs & Aggarwal, 1998). The primary hypothesis for its action is that it modulates the activity of basal forebrain cholinergic neurons that project widely to the hippocampus and cortex (Gibbs & Aggarwal, 1998; Bora et al., 2005), illustrating how neuromodulators may work in concert under normal conditions to facilitate plasticity. In ethological animal models, work on maternal rodents, which have altered levels of estrogen and other sex hormones compared to naïve females, has shown that the former are more adept at spatial learning tasks (Kinsley et al., 1999; Lambert et al., 2005), presumably owing to enhanced spine formation in the hippocampus (Woolley & McEwen, 1993; Kinsley et al., 2006). This is believed to subserve more efficient foraging behaviors at a time when the survival of many individuals is in the hands of the maternal female. Estrogen-facilitated improvement in learning during the postpartum period could benefit sensory systems as well, as litter survival depends on the female's ability to perceive and respond properly to the stimuli her young emit.

Though it has not yet been explored in depth in mammalian systems, evidence for estrogenic modulation of sensory plasticity has been found in the songbird. Female sparrows under a chronic estrogen stimulation regimen exhibit enhanced expression of the immediate early gene (IEG) *EGR1* (zenk/zif268/NGF-1A/Krox-24) in the auditory forebrain only in response to male song (Maney et al., 2006). The *EGR1* response to frequency-matched tones, on the other hand, is not affected by estrogen treatment. Because *EGR1* expression scales with neural activity, it can be inferred that estrogen selectively potentiates the neural response to birdsong in the avian auditory forebrain. Recent electrophysiological studies further suggest that estrogens can produce rapid effects on neuronal spiking in response to song (Tremere et al., 2009; Remage-Healey et al., 2010), and not just a slow genomic effect. It is not known whether the mammalian sensory cortex is similarly responsive to estrogen, though there is evidence for estrogen receptor expression in the infragranular layers of auditory cortex during development (Yokosuka et al., 1995) and adulthood (Charitidi & Canlon, 2010). Suggestively, a comparison of auditory cortical responses in virgin female mice and mothers reveals evidence of long-term plastic changes (Liu et al., 2006; Liu & Schreiner, 2007; Galindo-Leon et al., 2009) that may be partly attributable to differences in the hormonal environment between these females (Miranda & Liu, 2009). Future investigations will clarify the contribution of estrogen and other hormones to such plasticity. For further discussion of the common neuromodulatory mechanisms of auditory plasticity, see Section 10.4.3.

10.4.2 Genomic Mechanisms of Auditory Cortical Plasticity

In contrast to the large body of research on the electrophysiological manifestations and neuromodulatory mechanisms of cortical plasticity, there has been less work on the genomic mechanisms within auditory cortical neurons that enable long-term plasticity. IEGs such as transcription factors *FOS* and *EGR1* are obvious choices to investigate, based on extensive work looking at their expression in the mammalian auditory brain stem (Ehret & Fischer, 1991; Rouiller et al., 1992; Brown & Liu, 1995), and the auditory forebrain and song learning circuit of songbirds (Jarvis et al., 1995; Mello et al., 1995). As mentioned in Section 10.4.1.5 above, expression of these IEGs is often considered to be a "marker" for neural activity, and *FOS* activation in auditory cortex has indeed been found to follow a tonotopic pattern of expression (Zuschratter et al., 1995). However, caution should be applied to such an "activation" interpretation because a neuron can be electrophysiologically active without generating IEG expression. The classic auditory illustration of this is the absence of sound-driven *FOS* or *EGR1* expression in songbird Field L, despite this nucleus' role as the analog of the thalamorecipient layer of mammalian auditory cortex, and its robust sound-driven spiking (Sen et al., 2001). Hence, these activity-triggered transcription factors are instead thought to initiate cellular programs that underlie plasticity and the neural basis for acquiring a signal's behavioral relevance (Scheich & Zuschratter, 1995; Carretta et al., 1999).

More recently, "effector" IEGs that play direct roles in synaptic plasticity have gained attention. Among these, the gene for activity-regulated cytoskeleton-associated protein Arc (also known as Arg3.1) is particularly promising because it acts as a key

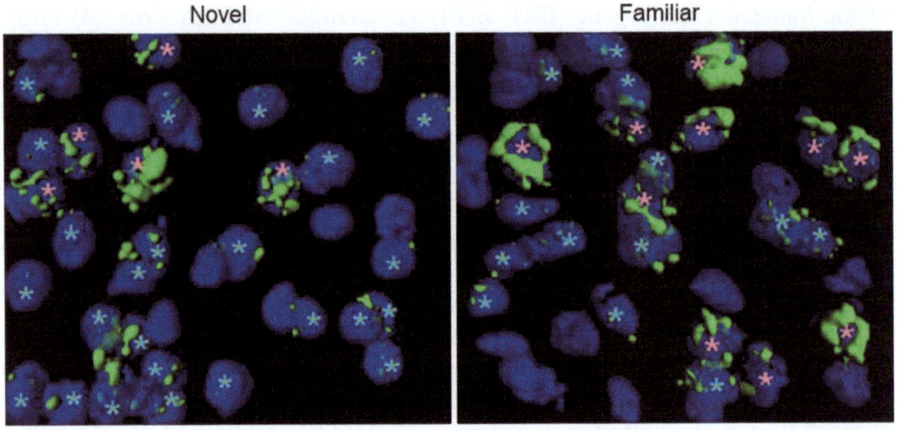

Fig. 10.8 Arc mRNA expression in mouse primary auditory cortex after passive exposure to novel and familiar sounds. Though the total proportion of cells expressing Arc (asterisks) is similar for novel and familiar sounds, the subcellular location of this expression changes. A novel sound induces more cells with foci of Arc (cyan asterisks) inside the nucleus. This Arc is quickly targeted into the cytoplasm (pink asterisks). This shift in the subcellular localization of expression may provide a prolonged molecular trace of the sound experience that facilitates ongoing synaptic plasticity. Images are three-dimensional reconstructions of cells labeled via fluorescent in situ hybridization. Blue shows 4′,6-diamidino-2-phenylindole (DAPI) staining of cell nuclei; green shows Arc mRNA. (Reprinted from Ivanova et al. [2011] with permission from Elsevier)

regulator of translation-dependent synaptic plasticity in the hippocampus (reviewed in Bramham et al., 2010). Arc mRNA is rapidly induced by synaptic activity (Link et al., 1995; Lyford et al., 1995) and transported into dendrites (Dynes & Steward, 2007), accumulating near activated synapses (Moga et al., 2004). Arc protein regulates glutamate receptor trafficking, postsynaptic density remodeling, and spine morphology (Chowdhury et al., 2006; Messaoudi et al., 2007; Peebles et al., 2010). Arc's role in hippocampal-based memory consolidation (Guzowski et al., 2000; Plath et al., 2006) has therefore made it a promising target for investigating molecular mechanisms of long-term auditory cortical plasticity (Mahlke & Wallhäusser-Franke, 2004; Carpenter-Hyland et al., 2010b; Ivanova et al., 2011). Recent data even suggest it may provide a common molecular mechanism for adult plasticity in both novel-sound, passive listening (Ivanova et al., 2011) and operant learning contexts (Carpenter-Hyland et al., 2010a) (Fig. 10.8). Arc should be considered only one of many interesting molecular targets modulated by the genomic response to sounds, as studies in other sensory modalities have demonstrated that a plethora of genes are up- or down-regulated by experience (Valles et al., 2011).

10.4.3 Common Mechanisms for Plasticity Throughout Life?

Much of the evidence presented in Sections 10.4.1 and 10.4.2 was taken from studies that used learning paradigms to induce plasticity. The relative lack of evidence from

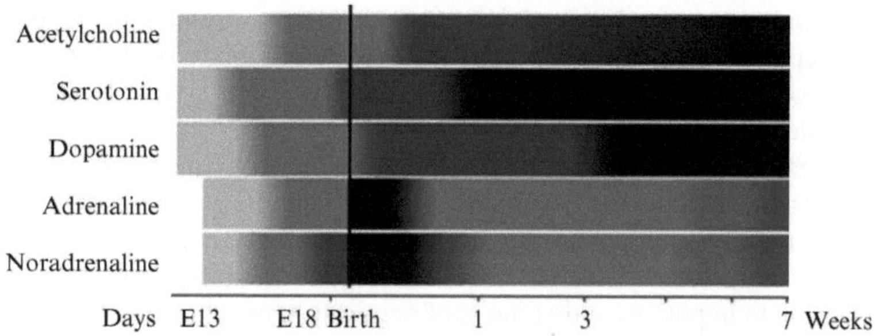

Fig. 10.9 Changing neuromodulator levels during early life. Grayscale gradient shows arbitrary whole-brain levels of neuromodulators in the rat as a function of age (logarithmic scale), with highest levels represented by black shading. (Data from Olson and Sieger [1972], Herregodts et al. [1990], Naeff et al. [1992], Sundstrom et al. [1993], and Almqvist et al. [1996] and used with permission from the publisher. Reprinted from Herlenius & Lagercrantz [2004] with permission from Elsevier)

studies of exposure-induced plasticity— whether in developing animals or adults—is the result of a comparative dearth of mechanistic studies in that area, perhaps stemming from the fact that exposure-induced plasticity is still a relatively new phenomenon in auditory cortex research. To understand fully auditory cortical plasticity and its place in learning processes, this is a gap that should be addressed. The aim of this section is to discuss a framework in which this matter can be investigated.

Whether plasticity is induced through exposure or processes under behavioral control, it is likely to be gated by the same neuromodulatory and genomic signaling events. That entirely distinct systems control auditory cortical retuning under different circumstances would be arguably inefficient, redundant, and ultimately disadvantageous to the animal. Rather, the specific manifestation of plasticity in different behavioral scenarios may be modulated by differential activation of the same neuromodulatory systems. In adulthood, specific behavioral demands elicit particular neuromodulatory events. For instance, an aversive event may trigger release of NE and 5-HT. A rewarding event facilitates dopamine release. These signaling events then go on to influence the likelihood of plasticity, depending on whether the original behavioral events are salient enough to warrant memory formation.

In development, however, plasticity is more easily induced. Whether a representation of a sound undergoes plasticity is not contingent upon the context surrounding its presentation, perhaps because the sensory environment into which an animal is born is inherently relevant and therefore worthy of neural encoding. Without the need for specific behavioral events to signal the need for plasticity, temporally controlled neuromodulatory events are unnecessary to induce it. Accordingly, plasticity may be essentially constitutive in young animals. This may be enabled by chronically elevated neuromodulator or protein expression levels in developing individuals compared to adults.

Neuromodulatory tone indeed changes throughout the lifetime. In many cases, neurotransmitter levels, receptor expression densities, or efferent innervation densities peak in cortex in early life around the sensitive period (Fig. 10.9). Studies in primary visual cortex, for instance, have revealed that expression of GABA, ACh, and NE receptors peaks there during the sensitive period (Shaw et al., 1985), as does the density of cholinergic efferents to its thalamorecipient layers (Bear & Singer, 1986). Similar changes have been observed in dopamine receptor levels throughout the whole brain (Nair & Mishra, 1995) and the serotonergic innervation of sensory cortex (D'Amato et al., 1987). Regardless of the details, greater access to neuromodulators during development might allow the auditory cortex to sustain a high level of ongoing plasticity, such as that seen in the sensitive period. For example, there is reason to hypothesize that the cholinergic system actually helps enable sensitive period plasticity during developmental sound exposure through nicotinic acetylcholine receptors (nAChR) (Metherate, 2004; Metherate & Hsieh, 2004). Near hearing onset, both nAChR and acetylcholinesterase (cholinergic enzyme) transiently increase their expression within sensory cortex, coinciding with a period when nicotine enhances *N*-methyl-D-aspartate (NMDA)-mediated excitatory postsynaptic potentials in auditory cortex (Aramakis & Metherate, 1998).

In addition to changes in wiring that may support enhanced plasticity, changes in response dynamics in the neuromodulatory nuclei themselves may help lower the threshold for plasticity induction. For example, the noradrenergic locus ceruleus, which is activated by arousing stimuli, is more easily excited during development than in adulthood, and its activation is longer lasting, as well (Nakamura et al., 1987). This effect is believed to be mediated by an immature pattern of inhibitory autoreceptor expression on locus ceruleus neurons, which serve to limit the period of excitation in adults (Nakamura et al., 1988). Whether through enhanced receptivity at the level of the target tissue or increased output from neuromodulatory centers, evidence abounds that plasticity is more readily induced in young animals through increased access to neuromodulators. Contrast this with adult animals, whose access to the same neuromodulatory systems is limited by greater temporal control of release.

Studies of plasticity mechanisms in developing animals will be particularly important in validating the hypothesis that learning- and exposure-induced plasticity stem from differential access to the same neuromodulatory events. To gain a complete understanding of the significance and purpose of auditory cortical plasticity, the relationships between neuromodulatory tone, neuromodulator control over plasticity, and the behavioral events that elicit neuromodulator release would need to be elucidated. A better understanding of neuromodulatory control over plasticity will ultimately lead to better understanding of the behavioral control of plasticity, as links between behavior and neuromodulatory signaling events have already been identified in other fields.

10.5 Summary

In the past several decades, research on auditory cortical plasticity has challenged views on brain structure and function that had persisted for hundreds of years. There is now no doubt that the adult brain can be reshaped—under the right circumstances—thanks to studies of plasticity in sensory cortex. Moreover, the notion that sensory processing areas can be cleanly separated from higher-level areas that handle cognitive functions has been dispelled by recent work that highlights the cognitive capabilities of auditory cortex.

Despite the volume of work that has focused on auditory cortical plasticity in the last several years, however, much about plasticity and its place in auditory processing remains to be understood, and this chapter has tried to draw attention to such open issues. One such issue is the segmented way in which plasticity is studied and, in turn, conclusions are drawn. This has left the research community with two separate bodies of literature on auditory plasticity: one largely featuring studies of learning-induced plasticity in adults and the other focused on exposure-induced plasticity in young animals. Recent forays into exposure-induced plasticity in adults have begun to bridge this gap, but additional work will be necessary to close it completely. Further, the question of what purpose auditory plasticity serves has yet to be answered definitively. Progress has been made on this front as well, though the case is made here that ethological paradigms, which attempt to replicate the environment in which sound acquisition naturally occurs, represent an underutilized but potentially fruitful means of addressing this question.

Finally, the benefits of research on cortical plasticity extend beyond just an improved understanding of sensation, perception and memory; it may also hold a key to new therapies for disorders of sensory processing. For example, tinnitus, a central auditory disorder wherein sufferers perceive background auditory sensations (e.g., ringing) in the absence of such a stimulus, has been linked to changes in neural synchrony and spontaneous activity in auditory cortex, as well as a host of subcortical nuclei (Eggermont & Roberts, 2004). Hypothesizing that the auditory cortex represents the final common pathway for the phantom percept and reversing neural abnormalities there would alleviate the primary symptom, Engineer et al. (2011) induced plasticity in rat auditory cortex by pairing tones at the non-tinnitus frequencies with stimulation of the vagus nerve, which triggers widespread release of neuromodulators. This approach was successful in eliminating the tinnitus percept in rats. This is of particular interest to human sufferers, as vagus nerve stimulation is a relatively noninvasive procedure already in widespread use for treatment of epilepsy and depression, and may someday constitute a novel treatment for tinnitus. Along these lines, a greater understanding of the mechanisms of plasticity could point toward further therapeutic insights, both in auditory cortex and beyond.

References

Ahissar, E., Abeles, M., Ahissar, M., Haidarliu, S., & Vaadia, E. (1998). Hebbian-like functional plasticity in the auditory cortex of the behaving monkey. *Neuropharmacology*, 37(4–5), 633–655.

Almqvist, P. M., Åkesson, E., Wahlberg, L. U., Pschera, H., Seiger, Å., & Sundström, E. (1996). First trimester development of the human nigrostriatal dopamine system. *Experimental Neurology*, 139, 227–237.

Aramakis, V. B., & Metherate, R. (1998). Nicotine selectively enhances NMDA receptor-mediated synaptic transmission during postnatal development in sensory neocortex. *Journal of Neuroscience*, 18(20), 8485–8495.
Aston-Jones, G., Chiang, C., & Alexinsky, T. (1991). Discharge of noradrenergic locus coeruleus neurons in behaving rats and monkeys suggests a role in vigilance. *Progress in Brain Research*, 88, 501–520.
Bakin, J. S., & Weinberger, N. M. (1990). Classical conditioning induces CS-specific receptive field plasticity in the auditory cortex of the guinea pig. *Brain Research*, 536(1–2), 271–286.
Bakin, J. S., & Weinberger, N. M. (1996). Induction of a physiological memory in the cerebral cortex by stimulation of the nucleus basalis. *Proceedings of the National Academy of Sciences of the USA*, 93(20), 11219–11224.
Bakin, J. S., South, D. A., & Weinberger, N. M. (1996). Induction of receptive field plasticity in the auditory cortex of the guinea pig during instrumental avoidance conditioning. *Behavioral Neuroscience*, 110(5), 905–913.
Bao, S., Chan, V. T., & Merzenich, M. M. (2001). Cortical remodeling induced by activity of ventral tegmental dopamine neurons. *Nature*, 412(6842), 79–83.
Baranyi, A., & Feher, O. (1981). Synaptic facilitation requires paired activation of convergent pathways in the neocortex. *Nature*, 290(5805), 413–415.
Bear, M. F., & Daniels, J. D. (1983). The plastic response to monocular deprivation persists in kitten visual cortex after chronic depletion of norepinephrine. *Journal of Neuroscience*, 3(2), 407–416.
Bear, M. F., & Singer, W. (1986). Modulation of visual cortical plasticity by acetylcholine and noradrenaline. *Nature*, 320(6058), 172–176.
Berlau, K. M., & Weinberger, N. M. (2008). Learning strategy determines auditory cortical plasticity. *Neurobiology of Learning and Memory*, 89(2), 153–166.
Bickerdike, M. J., Wright, I. K., & Marsden, C. A. (1993). Social isolation attenuates rat forebrain 5-HT release induced by KCI stimulation and exposure to a novel environment. *Behavioural Pharmacology*, 4(3), 231–236.
Bieszczad, K. M., & Weinberger, N. M. (2010a). Learning strategy trumps motivational level in determining learning-induced auditory cortical plasticity. *Neurobiology of Learning and Memory*, 93(2), 229–239.
Bieszczad, K. M., & Weinberger, N. M. (2010b). Remodeling the cortex in memory: Increased use of a learning strategy increases the representational area of relevant acoustic cues. *Neurobiology of Learning and Memory*, 94(2), 127–144.
Bollinger, J. J. (2006). *Adult auditory cortical plasticity modulated by locus coeruleus activity*. Doctoral dissertation, University of California, San Francisco, San Francisco.
Bora, S. H., Liu, Z., Kecojevic, A., Merchenthaler, I., & Koliatsos, V. E. (2005). Direct, complex effects of estrogens on basal forebrain cholinergic neurons. *Experimental Neurology*, 194(2), 506–522.
Boutelle, M. G., Zetterstrom, T., Pei, Q., Svensson, L., & Fillenz, M. (1990). In vivo neurochemical effects of tail pinch. *Journal of Neuroscience Methods*, 34(1–3), 151–157.
Bramham, C. R., Alme, M. N., Bittins, M., Kuipers, S. D., Nair, R. R., Pai, B., et al. (2010). The Arc of synaptic memory. *Experimental Brain Research*, 200(2), 125–140.
Brown, M., Irvine, D. R., & Park, V. N. (2004). Perceptual learning on an auditory frequency discrimination task by cats: Association with changes in primary auditory cortex. *Cerebral Cortex*, 14(9), 952–965.
Brown, M. C., & Liu, T. S. (1995). Fos-like immunoreactivity in central auditory neurons of the mouse. *Journal of Comparative Neurology*, 357(1), 85–97.
Brunelli, M., Castellucci, V., & Kandel, E. R. (1976). Synaptic facilitation and behavioral sensitization in *Aplysia*: Possible role of serotonin and cyclic AMP. *Science*, 194(4270), 1178–1181.
Carpenter-Hyland, E. P., Bunting, K., Blake, D. T., & Vazdarjanova, A. (2010a). Arc anti-sense in primary auditory cortex blocks perceptual learning. Abstract presented at the Society for Neuroscience, San Diego, CA, November 13–17. Abstract 406.15.
Carpenter-Hyland, E. P., Plummer, T. K., Vazdarjanova, A., & Blake, D. T. (2010b). Arc expression and neuroplasticity in primary auditory cortex during initial learning are inversely related to neural activity. *Proceedings of the National Academy of Sciences*, 107(33), 14828–14832.

Carretta, D., Herve-Minvielle, A., Bajo, V. M., Villa, A. E., & Rouiller, E. M. (1999). c-*Fos* expression in the auditory pathways related to the significance of acoustic signals in rats performing a sensory-motor task. *Brain Research*, 841(1–2), 170–183.
Chang, E. F., & Merzenich, M. M. (2003). Environmental noise retards auditory cortical development. *Science*, 300(5618), 498–502.
Charitidi, K., & Canlon, B. (2010). Estrogen receptors in the central auditory system of male and female mice. *Neuroscience*, 165(3), 923–933.
Chavez, C. M., McGaugh, J. L., & Weinberger, N. M. (2009). The basolateral amygdala modulates specific sensory memory representations in the cerebral cortex. *Neurobiology of Learning and Memory*, 91(4), 382–392.
Chowdhury, S., Shepherd, J. D., Okuno, H., Lyford, G., Petralia, R. S., Plath, N., et al. (2006). Arc/Arg3.1 interacts with the endocytic machinery to regulate AMPA receptor trafficking. *Neuron*, 52(3), 445–459.
Condon, C. D., & Weinberger, N. M. (1991). Habituation produces frequency-specific plasticity of receptive fields in the auditory cortex. *Behavioral Neuroscience*, 105(3), 416–430.
D'Amato, R. J., Blue, M. E., Largent, B. L., Lynch, D. R., Ledbetter, D. J., Molliver, M. E., & Snyder, S. H. (1987). Ontogeny of the serotonergic projection to rat neocortex: Transient expression of a dense innervation to primary sensory areas. *Proceedings of the National Academy of Sciences of the USA*, 84(12), 4322–4326.
de Villers-Sidani, E., Chang, E. F., Bao, S., & Merzenich, M. M. (2007). Critical period window for spectral tuning defined in the primary auditory cortex (A1) in the rat. *Journal of Neuroscience*, 27(1), 180–189.
Diamond, D. M., & Weinberger, N. M. (1984). Physiological plasticity of single neurons in auditory cortex of the cat during acquisition of the pupillary conditioned response: II. Secondary field (AII). *Behavioral Neuroscience*, 98(2), 189–210.
Diamond, D. M., & Weinberger, N. M. (1986). Classical conditioning rapidly induces specific changes in frequency receptive fields of single neurons in secondary and ventral ectosylvian auditory cortical fields. *Brain Research*, 372(2), 357–360.
Dorrn, A. L., Yuan, K., Barker, A. J., Schreiner, C. E., & Froemke, R. C. (2010). Developmental sensory experience balances cortical excitation and inhibition. *Nature*, 465(7300), 932–936.
Dynes, J. L., & Steward, O. (2007). Dynamics of bidirectional transport of Arc mRNA in neuronal dendrites. *Journal of Comparative Neurology*, 500(3), 433–447.
Edeline, J. M., & Weinberger, N. M. (1993). Receptive field plasticity in the auditory cortex during frequency discrimination training: Selective retuning independent of task difficulty. *Behavioral Neuroscience*, 107(1), 82–103.
Edeline, J. M., Pham, P., & Weinberger, N. M. (1993). Rapid development of learning-induced receptive field plasticity in the auditory cortex. *Behavioral Neuroscience*, 107(4), 539–551.
Edeline, J. M., Manunta, Y., & Hennevin, E. (2011). Induction of selective plasticity in the frequency tuning of auditory cortex and auditory thalamus neurons by locus coeruleus stimulation. *Hearing Research*, 274(1–2), 75–84.
Eggermont, J. J., & Roberts, L. E. (2004). The neuroscience of tinnitus. *Trends in the Neurosciences*, 27(11), 676–682.
Ehret, G., & Fischer, R. (1991). Neuronal activity and tonotopy in the auditory system visualized by c-Fos gene expression. *Brain Research*, 567(2), 350–354.
Engineer, N. D., Riley, J. R., Seale, J. D., Vrana, W. A., Shetake, J. A., Sudanagunta, S. P., et al. (2011). Reversing pathological neural activity using targeted plasticity. *Nature*, 470(7332), 101–104.
España, R. A., & Berridge, C. W. (2006). Organization of noradrenergic efferents to arousal-related basal forebrain structures. *Journal of Comparative Neurology*, 496(5), 668–683.
Fichtel, I., & Ehret, G. (1999). Perception and recognition discriminated in the mouse auditory cortex by c-Fos labeling. *NeuroReport*, 10(11), 2341–2345.
Foote, S. L., Aston-Jones, G., & Bloom, F. E. (1980). Impulse activity of locus coeruleus neurons in awake rats and monkeys is a function of sensory stimulation and arousal. *Proceedings of the National Academy of Sciences of the USA*, 77(5), 3033–3037.
Freedman, D. G., King, J. A., & Elliot, O. (1961). Critical period in the social development of dogs. *Science*, 133(3457), 1016–1017.

Fregnac, Y., Shulz, D., Thorpe, S., & Bienenstock, E. (1988). A cellular analogue of visual cortical plasticity. *Nature*, 333(6171), 367–370.

Galambos, R., Sheatz, G., & Vernier, V. G. (1956). Electrophysiological correlates of a conditioned response in cats. *Science*, 123(3192), 376–377.

Galindo-Leon, E. E., Lin, F. G., & Liu, R. C. (2009). Inhibitory plasticity in a lateral band improves cortical detection of natural vocalizations. *Neuron*, 62(5), 705–716.

Gao, E., & Suga, N. (2000). Experience-dependent plasticity in the auditory cortex and the inferior colliculus of bats: Role of the corticofugal system. *Proceedings of the National Academy of Sciences of the USA*, 97(14), 8081–8086.

Garg, M. (1969). The effect of nicotine on two different types of learning. *Psychopharmacologia*, 15(5), 408–414.

Geissler, D. B., & Ehret, G. (2004). Auditory perception vs. recognition: Representation of complex communication sounds in the mouse auditory cortical fields. *European Journal of Neuroscience*, 19(4), 1027–1040.

Gibbs, R. B., & Aggarwal, P. (1998). Estrogen and basal forebrain cholinergic neurons: Implications for brain aging and Alzheimer's disease-related cognitive decline. *Hormones and Behavior*, 34(2), 98–111.

González, J. C., Albinana, E., Baldelli, P., Garcia, A. G., & Hernandez-Guijo, J. M. (2011). Presynaptic muscarinic receptor subtypes involved in the enhancement of spontaneous GABAergic postsynaptic currents in hippocampal neurons. *European Journal of Neuroscience*, 33(1), 69–81.

Guzowski, J. F., Lyford, G. L., Stevenson, G. D., Houston, F. P., McGaugh, J. L., Worley, P. F., & Barnes, C. A. (2000). Inhibition of activity-dependent Arc protein expression in the rat hippocampus impairs the maintenance of long-term potentiation and the consolidation of long-term memory. *Journal of Neuroscience*, 20(11), 3993–4001.

Hagan, J. J., Jansen, J. H., & Broekkamp, C. L. (1987). Blockade of spatial learning by the M_1 muscarinic antagonist pirenzepine. *Psychopharmacology*, 93(4), 470–476.

Han, Y. K., Kover, H., Insanally, M. N., Semerdjian, J. H., & Bao, S. (2007). Early experience impairs perceptual discrimination. *Nature Neuroscience*, 10(9), 1191–1197.

Haring, J. H., & Wang, R. Y. (1986). The identification of some sources of afferent input to the rat nucleus basalis magnocellularis by retrograde transport of horseradish peroxidase. *Brain Research*, 366(1–2), 152–158.

Harlow, H. F., Dodsworth, R. O., & Harlow, M. K. (1965). Total social isolation in monkeys. *Proceedings of the National Academy of Sciences of the USA*, 54(1), 90–97.

Harper, M. S., & Wallace, M. N. (1995). Changes in density of brainstem afferents in ferret primary auditory cortex (AI) during postnatal development. *Journal of Anatomy*, 186 (Pt 2), 373–382.

Hasselmo, M. E., & Bower, J. M. (1992). Cholinergic suppression specific to intrinsic not afferent fiber synapses in rat piriform (olfactory) cortex. *Journal of Neurophysiology*, 67(5), 1222–1229.

Hasselmo, M. E., & Bower, J. M. (1993). Acetylcholine and memory. *Trends in the Neurosciences*, 16(6), 218–222.

Hasselmo, M. E., Anderson, B. P., & Bower, J. M. (1992). Cholinergic modulation of cortical associative memory function. *Journal of Neurophysiology*, 67(5), 1230–1246.

Herregodts, P., Velkeniers, B., Ebinger, G., Michotte, Y., Vanhaelst, L., & Hooghe-Peters, E. (1990). Development of monoaminergic neurotransmitters in fetal and postnatal rat brain: Analysis by HPLC with electrochemical detection. *Journal of Neurochemistry*, 55, 774–779.

Hu, H., Real, E., Takamiya, K., Kang, M. G., Ledoux, J., Huganir, R. L., & Malinow, R. (2007). Emotion enhances learning via norepinephrine regulation of AMPA-receptor trafficking. *Cell*, 131(1), 160–173.

Hubel, D. H., & Wiesel, T. N. (1970). The period of susceptibility to the physiological effects of unilateral eye closure in kittens. *Journal of Physiology*, 206(2), 419–436.

Hui, G. K., Wong, K. L., Chavez, C. M., Leon, M. I., Robin, K. M., & Weinberger, N. M. (2009). Conditioned tone control of brain reward behavior produces highly specific representational gain in the primary auditory cortex. *Neurobiology of Learning and Memory*, 92(1), 27–34.

Hurley, L.M. & Hall, I.C. (2011). Context-dependent modulation of auditory processing by serotonin. *Hearing Research*, 279(1–2), 74–84.

Hurley, L. M., & Pollak, G. D. (2001). Serotonin effects on frequency tuning of inferior colliculus neurons. *Journal of Neurophysiology*, 85(2), 828–842.

Insanally, M. N., Kover, H., Kim, H., & Bao, S. (2009). Feature-dependent sensitive periods in the development of complex sound representation. *Journal of Neuroscience*, 29(17), 5456–5462.

Ivanova, T., Matthews, A., Gross, C., Mappus, R. C., Gollnick, C., Swanson, A., et al. (2011). Arc/Arg3.1 mRNA expression reveals a sub-cellular trace of prior sound exposure in adult primary auditory cortex. *Neuroscience*, 181, 117–126.

Jacobs, B. L., & Azmitia, E. C. (1992). Structure and function of the brain serotonin system. *Physiological Reviews*, 72(1), 165–229.

Jarvis, E. D., Mello, C. V., & Nottebohm, F. (1995). Associative learning and stimulus novelty influence the song-induced expression of an immediate early gene in the canary forebrain. *Learning and Memory*, 2(2), 62–80.

Ji, W., & Suga, N. (2003). Development of reorganization of the auditory cortex caused by fear conditioning: Effect of atropine. *Journal of Neurophysiology*, 90(3), 1904–1909.

Ji, W., & Suga, N. (2007). Serotonergic modulation of plasticity of the auditory cortex elicited by fear conditioning. *Journal of Neuroscience*, 27(18), 4910–4918.

Ji, W., Gao, E., & Suga, N. (2001). Effects of acetylcholine and atropine on plasticity of central auditory neurons caused by conditioning in bats. *Journal of Neurophysiology*, 86(1), 211–225.

Kamke, M. R., Brown, M., & Irvine, D. R. (2005). Basal forebrain cholinergic input is not essential for lesion-induced plasticity in mature auditory cortex. *Neuron*, 48(4), 675–686.

Kasamatsu, T., & Pettigrew, J. D. (1976). Depletion of brain catecholamines: Failure of ocular dominance shift after monocular occlusion in kittens. *Science*, 194(4261), 206–209.

Kilgard, M. P., & Merzenich, M. M. (1998a). Plasticity of temporal information processing in the primary auditory cortex. *Nature Neuroscience*, 1(8), 727–731.

Kilgard, M. P., & Merzenich, M. M. (1998b). Cortical map reorganization enabled by nucleus basalis activity. *Science*, 279(5357), 1714–1718.

Kilgard, M. P., Pandya, P. K., Vazquez, J. L., Rathbun, D. L., Engineer, N. D., & Moucha, R. (2001). Spectral features control temporal plasticity in auditory cortex. *Audiology and Neurotology*, 6(4), 196–202.

Kilgard, M. P., Vazquez, J. L., Engineer, N. D., & Pandya, P. K. (2007). Experience dependent plasticity alters cortical synchronization. *Hearing Research*, 229(1–2), 171–179.

Kim, H., & Bao, S. (2009). Selective increase in representations of sounds repeated at an ethological rate. *Journal of Neuroscience*, 29(16), 5163–5169.

Kinsley, C. H., Madonia, L., Gifford, G. W., Tureski, K., Griffin, G. R., Lowry, C., et al. (1999). Motherhood improves learning and memory. *Nature*, 402(6758), 137–138.

Kinsley, C. H., Trainer, R., Stafisso-Sandoz, G., Quadros, P., Marcus, L. K., Hearon, C., et al. (2006). Motherhood and the hormones of pregnancy modify concentrations of hippocampal neuronal dendritic spines. *Hormones and Behavior*, 49(2), 131–142.

Kirkwood, A., Rozas, C., Kirkwood, J., Perez, F., & Bear, M. F. (1999). Modulation of long-term synaptic depression in visual cortex by acetylcholine and norepinephrine. *Journal of Neuroscience*, 19(5), 1599–1609.

Kisley, M. A., & Gerstein, G. L. (2001). Daily variation and appetitive conditioning-induced plasticity of auditory cortex receptive fields. *European Journal of Neuroscience*, 13(10), 1993–2003.

Knudsen, E. I. (2004). Sensitive periods in the development of the brain and behavior. *Journal of Cognitive Neuroscience*, 16(8), 1412–1425.

Korzan, W. J., Summers, T. R., & Summers, C. H. (2000). Monoaminergic activities of limbic regions are elevated during aggression: Influence of sympathetic social signaling. *Brain Research*, 870(1–2), 170–178.

Kraus, M., Schicknick, H., Wetzel, W., Ohl, F., Staak, S., & Tischmeyer, W. (2002). Memory consolidation for the discrimination of frequency-modulated tones in mongolian gerbils is sensitive to protein-synthesis inhibitors applied to the auditory cortex. *Learning and Memory*, 9(5), 293–303.

Kuhl, P. K. (2004). Early language acquisition: Cracking the speech code. *Nature Reviews Neuroscience*, 5(11), 831–843.
Lambert, K. G., Berry, A. E., Griffins, G., Amory-Meyers, E., Madonia-Lomas, L., Love, G., & Kinsley, C. H. (2005). Pup exposure differentially enhances foraging ability in primiparous and nulliparous rats. *Physiology and Behavior*, 84(5), 799–806.
Link, W., Konietzko, U., Kauselmann, G., Krug, M., Schwanke, B., Frey, U., & Kuhl, D. (1995). Somatodendritic expression of an immediate early gene is regulated by synaptic activity. *Proceedings of the National Academy of Sciences of the USA*, 92(12), 5734–5738.
Liu, R. C., & Schreiner, C. E. (2007). Auditory cortical detection and discrimination correlates with communicative significance. *PLoS Biology*, 5(7), e173.
Liu, R. C., Linden, J. F., & Schreiner, C. E. (2006). Improved cortical entrainment to infant communication calls in mothers compared with virgin mice. *European Journal of Neuroscience*, 23(11), 3087–3097.
Lyford, G. L., Yamagata, K., Kaufmann, W. E., Barnes, C. A., Sanders, L. K., Copeland, N. G., et al. (1995). Arc, a growth factor and activity-regulated gene, encodes a novel cytoskeleton-associated protein that is enriched in neuronal dendrites. *Neuron*, 14(2), 433–445.
Mahlke, C., & Wallhäusser-Franke, E. (2004). Evidence for tinnitus-related plasticity in the auditory and limbic system, demonstrated by Arg3.1 and c-Fos immunocytochemistry. *Hearing Research*, 195(1–2), 17–34.
Maney, D. L., Cho, E., & Goode, C. T. (2006). Estrogen-dependent selectivity of genomic responses to birdsong. *European Journal of Neuroscience*, 23(6), 1523–1529.
Mangan, G. L. (1983). The effects of cigarette smoking on verbal learning and retention. *Journal of General Psychology*, 108, 203–210.
Manunta, Y., & Edeline, J. M. (2004). Noradrenergic induction of selective plasticity in the frequency tuning of auditory cortex neurons. *Journal of Neurophysiology*, 92(3), 1445–1463.
McGaugh, J. L. (1966). Time-dependent processes in memory storage. *Science*, 153(742), 1351–1358.
Mello, C., Nottebohm, F., & Clayton, D. (1995). Repeated exposure to one song leads to a rapid and persistent decline in an immediate early gene's response to that song in zebra finch telencephalon. *Journal of Neuroscience*, 15(10), 6919–6925.
Messaoudi, E., Kanhema, T., Soule, J., Tiron, A., Dagyte, G., da Silva, B., & Bramham, C. R. (2007). Sustained Arc/Arg3.1 synthesis controls long-term potentiation consolidation through regulation of local actin polymerization in the dentate gyrus in vivo. *Journal of Neuroscience*, 27(39), 10445–10455.
Metherate, R. (2004). Nicotinic acetylcholine receptors in sensory cortex. *Learning and Memory*, 11(1), 50–59.
Metherate, R., & Hsieh, C. Y. (2004). Synaptic mechanisms and cholinergic regulation in auditory cortex. *Progress in Brain Research*, 145, 143–156.
Miasnikov, A. A., Chen, J. C., & Weinberger, N. M. (2006). Rapid induction of specific associative behavioral memory by stimulation of the nucleus basalis in the rat. *Neurobiology of Learning and Memory*, 86(1), 47–65.
Miasnikov, A. A., Chen, J. C., & Weinberger, N. M. (2011). Consolidation and long-term retention of an implanted behavioral memory. *Neurobiology of Learning and Memory*, 95(3), 286–295.
Miranda, J. A., & Liu, R. C. (2009). Dissecting natural sensory plasticity: Hormones and experience in a maternal context. *Hearing Research*, 252(1–2), 21–28.
Moga, D. E., Calhoun, M. E., Chowdhury, A., Worley, P., Morrison, J. H., & Shapiro, M. L. (2004). Activity-regulated cytoskeletal-associated protein is localized to recently activated excitatory synapses. *Neuroscience*, 125(1), 7–11.
Moucha, R., Pandya, P. K., Engineer, N. D., Rathbun, D. L., & Kilgard, M. P. (2005). Background sounds contribute to spectrotemporal plasticity in primary auditory cortex. *Experimental Brain Research*, 162(4), 417–427.
Naeff, B., Schlumpf, M., & Lichtensteiger, W. (1992). Pre- and postnatal development of high-affinity [^{3}H]nicotine binding sites in rat brain regions: An autoradiographic study. *Developmental Brain Research*, 68, 163–174.

Nair, V. D., & Mishra, R. K. (1995). Ontogenic development of dopamine D4 receptor in rat brain. *Developmental Brain Research*, 90(1–2), 180–183.

Nakamura, S., Kimura, F., & Sakaguchi, T. (1987). Postnatal development of electrical activity in the locus ceruleus. *Journal of Neurophysiology*, 58(3), 510–524.

Nakamura, S., Sakaguchi, T., Kimura, F., & Aoki, F. (1988). The role of alpha 1-adrenoceptor-mediated collateral excitation in the regulation of the electrical activity of locus coeruleus neurons. *Neuroscience*, 27(3), 921–929.

Noreña, A. J., Gourevitch, B., Aizawa, N., & Eggermont, J. J. (2006). Spectrally enhanced acoustic environment disrupts frequency representation in cat auditory cortex. *Nature Neuroscience*, 9(7), 932–939.

Ohl, F. W., & Scheich, H. (1996). Differential frequency conditioning enhances spectral contrast sensitivity of units in auditory cortex (field Al) of the alert Mongolian gerbil. *European Journal of Neuroscience*, 8(5), 1001–1017.

Olson, L., & Sieger, A. (1972). Early prenatal ontogeny of central monoamine neurons in the rat: Fluorescence histochemical observations. *Anatomy and Embryology*, 137, 301–316.

Packard, M. G. (1998). Posttraining estrogen and memory modulation. *Hormones and Behavior*, 34(2), 126–139.

Peebles, C. L., Yoo, J., Thwin, M. T., Palop, J. J., Noebels, J. L., & Finkbeiner, S. (2010). Arc regulates spine morphology and maintains network stability in vivo. *Proceedings of the National Academy of Sciences of the USA*, 107(42), 18173–18178.

Percaccio, C. R., Engineer, N. D., Pruette, A. L., Pandya, P. K., Moucha, R., Rathbun, D. L., & Kilgard, M. P. (2005). Environmental enrichment increases paired-pulse depression in rat auditory cortex. *Journal of Neurophysiology*, 94(5), 3590–3600.

Percaccio, C. R., Pruette, A. L., Mistry, S. T., Chen, Y. H., & Kilgard, M. P. (2007). Sensory experience determines enrichment-induced plasticity in rat auditory cortex. *Brain Research*, 1174, 76–91.

Pettigrew, J. D., & Kasamatsu, T. (1978). Local perfusion of noradrenaline maintains visual cortical plasticity. *Nature*, 271(5647), 761–763.

Picciotto, M. R., Zoli, M., Lena, C., Bessis, A., Lallemand, Y., Le Novere, N., et al. (1995). Abnormal avoidance learning in mice lacking functional high-affinity nicotine receptor in the brain. *Nature*, 374(6517), 65–67.

Pienkowski, M., & Eggermont, J. J. (2009). Long-term, partially-reversible reorganization of frequency tuning in mature cat primary auditory cortex can be induced by passive exposure to moderate-level sounds. *Hearing Research*, 257(1–2), 24–40.

Pienkowski, M., & Eggermont, J. J. (2010a). Passive exposure of adult cats to moderate-level tone pip ensembles differentially decreases AI and AII responsiveness in the exposure frequency range. *Hearing Research*, 268(1–2), 151–162.

Pienkowski, M., & Eggermont, J. J. (2010b). Intermittent exposure with moderate-level sound impairs central auditory function of mature animals without concomitant hearing loss. *Hearing Research*, 261(1–2), 30–35.

Pilley, J. W., & Reid, A. K. (2011). Border collie comprehends object names as verbal referents. *Behavioural Processes*, 86(2), 184–195.

Plath, N., Ohana, O., Dammermann, B., Errington, M. L., Schmitz, D., Gross, C., et al. (2006). Arc/Arg3.1 is essential for the consolidation of synaptic plasticity and memories. *Neuron*, 52(3), 437–444.

Polley, D. B., Heiser, M. A., Blake, D. T., Schreiner, C. E., & Merzenich, M. M. (2004). Associative learning shapes the neural code for stimulus magnitude in primary auditory cortex. *Proceedings of the National Academy of Sciences of the USA*, 101(46), 16351–16356.

Polley, D. B., Steinberg, E. E., & Merzenich, M. M. (2006). Perceptual learning directs auditory cortical map reorganization through top-down influences. *Journal of Neuroscience*, 26(18), 4970–4982.

Popescu, M. V., & Polley, D. B. (2010). Monaural deprivation disrupts development of binaural selectivity in auditory midbrain and cortex. *Neuron*, 65(5), 718–731.

Puckett, A. C., Pandya, P. K., Moucha, R., Dai, W., & Kilgard, M. P. (2007). Plasticity in the rat posterior auditory field following nucleus basalis stimulation. *Journal of Neurophysiology*, 98(1), 253–265.

Radwanska, K., Nikolaev, E., & Kaczmarek, L. (2010). Central noradrenergic lesion induced by DSP-4 impairs the acquisition of avoidance reactions and prevents molecular changes in the amygdala. *Neurobiology of Learning and Memory*, 94(3), 303–311.

Ramsey, L. C., Sinha, S. R., & Hurley, L. M. (2010). 5-HT1A and 5-HT1B receptors differentially modulate rate and timing of auditory responses in the mouse inferior colliculus. *European Journal of Neuroscience*, 32(3), 368–379.

Razak, K. A., Richardson, M. D., & Fuzessery, Z. M. (2008). Experience is required for the maintenance and refinement of FM sweep selectivity in the developing auditory cortex. *Proceedings of the National Academy of Sciences of the USA*, 105(11), 4465–4470.

Recanzone, G. H., Schreiner, C. E., & Merzenich, M. M. (1993). Plasticity in the frequency representation of primary auditory cortex following discrimination training in adult owl monkeys. *Journal of Neuroscience*, 13(1), 87–103.

Reed, A., Riley, J., Carraway, R., Carrasco, A., Perez, C., Jakkamsetti, V., & Kilgard, M. P. (2011). Cortical map plasticity improves learning but is not necessary for improved performance. *Neuron*, 70(1), 121–131.

Remage-Healey, L., Coleman, M. J., Oyama, R. K., & Schlinger, B. A. (2010). Brain estrogens rapidly strengthen auditory encoding and guide song preference in a songbird. *Proceedings of the National Academy of Sciences of the USA*, 107(8), 3852–3857.

Rouiller, E. M., Wan, X. S. T., Moret, V., & Liang, F. (1992). Mapping of c-*Fos* expression elicited by pure tones stimulation in the auditory pathways of the rat, with emphasis on the cochlear nucleus. *Neuroscience Letters*, 144(1–2), 19–24.

Rutkowski, R. G., & Weinberger, N. M. (2005). Encoding of learned importance of sound by magnitude of representational area in primary auditory cortex. *Proceedings of the National Academy of Sciences of the USA*, 102(38), 13664–13669.

Scheich, H., & Zuschratter, W. (1995). Mapping of stimulus features and meaning in gerbil auditory cortex with 2-deoxyglucose and c-*Fos* antibodies. *Behavioural Brain Research*, 66(1–2), 195–205.

Sen, K., Theunissen, F. E., & Doupe, A. J. (2001). Feature analysis of natural sounds in the songbird auditory forebrain. *Journal of Neurophysiology*, 86(3), 1445–1458.

Shaw, C., Needler, M. C., Wilkinson, M., Aoki, C., & Cynader, M. (1985). Modification of neurotransmitter receptor sensitivity in cat visual cortex during the critical period. *Brain Research*, 22(1), 67–73.

Sherwin, B. B. (2003). Estrogen and cognitive functioning in women. *Endocrine Reviews*, 24(2), 133–151.

Siegelbaum, S. A., Camardo, J. S., & Kandel, E. R. (1982). Serotonin and cyclic AMP close single K^+ channels in Aplysia sensory neurones. *Nature*, 299(5882), 413–417.

Sundström, E., Kölare, S., Souverbic, F., Samuelsson, E. B., Pschera, H., Lunell, N. O., & Seiger, Å. (1993). Neurochemical differentiation of human bulbospinal monoaminergic neurons during the first trimester. *Developmental Brain Research*, 75, 1–12.

Taffe, M. A., Weed, M. R., Gutierrez, T., Davis, S. A., & Gold, L. H. (2002). Differential muscarinic and NMDA contributions to visuo-spatial paired-associate learning in rhesus monkeys. *Psychopharmacology*, 160(3), 253–262.

Tremere, L. A., Jeong, J. K., & Pinaud, R. (2009). Estradiol shapes auditory processing in the adult brain by regulating inhibitory transmission and plasticity-associated gene expression. *Journal of Neuroscience*, 29(18), 5949–5963.

Valles, A., Boender, A. J., Gijsbers, S., Haast, R. A., Martens, G. J., & de Weerd, P. (2011). Genomewide analysis of rat barrel cortex reveals time- and layer-specific mRNA expression changes related to experience-dependent plasticity. *Journal of Neuroscience*, 31(16), 6140–6158.

van Praag, H., Kempermann, G., & Gage, F. H. (2000). Neural consequences of environmental enrichment. *Nature Reviews Neuroscience*, 1(3), 191–198.

van Wassenhove, V., & Nagarajan, S. S. (2007). Auditory cortical plasticity in learning to discriminate modulation rate. *Journal of Neuroscience*, 27(10), 2663–2672.

Watanabe, T., Taguchi, Y., Shiosaka, S., Tanaka, J., Kubota, H., Terano, Y., et al. (1984). Distribution of the histamingergic neuron system in the central nervous system of rats: A fluorescent immunohistochemical analysis with histidine decarboxylase as a marker. *Brain Research*, 295, 13–25.

Weinberger, N. M. (2004). Specific long-term memory traces in primary auditory cortex. *Nature Reviews Neuroscience*, 5(4), 279–290.

Weinberger, N. M., Hopkins, W., & Diamond, D. M. (1984). Physiological plasticity of single neurons in auditory cortex of the cat during acquisition of the pupillary conditioned response: I. Primary field (AI). *Behavioral Neuroscience*, 98(2), 171–188.

Weinberger, N. M., Javid, R., & Lepan, B. (1993). Long-term retention of learning-induced receptive-field plasticity in the auditory cortex. *Proceedings of the National Academy of Sciences of the USA*, 90(6), 2394–2398.

Weinberger, N. M., Miasnikov, A. A., & Chen, J. C. (2006). The level of cholinergic nucleus basalis activation controls the specificity of auditory associative memory. *Neurobiology of Learning and Memory*, 86(3), 270–285.

Woolley, C. S., & McEwen, B. S. (1993). Roles of estradiol and progesterone in regulation of hippocampal dendritic spine density during the estrous cycle in the rat. *Journal of Comparative Neurology*, 336(2), 293–306.

Yokosuka, M., Okamura, H., & Hayashi, S. (1995). Transient expression of estrogen receptor-immunoreactivity (ER-IR) in the layer V of the developing rat cerebral cortex. *Developmental Brain Research*, 84(1), 99–108.

Zhang, L. I., Bao, S., & Merzenich, M. M. (2001). Persistent and specific influences of early acoustic environments on primary auditory cortex. *Nature Neuroscience*, 4(11), 1123–1130.

Zhang, L. I., Bao, S., & Merzenich, M. M. (2002). Disruption of primary auditory cortex by synchronous auditory inputs during a critical period. *Proceedings of the National Academy of Sciences of the USA*, 99(4), 2309–2314.

Zhou, X., & Merzenich, M. M. (2007). Intensive training in adults refines A1 representations degraded in an early postnatal critical period. *Proceedings of the National Academy of Sciences of the USA*, 104(40), 15935–15940.

Zhou, X., & Merzenich, M. M. (2008). Enduring effects of early structured noise exposure on temporal modulation in the primary auditory cortex. *Proceedings of the National Academy of Sciences of the USA*, 105(11), 4423–4428.

Zhou, X., Nagarajan, N., Mossop, B. J., & Merzenich, M. M. (2008). Influences of un-modulated acoustic inputs on functional maturation and critical-period plasticity of the primary auditory cortex. *Neuroscience*, 154(1), 390–396.

Zhou, X., Panizzutti, R., de Villers-Sidani, E., Madeira, C., & Merzenich, M. M. (2011). Natural restoration of critical period plasticity in the juvenile and adult primary auditory cortex. *Journal of Neuroscience*, 31(15), 5625–5634.

Zuschratter, W., Gass, P., Herdegen, T., & Scheich, H. (1995). Comparison of frequency-specific c-*Fos* expression and fluoro-2-deoxyglucose uptake in auditory cortex of gerbils (*Meriones unguiculatus*). *European Journal of Neuroscience*, 7(7), 1614–1626.

MIX
Papier aus verantwortungsvollen Quellen
Paper from responsible sources
FSC® C105338

If you have any concerns about our products,
you can contact us on
ProductSafety@springernature.com

In case Publisher is established outside the EU,
the EU authorized representative is:
Springer Nature Customer Service Center GmbH
Europaplatz 3, 69115 Heidelberg, Germany

Printed by Libri Plureos GmbH
in Hamburg, Germany